A Publication of Cahners Publishing Company, Inc.

W0230436

ADVANCES IN
Volume 5
ELECTRONIC CIRCUIT PACKAGING

Proceedings of the Fifth International Electronic Circuit Packaging Symposium sponsored by the University of Colorado,

EDN (Electrical Design News), and Design News, held at Boulder, Colorado, August 19-21, 1964

Edited by Lawrence L. Rosine, Editor, EDN

SPRINGER SCIENCE+BUSINESS MEDIA, LLC 1965

Library of Congress Catalog Card Number: 62-2203

© *Copyright 1965 Springer Science+Business Media New York*
Originally published by Rogers Publishing Company, Inc. in 1965
Softcover reprint of the hardcover 1st edition 1965
ISBN 978-1-4899-7295-8   ISBN 978-1-4899-7307-8 (eBook)
DOI 10.1007/978-1-4899-7307-8

Foreword

This fifth volume of *Advances in Electronic Circuit Packaging* contains information assembled by specialists who are pushing to new frontiers in the field of packaging. The firms represented by these gentlemen are typical of the major electronic industries. Coverage of all the facets of electronic packaging, earthbound hardware to orbiting systems, is included in these articles.

While the success of such a conference as the International Electronic Circuit Packaging Symposium is certainly due in great part to the excellence of the papers presented during the sessions, credit must also be given to the untiring efforts of a group that receives little prominence. They are the members of the Program Selection Committee. These experts in the field, most of whom have presented papers at previous symposia, spend many hours of their own time discussing, selecting, rejecting, and rediscussing the papers submitted for possible use. This is no easy task. Almost all of the papers submitted are good. Some good papers were undoubtedly rejected so that the sessions could be kept to the allotted time. Some of the rejected papers have appeared over the past months in the packaging section of EDN. Others have appeared, in whole or part, at other technical sessions.

The process of selection of these papers is not a simple one. Abstracts of the papers are reviewed by each member of the selection committee. The paper is rated individually by a numbering system. The numbers for each paper are totaled, and a meeting of the committee is called. As the committee members are active in the field, they are able to objectively review each paper on its merit. A final decision is then reached based on this discussion and the number of points each paper receives. The man-hours spent on this selection far exceed the total hours of the symposium. This is done to assure the attendees of the symposium — and the readers of the proceedings — that they are obtaining the very latest, comprehensive information available in the field of electronic circuit packaging.

Therefore, we deem it fitting that the members of the Program Selection Committee should be listed as the true editors of *Advances in Electronic Circuit Packaging*.

Glen Boe, *Cahners Publishing Company*

D. A. Beck, *Bendix Research Laboratories*

G. E. Gless, *University of Colorado*

J. R. Goodykoontz, *Space Technology Laboratories, Inc.*

E. J. Lorenz, *IBM Corporation*

*R. C. Mayne, *Jet Propulsion Laboratory*

*E. C. Neidel, *Sandia Corporation*

W. J. Prise, *Lockheed Missiles & Space Co.*

*M. I. Ross, *The Milton Ross Company*

J. C. Rubin, *Eastman Kodak Co.*

H. J. Scagnelli, *Bell Telephone Laboratories, Inc.*

L. S. Shuey, *Sprague Electric Co.*

T. A. Telfer, *General Electric Co.*

Lawrence L. Rosine
Editor, EDN

*Indicates new members since the fifth symposium.

Contents

Encapsulants for Electronic Packaging

GEORGE R. DALLIMORE

Lockheed Missile and Space Company
Sunnyvale, California

Many types and classes of embedding compounds are available for encapsulating electronic packages. For a particular application, an embedding compound must be selected that meets the design parameters, the electrical, mechanical, thermal, and environmental requirements. The factual property data necessary for such selection and the comparative advantages and disadvantages of the available encapsulating and potting compounds are tabulated and discussed. The problem of selecting a specific encapsulant for a particular electronic package currently being used in an aerospace application is presented. The matching of the design parameter, the electrical, mechanical, thermal, and environmental requirements to the electrical, physical, and thermal properties of the available embedding materials is described.

INTRODUCTION

MANY TYPES OF EPOXY, silicone, and polyurethane casting, coating, and foam-in-place materials are available for the protective encapsulation of electronic circuits. Little information is available on the comparative advantages and disadvantages of these encapsulants for electronic circuit packaging. Which encapsulating material to use for a particular electronic device is a problem facing many electronic packaging engineers. Technical property data on encapsulants are widely scattered among vendor's literature, military specifications, company specifications, technical journals, and technical reports by the various users of these materials. It is the intent of this paper to tabulate and classify the technical property data and the comparative advantages and disadvantages of the available encapsulating materials and to show how this information was used in the selection of an encapsulant for the environmental protection of a particular electronic package.

SELECTION OF ENCAPSULANT

In most electronic packaging applications the encapsulant has the triple function of a structural member, electrical insulator, and heat-conducting medium, and must maintain these properties through many adverse environments. The selection of encapsulants to perform these important functions is an important part of electronic packaging. Encapsulants should be selected on the basis of a detailed technical analysis of the electrical, mechanical, thermal, and environmental requirements. The technical analysis should be conducted, and the encapsulant selected, during the initial stages of the packaging design. The actual layout of the package should be begun with the electrical, mechanical, thermal, and physical properties of the selected encapsulant firmly in mind.

Since the electrical, mechanical, thermal, and physical properties of the encapsulants are all temperature dependent, a detailed thermal analysis of the circuit is especially important. The results of this thermal calculation or measurement will determine whether highly filled

1

high-thermal-conductivity encapsulants are to be considered, or if lighter-weight materials such as syntactic foams, unfilled resins, or foam-in-place materials can be used. The thermal environments in which the electronic package must operate and be stored are also important considerations when selecting encapsulants. Shrinkage of the resin during cure along with differences in coefficients of expansion of the resin and components cause severe internal stresses to be set up in electronic packages during thermal cycling and/or storage. The resultant stresses have been known to cause cracking of the encapsulant material or cracking of sensitive components such as diodes, capacitors, and resistors encased in glass. Thermal cracking of the encapsulant can usually be solved by eliminating metal inserts and hardware or by reinforcing the resin in the stressed area or by using more resilient encapsulating materials.

The mechanical properties of the encapsulant are also quite important since the encapsulant is usually used as the only structural support for the components. The encapsulant must be rigid and strong enough to provide this support during handling and installation, and under vibration and mechanical shock environments. Because of their high mechanical and structural properties, epoxy resins have been used almost exclusively for the protective encapsulation of electronic devices for the aerospace industries.

At temperatures below 85°C the electrical properties of most encapsulants are sufficient for the majority of low-voltage electronic devices operating below one megacycle. Above 85°C

TABLE I

Epoxy Resin—Unfilled, Rigid

Properties	Unit	Typical value
Physical		
Specific gravity	—	1.0–1.2
Hardness	Shore D	>80
Moisture absorption	%	0.1–0.3
Electrical		
Volume resistivity	ohm-cm	10^{15}–10^{18}
Dielectric strength	volts/mil	450
Dielectric constant	—	2.8–3.5
Dissipation factor	—	0.005–0.010
Mechanical		
Flexural strength	psi	18×10^3
Tensile strength	psi	9–12×10^3
Elongation	%	5
Compressive strength	psi	18–20×10^3
Modulus of elasticity	psi	4–5×10^5
Thermal		
Conductivity	cal/sec/cm/°C/cm	4×10^{-4}
Expansion	in./in./°C	65×10^{-6}
Heat resistance	°C	up to 200

Relative Appraisal of Properties

Advantages	Disadvantages
Transparent	Poor thermal shock resistance
Excellent electricals	High stress on components
High mechanicals	High shrinkage
Excellent adhesion	Poor impact resistance
Low moisture absorption	Medium coefficient of expansion
Fair thermal conductivity	Not repairable

particular attention must be paid to the degradation of electrical properties with temperature, especially with the semiresilient and flexible epoxies and polyurethane foams. With high-gain and/or high-frequency circuits, particular attention must be paid to possible interelement coupling effects caused by high-dielectric-constant encapsulants. Low-dielectric-constant foams have had widespread application in high-frequency circuitry. In the encapsulation of high-voltage devices, particular attention must be paid not only to the dielectric strength of the encapsulant but also to its adhesion to and compatibility with other insulating materials such as sleeving and wire insulations in the package. Encapsulants for high-voltage applications must also be capable of being vacuum cast to eliminate internal voids which can cause breakdown or corona at high altitudes.

Consideration must also be given to the various environments to which the finished package may be exposed. Most electronic packages have to contend with water in the form of high humidity, actual immersion, or salt-spray resistance. Fortunately, practically all the solid encapsulants have sufficient moisture resistance so that seldom is a choice of encapsulant based solely on the differences in moisture absorption. Foamed plastics are not inherently resistant to moisture, and special coatings or containers should be used. Vacuum resistance is very dependent upon temperature and length of exposure. Encapsulants containing solvents, plasticizers, and nonreactive diluents have very poor vacuum and radiation resistance. Rigid highly filled (mineral filler) resins and silicone rubbers exhibit the best radiation resistance.

TABLE II

Epoxy Resin—Unfilled, Semiflexible

Properties	Unit	Typical value
Physical		
Specific gravity	—	1.0–1.2
Hardness	Shore D	50–80
Moisture absorption	%	0.75
Electrical		
Volume resistivity	ohm-cm	10^{12}–10^{14}
Dielectric strength	volts/mil	350
Dielectric constant	—	4.0–5.0
Dissipation factor	—	
Mechanical		
Flexural strength	psi	12×10^3
Tensile strength	psi	3–4×10^3
Elongation	%	25
Compressive strength	psi	7–10×10^3
Modulus of elasticity	psi	1×10^4
Thermal		
Conductivity	cal/sec/cm/°C/cm	4×10^{-4}
Expansion	in./in./°C	90–100×10^{-6}
Heat resistance	°C	100 max

Relative Appraisal of Properties

Advantages	Disadvantages
Good thermal shock resistance	Poor electricals (above 75°C)
Transparent	Poor mechanicals (above 75°C)
Fair electrical (to 75°C)	High coefficient of expansion
Fair mechanical (to 75°C)	
Low moisture absorption	
Some repairability	
Fair thermal conductivity	

SPECIFIC ENCAPSULATING MATERIALS

Epoxy Resins

Epoxy resins, because of their overall excellent electrical, mechanical, and physical properties, are the most widely used encapsulating resins for electronic packaging. The ability to modify almost any property through the use of selected fillers, flexibilizers, modifiers, and curing agents, have given epoxies the versatility to meet almost every environmental requirement imposed on electronic packaging materials. The various epoxy resin systems can be subdivided into seven major classes, each class being generally characterized by its specific gravity and hardness. Tables I–VII present typical property data and generalized advantages and disadvantages of these materials. Handling and curing properties such as viscosity, pot life, cure temperature, cure time, and exotherm have been deliberately left out of the tables since there are available within each class resin systems with varying handling properties and cure schedules.

Foams

Rigid polyurethane foams with densities from 2 to 20 lb/ft^3 are used extensively for encapsulating low-voltage, low-proper-dissipating electronic devices. For most applications

TABLE III

Epoxy Resin—Unfilled, Flexible

Properties	Unit	Typical value
Physical		
Specific gravity	—	1.0–1.2
Hardness	Shore	65A–50D
Moisture absorption	%	—
Electrical		
Volume resistivity	ohm-cm	10^{10}–10^{12}
Dielectric strength	volts/mil	275–350
Dielectric constant	—	5.4–9.4
Dissipation factor	—	0.15–0.3
Mechanical		
Flexural strength	psi	—
Tensile strength	psi	4×10^3
Elongation	%	100
Compressive strength	psi	—
Modulus of elasticity	psi	—
Thermal		
Conductivity	cal/sec/cm/°C/cm	4×10^{-4}
Expansion	in./in./°C	150×10^{-6}
Heat resistance	°C	75° max

Relative Appraisal of Properties

Advantages	Disadvantages
Excellent thermal shock resistance	Poor electricals (above 75°C)
Transparent	Low mechanicals
Fair electricals (to 75°C)	Poor heat resistance
Fair thermal conductivity	Highest coefficient of expansion
Repairable	High dielectric constant
Low stress on components	High dissipation factor

involving miniature components and small modules, the $2 \, lb/ft^3$ foam has had the most widespread usage. It has adequate mechanical properties to protect miniature components and is the lightest in weight of the available encapsulating materials. The low values of the dielectric constants and dissipation factors of these foams are used advantageously in high-frequency applications and where interelement coupling effects are critical. In addition, modules encapsulated in foams are more easily repaired than those encapsulated in solid encapsulants. The poor thermal conductivity of foams may be overcome by designing suitable heat-flow paths to a heat sink. Moisture sensitivity can be eliminated by providing a barrier material over the foam, or by foaming the device in a suitable metal or plastic container. Table VIII outlines the properties, advantages, and disadvantages of foams.

Silicone Rubbers and Flexible Resins

RTV silicone rubbers and flexible resins which have excellent electrical properties and high temperature resistance are not as widely used as epoxy resins because of their poor mechanical properties and their high cost. The low mechanical strengths of these materials limit their usefulness as encapsulants for electronic packaging subject to severe vibration and mechanical shock. Silicones have been successfully used for their electrical and thermal properties

TABLE IV

Epoxy Resin—Filled, Rigid (Solid Filler)

Properties	Unit	Typical value
Physical		
Specific gravity	—	1.4–2
Hardness	Shore D	>80
Moisture absorption	%	0.1–0.3
Electrical		
Volume resistivity	ohm-cm	10^{14}–10^{15}
Dielectric strength	volts/mil	400–500
Dielectric constant	—	4.0–4.7
Dissipation factor	—	0.01–0.04
Mechanical		
Flexural strength	psi	15–20×10^3
Tensile strength	psi	9–12×10^3
Elongation	%	1–2
Compressive strength	psi	20–30×10^3
Modulus of elasticity	psi	0.7–1.2×10^6
Thermal		
Conductivity	cal/sec/cm/°C/cm	8–24×10^{-4}
Expansion	in./in./°C	20–40×10^{-6}
Heat resistance	°C	up to 200

Relative Appraisal of Properties

Advantages	Disadvantages
High mechanicals	Not repairable
Low moisture absorption	High dielectric constant
Low coefficient of expansion	Highest weight
Good thermal conductivity	Opaque
Excellent adhesion	
Fair thermal shock resistance	
Low shrinkage	

by using other materials to provide structural rigidity. Tables IX and X outline the properties, advantages, and disadvantages of RTV silicone rubbers and flexible silicone resins, respectively.

Polyurethane Elastomers

Solid polyurethane materials have had limited usage in electronic packaging. These materials have excellent thermal-shock characteristics, but their poor adhesion, high dielectric constants, high coefficients of expansion, in addition to the fact that they are not easily repaired, limit their usefulness as electronic encapsulating resins. Table XI outlines the properties, advantages, and disadvantages of these materials.

SELECTION OF ENCAPSULANTS FOR FLIGHT-CONTROL WELDED MODULES

The Flight Control Electronics Package for the Polaris Missile consists of a number of welded electronic modules (Fig. 1) which control and direct the in-flight movements of the missile. The resin material selected to encapsulate these modules must be capable of protecting the electronic components from the effects of humidity, high- and low-temperature storage, vibration, acceleration, launching and stage-separation shocks, pressure, and vacuum exposure. In addition, the encapsulant must be light in weight, an excellent insulator, and dielectric material with sufficient thermal conductivity to dissipate up to 2 W when heat-sinked to the chassis.

TABLE V

Epoxy Resin—Filled, Rigid (Hollow Filler, "Syntactic Foam")

Properties	Unit	Typical value
Physical		
Specific gravity	—	0.8
Hardness	Shore D	>80
Moisture absorption	%	1.2
Electrical		
Volume resistivity	ohm-cm	10^{13}
Dielectric strength	volts/mil	300
Dielectric constant	—	3.8–4.5
Dissipation factor	—	0.020
Mechanical		
Flexural strength	psi	4×10^3
Tensile strength	psi	—
Elongation	%	—
Compressive strength	psi	$>9 \times 10^3$
Modulus of elasticity	psi	—
Thermal		
Conductivity	cal/sec/cm/°C/cm	3×10^{-4}
Expansion	in./in./°C	38×10^{-6}
Heat resistance	°C	up to 200

Relative Appraisal of Properties

Advantages	Disadvantages
Low weight	Opaque
Low coefficient of expansion	Not repairable
Excellent electricals	
High mechanicals	
Excellent adhesion	
Fair thermal shock resistance	
Fair thermal conductivity	

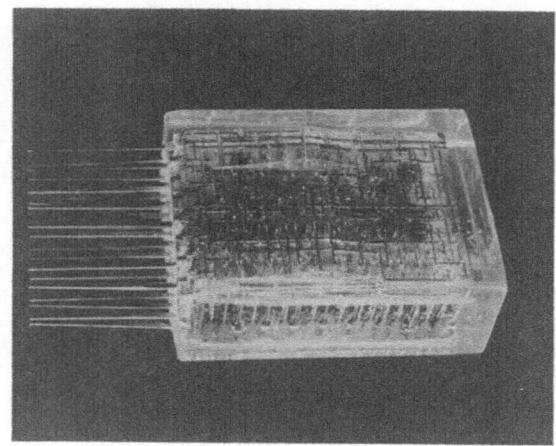

Fig. 1. Welded cordwood module.

TABLE VI

Epoxy Resin—Filled, Semiresilient (Solid Filler)

Properties	Unit	Typical value
Physical		
Specific gravity	—	1.4–1.8
Hardness	Shore D	50–80
Moisture absorption	%	0.5
Electrical		
Volume resistivity	ohm-cm	10^{14}
Dielectric strength	volts/mil	375–450
Dielectric constant	—	3.8–5.5
Dissipation factor	—	0.02–0.09
Mechanical		
Flexural strength	psi	—
Tensile strength	psi	3.9×10^3
Elongation	%	20
Compressive strength	psi	20×10^3
Modulus of elasticity	psi	—
Thermal		
Conductivity	cal/sec/cm/°C/cm	8×10^{-4}
Expansion	in./in./°C	100×10^{-6}
Heat resistance	°C	up to 130

Relative Appraisal of Properties

Advantages	Disadvantages
Excellent thermal shock resistance	High dielectric constant
Low moisture absorption	High dissipation factor
Good adhesion	High coefficient of expansion
Good thermal conductivity	High weight
Some repairability	Low mechanicals (above 75°C)
Low stress on components	Low electricals (above 75°C)
Good mechanicals (to 75°C)	
Good electricals (to 75°C)	

Fig. 2. Encapsulated modules.

TABLE VII

Epoxy Resin—Filled, Semiresilient (Hollow Filler, "Syntactic Foam")

Properties	Unit	Typical value
Physical		
Specific gravity	—	0.65–0.8
Hardness	Shore D	65–75
Moisture absorption	%	—
Electrical		
Volume resistivity	ohm-cm	10^{13}–10^{14}
Dielectric strength	volts/mil	—
Dielectric constant	—	2.6–3.2
Dissipation factor	—	0.11–0.13
Mechanical		
Flexural strength	psi	—
Tensile strength	psi	—
Elongation	%	—
Compressive strength	psi	4–7×10^3
Modulus of elasticity	psi	8.5–10.5×10^4
Thermal		
Conductivity	cal/sec/cm/°C/cm	3×10^{-4}
Expansion	in./in./°C	—
Heat resistance	°C	up to 130

Relative Appraisal of Properties

Advantages	Disadvantages
Low weight	Opaque
Low coefficient of expansion	Poor electricals (above 75°C)
Excellent thermal shock	Poor mechanicals (above 75°C)
Good adhesion	High coefficient of expansion
Some repairability	
Low stress on components	
Good mechanicals (to 75°C)	
Excellent electricals (to 75°C)	

In order to provide shock and vibration resistance, a rigid, high-modulus epoxy resin system was selected that could provide the necessary mechanical support for components and eliminate as much structural hardware as possible to conserve weight (Fig. 2). Modules were designed to be held in the chassis by a compressive wedge which bears against wedges on the module, which not only provides positive lock-down of modules but also provides a means of maintaining intimate contact between the module side and the chassis web for heat-sinking purposes (Fig. 3). The requirement that the encapsulant be capable of withstanding thermal cycling indicated that a filled resin system be used. The lower coefficient of expansion of the filled system would also be an advantage in the method of mounting. The requirement that weight be held to a minimum meant that conventional filled (solid filler) epoxy resins could not be used. A filled, low-weight epoxy resin requirement could only be met by a syntactic foam (microballoon filled epoxy resin). The lowered thermal conductivity, because of the insulating nature of the microballoons, did not prove to be a problem except for one module which required that a metallic heat sink be embedded.

PROCESSING OF THE SYNTACTIC FOAM

The high viscosity of the syntactic foams and the close spacing in the welded module require that a heat-cure system be selected so that the resin can be heated to lower the viscosity,

TABLE VIII

Rigid Polyurethane Foams

Properties	Unit	Typical value		
Physical				
Density	lb/ft	2	10	20
Water absorption	%	1.4	0.7	0.7
Electrical				
Volume resistivity	ohm-cm	10^{13}	10^{13}	10^{13}
Dielectric strength	volts/mil	200	250	300
Dielectric constant	—	1.0–1.1	1.2–1.3	1.4–1.5
Dissipation factor	—	0.001	0.003	0.006
Mechanical				
Flexural strength	psi	10–70	100–500	500–1100
Tensile strength	psi	45–65	250–400	1000–1200
Compressive strength	psi	20–40	180–350	800–1400
Modulus of elasticity	psi	450–700	11,000	38,000
Thermal				
Conductivity	cal/sec/cm/°C/cm	0.48×10^{-4}	1.1×10^{-4}	1.1×10^{-4}
Expansion	in./in./°C	18×10^{-6}	42×10^{-6}	70×10^{-6}
Heat resistance	°C		55–150	

Relative Appraisal of Properties

Advantages	Disadvantages
Lowest weight	Tendency to swell under humidity
Repairability	Poor thermal conductivity
Excellent strength to weight	Poor dielectric strength
Low dielectric constant	
Low dissipation factor	
Excellent thermal shock resistance	
Good adhesion	
Low coefficient of expansion	

Fig. 3. Modules assembled in chassis.

TABLE IX

RTV Silicone Rubbers

Properties	Unit	Typical value
Physical		
Specific gravity	—	1.15–1.45
Hardness	Shore A	30–75
Moisture absorption	%	0.5
Electrical		
Volume resistivity	ohm-cm	10^{12}–10^{14}
Dielectric strength	volts/mil	450–550
Dielectric constant	—	3.0–3.6
Dissipation factor	—	0.015–0.030
Mechanical		
Tensile strength	psi	275–500
Elongation	%	150
Thermal		
Conductivity	cal/sec/cm/°C/cm	4.3–5.7
Expansion	in./in./°C	250–280×10^{-6}
Heat resistance	°C	175–200

Relative Appraisal of Properties

Advantages	Disadvantages
Excellent electricals	Low mechanicals
Excellent thermal shock resistance	Poor shock and vibration resistance
High temperature resistance	Poor adhesion (must use primers)
Repairable	High coefficient of expansion
High vacuum resistance	
No exotherm	
No stress on components	

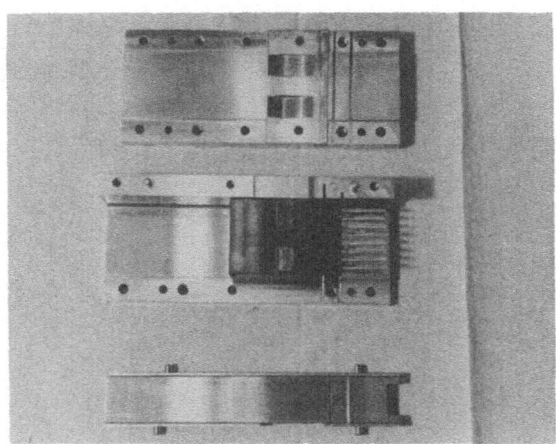

Fig. 4. Metal encapsulating mold.

TABLE X

Flexible Silicone Resins

Properties	Unit	Typical value
Physical		
Specific gravity	—	1.05–1.25
Hardness	Shore A	40–45
Moisture absorption	%	0.1
Electrical		
Volume resistivity	ohm-cm	10^{15}
Dielectric strength	volts/mil	550
Dielectric constant	—	2.7–3.2
Dissipation factor	—	0.001–0.008
Mechanical		
Tensile strength	psi	800–1000
Elongation	%	100
Thermal		
Conductivity	cal/sec/cm/°C/cm	$3.5–7.5 \times 10^{-4}$
Expansion	in./in./°C	$250–300 \times 10^{-6}$
Heat resistance	°C	200–250

Relative Appraisal of Properties

Advantages	Disadvantages
Excellent electricals	Low mechanicals
Excellent thermal shock resistance	Poor shock and vibration resistance
High temperature resistance	Poor adhesion (primers required)
Repairability	High coefficient of expansion
High vacuum resistance	Inhibition by epoxy curing agents
No exotherm	
Transparent (some)	
No stress on components	

TABLE XI

Polyurethane Elastomers

Properties	Unit	Typical value
Physical		
Specific gravity	—	1.2
Hardness	Shore A	65–90
Moisture absorption	%	—
Electrical		
Volume resistivity	ohm-cm	10^{12}–10^{13}
Dielectric strength	volts/mil	275–350
Dielectric constant	—	6.5
Dissipation factor	—	0.05
Mechanical		
Tensile strength	psi	2000–5000
Elongation	%	500–600
Thermal		
Conductivity	cal/sec/cm/°C/cm	3.75×10^{-4}
Expansion	in./in./°C	240×10^{-6}
Heat resistance	°C	150

Relative Appraisal of Properties

Advantages	Disadvantages
Excellent thermal shock resistance	Poor adhesion (primers required)
Transparent	Not repairable (does not bond to itself)
Low moisture absorption	High coefficient of expansion
Low stress on components	High dielectric constant
	High dissipation factor

and yet have sufficient pot life to be thoroughly degassed and poured under vacuum to ensure the absence of voids in the finished module. The tendency for the lightweight microballoons to float to the surface is quite pronounced and can only be controlled by careful control of the resin viscosity (temperature). Exothermic heat, which would lower the viscosity and allow the microballoons to pop to the surface, must be kept under control. The insulating nature of the microballoon makes exotherm control difficult, and necessitates the use of heavy steel molds which act as a heat sink (Fig. 4).

SUMMARY

The selection of an encapsulant for an electronic package is an engineering problem that can only be solved by careful technical analysis of the electrical, mechanical, thermal, physical, and environmental requirements. Tables I–XI provide typical properties on the various classes and types of encapsulants, as well as an appraisal of the relative advantages and disadvantages of each. It is hoped that this information may aid the packaging designer in his search for the right encapsulant for his design.

Use of Radiation-Cross-Linked Materials for Encapsulating and Terminating Devices

PAUL SHERLOCK

Raychem Corporation
Redwood City, California

Recent developments in heat-shrinkable radiation-cross-linked plastics have made possible a number of new devices for encapsulating components and terminating wires. Over the past several years irradiated heat-shrinkable tubings and molded products have been used in increasing quantities for harness coverings, for the strain relief of wire terminations and connectors, wire markers, and component covers. The new cross-linked heat-shrinkable plastics may be used in combination with (1) other thermoplastics and sealants for the encapsulation of components and (2) solders for wire terminations. This paper describes the recent developments in cross-linked materials and the application of these new developments in encapsulating and terminating devices.

NATURE OF MATERIAL

IT IS KNOWN that many polymers such as polyvinyl chloride, silicone rubber, nylon, natural rubber, etc., can be cross-linked by ionizing radiation. However, this paper will only concern itself with the two polymer systems used for most of the devices which will be described, namely: (1) high-strength stabilized polyolefins; and (2) polyvinylidene fluoride (Kynar*).

Both of these materials are capable of operation at relatively high temperatures. For continuous operation the polyolefins are rated at 135°C while Kynar is rated at 150°C. For shorter periods of time, both of these materials can be used up to a temperature of 300°C. Kynar is more expensive than the polyolefins; however, it has certain features which make its use in some specific applications worth the higher cost. For example, Kynar has approximately twice the tensile strength of most polyolefins; at the same time, it is transparent and well flame retarded, and can be formed with wall thicknesses as low as 0.002 in.

These polymers are composed of long chain-like molecules. At temperatures below 135°C (275°F) for the polyolefins and 175°C (347°F) for polyvinylidene fluoride, a portion of these molecules is arranged systematically into highly ordered crystalline structures, which accounts for their characteristic hardness and toughness. Above this temperature range the crystals melt and the polymers flow. This molecular structure can be altered, however, by high-energy beta radiation to form cross-links between chains such as occur during the vulcanization of rubber. These cross-links, or primary valence bonds, impart to the polymer a three-dimensional gel network which gives the polymer form stability at temperatures above its crystalline melting point. At these temperatures the irradiated polymer behaves as an elastomer, is quite rubbery, and can be deformed readily by an applied stress. Upon removal of the stress the cross-linked material will return to its original shape and size. If, however, the material is cooled below its crystalline melting point while being held in the stressed state, it will remain in this

---

* Registered trademark, Pennsalt.

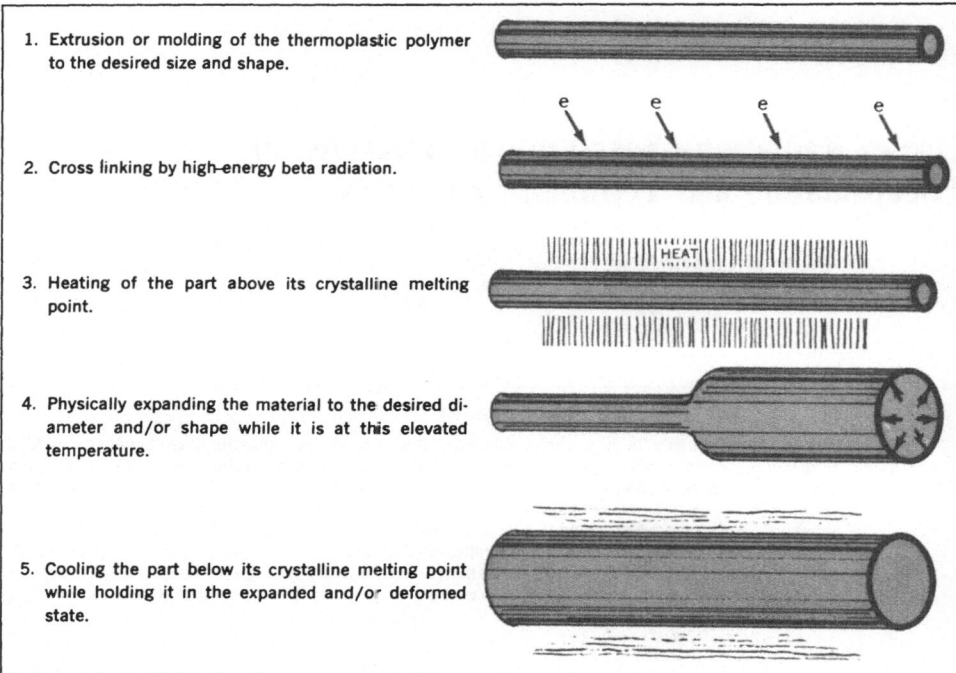

1. Extrusion or molding of the thermoplastic polymer to the desired size and shape.

2. Cross linking by high-energy beta radiation.

3. Heating of the part above its crystalline melting point.

4. Physically expanding the material to the desired diameter and/or shape while it is at this elevated temperature.

5. Cooling the part below its crystalline melting point while holding it in the expanded and/or deformed state.

Fig. 1.

configuration until it is again heated above the crystalline melting point. With this background in mind, the specific steps in the manufacture of cross-linked plastic parts may be as shown in Fig. 1.

Recent improvements, listed below, have made possible the encapsulating and terminating techniques to be described in this article.

1. High Shrinkage Ratios. Earlier heat-shrinkable tubings had shrinkage ratios of about 3 : 2. Now practically all standard tubings are at least 2 : 1, with specific types as high as 5 : 1.

2. Strength at Temperature. Perhaps the single most important characteristic of these cross-linked polymers is the fact that they do have some strength above their crystalline melting points in contrast to these materials before cross-linking. This permits their use at elevated temperatures, and, more specifically, wires can be soldered through these materials without destroying their physical integrity. Temperatures as high as 300°C are easily withstood for a period well in excess of the time required to solder.

3. Strengths of Polyolefin and Kynar. The vast spectra of polyolefins available have tensile strengths of the order of 2000–4000 psi. In comparison, the recently developed radiation-cross-linked Kynar has a tensile strength of approximately 7000 psi.

4. Selective Cross-Linking. By a process known as selective cross-linking it is now possible to produce a tubing with an outer wall which is heat shrinkable and a thermoplastic inner wall which melts and flows simultaneously with the shrinkage of the outer wall. Upon cooling, the entire mass becomes a hard homogeneous molding.

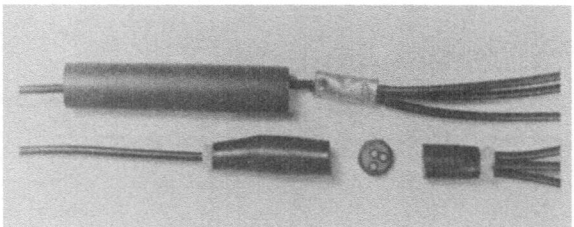

Fig. 2.

5. Thinner Walls. With the new irradiated Kynar materials, expanded walls as thin as 0.002 in. are possible. This provides a minimum size buildup with toughness and dielectric strengths in excess of 2500 V.

ENCAPSULATING DEVICES

With the high-strength, high-shrink-ratio polymers lined with thermoplastics or sealants, an automatic molding over irregularly shaped wire splices and components can be readily achieved. Upon brief heating to above the crystalline melting point, the outer tubing or molded part shrinks. Simultaneously, the inner wall melts and flows, and under the pressure of the shrinking outer wall is forced into the interstices of the component being covered. Some specific applications are shown to illustrate these devices.

1. Missile Wire-Splice Cover. To prevent corrosion of wire splices from nitrogen tetroxide and UDMH, in one of the major missile programs a part consisting of an irradiated polyolefin outer sleeve with a highly modified polyolefin liner was used. The liner was specifically formulated to provide moisture sealing as well as to resist attack from the fuel and oxidizer (Fig. 2).

2. Jacket-to-Wire Transitions. On a space capsule application, the requirement was to seal between the jacket and individual wires of a multconductor shielded cable. Wire insulation and jacket were irradiated modified polyolefin. Selectively cross-linked irradiated polyolefin tubing provided an encapsulation which withstood the water immersion test of MIL-C-26500 (Fig. 3).

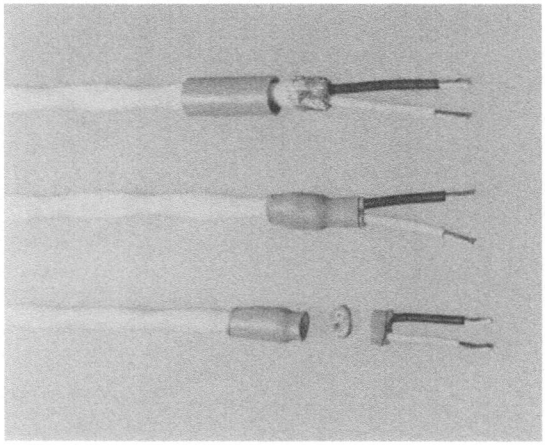

Fig. 3.

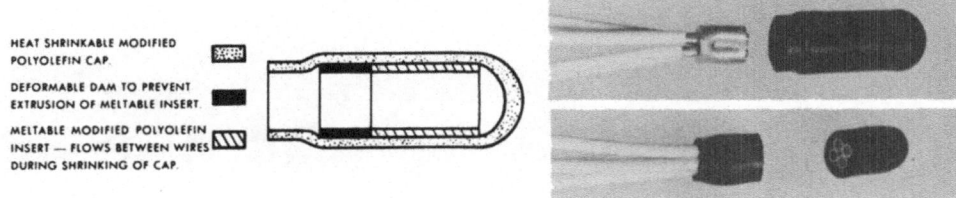

Fig. 4.

3. Stub-Splice Cap. In order to mold or "pot" crimp stub splices on a supersonic military aircraft, a heat-shrinkable molded cap with meltable inserts was developed. Since the splices had to withstand 400°F for short periods, and the melting temperature of the non-cross-linked polyolefin potting insert was 275°F, a partially cross-linked and, hence, deformable cylindrical insert was located near the opening to act as a dam, preventing extrusion of the potting insert. The cap withstood the qualification test consisting of 200 temperature cycles from room temperature to 400°F for five minutes each, followed by the water immersion test of MIL-C-26500 (Fig. 4).

4. Diode Cover. To replace the lacquer coating on glass diodes, and thus allow production-line automation, a black irradiated Kynar approximately 0.100 in. supplied diameter with a 0.002 in. wall thickness was developed. The thin wall allowed the diode to meet size specifications established on the lacquered design (Fig. 5).

TERMINATING DEVICES

Since the new cross-linked tubings can withstand temperatures well above that which will melt even 60/40 or 63/37 solder, a new concept for wire termination becomes possible. This concept involves a precisely designed and manufactured flux-coated solder preform inside a heat-shrinkable encapsulating sleeve. By a controlled heating technique the sleeve shrinks tightly, forcing the wires into the proper position for joining and forming a cavity to contain solder flow. Further heating causes the flux to act and the solder to melt to form a soldered, insulated, and encapsulated joint. The combination of these features eliminates the normal

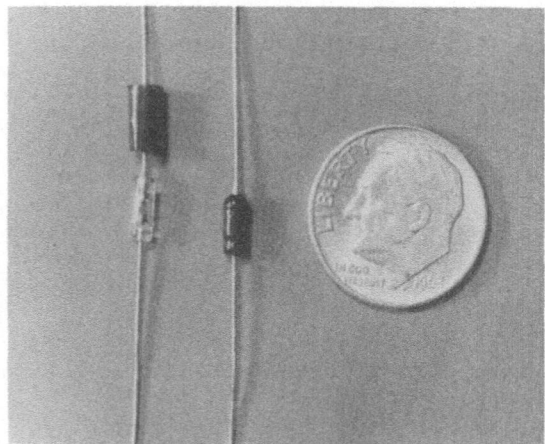

Fig. 5.

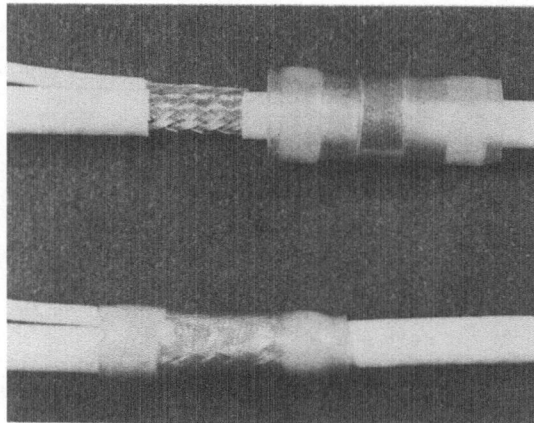

Fig. 6.

operator variables commonly associated with the soldered termination. This concept is important because it combines the outstanding electrical characteristics of a solder joint with a reproducibility that is equal to or better than crimp-type terminations. This reproducibility has not been possible before with solder.

Qualification testing has proven that the appearance of solder flow indicates a properly wetted joint, assuming that the proper type of wires are used. This relationship is true because of the elimination of the variables. Inspection for solder flow is a simple matter through the transparent sleeve.

Several configurations of this solder device are already being used. One version, known as a solder sleeve, is a shield terminating device consisting of a heat-shrinkable irradiated Kynar sleeve approximately $\frac{5}{8}$ in. long containing three inserts—a preform of flux-coated solder at the center, and a sealing ring at each end (see Fig. 6). The solder is Sn63 in the series used with irradiated or Teflon wires, and an alloy containing 25% indium, $37\frac{1}{2}$% tin, and $37\frac{1}{2}$% lead, in the design used with polyvinyl chloride (PVC) insulated wires. When heated by a hot-air source of approximately 600°F evenly directed around the surface of the sleeve, an insulated, encapsulated solder joint is formed in 8–20 sec, depending upon the type of wire. Induction-heating techniques for heating the solder, with auxiliary hot-air heating for shrinking the sleeve, can solder as many as fifteen assemblies simultaneously in a total of 8–10 sec.

The characteristics of this shield terminating system compared with the crimp ferrule techniques are as follows:

1. Size and Weight. Table I compares the sizes and weights of solder sleeve terminations to the three conventional crimp-shield terminations when installed over a 20 AWG shielded wire with a 22 AWG ground lead. Figure 6 illustrates the small-diameter buildup.

TABLE I

Device	Maximum O.D. of insulation, in.	Weight, lb/1000 parts
Solder sleeve	0.175	0.53
Crimp ferrule A	0.200	1.03
Crimp ferrule B	0.250	1.45
Crimp ferrule C	0.225	1.60

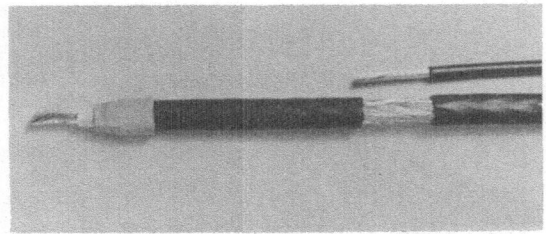

Fig. 7.

2. RFI Shielding. The fact that the solder sleeve can be readily installed at any point along the cable allows shielding to extend close to the connector pins. Figure 7 shows the cable preparation for this type of installation. Tests indicate that an increase in effectiveness of approximately 20% (an average of 6 db) is achieved by taking advantage of this feature on a 28-in. cable length. The shielding efficiency was obtained by measuring the coupling between two unshielded cables and using this as the zero db reference. This test was conducted over frequencies from 500 kc/sec to 5 Mc/sec.

3. Encapsulation and Strain Relief. The sealing rings of the solder sleeve tightly mold around the wire insulation on both sides of the solder joint so all flexing action is outside the solder area. This provides vibration resistance and protection from handling damage.

4. Inspectability. The transparency of the solder sleeve insulation allows inspection of the joint along its entire length so that improper stripping and bunched or nicked wires can be easily observed.

5. Cost Savings. The experience of users indicates a reduction of 20–40% in assembly time with the solder sleeves applied with the hot-air blower when compared with crimp-type ferrules. With multiple-induction heating, the time saving becomes greater.

In addition to time savings, the need for a precise crimp-tool die for each size is eliminated. A single heating tool readily handles all sizes. Defects in the heating device are easily detected by an increase in the heating times required, so that improper adjustments of heaters do not go undetected.

6. Inventory of Parts. One size solder sleeve fits an estimated 80% of all commonly used shielded hookup wires, covering all diameters up to 0.175 in. Two additional sizes have been developed, one for compactness on miniature wires, and another for cables up to 0.280 in. in diameter.

7. Performance. Table II compares the solder sleeve terminations with the requirements of MIL-F-21608A, Ferrule, Shield, Grounding, Insulated, Crimp-Style, Brass.

TABLE II

	Requirements of MIL-F-21608A	Solder sleeve performance, typical data
Allowable increase in voltage drop after temperature cycling (millivolts)	3.00	0.04
Allowable increase in voltage drop after vibration testing (millivolts)	3.00 (10 to 55 cycles/sec)	0.07 (10 to 55 plus 55 to 2000 cycles/sec)
Minimum pull strength 22 AWG ground lead (pounds)	15	26 (caused by lead breakage)
Dielectric strength of insulation (volts)	500	7000

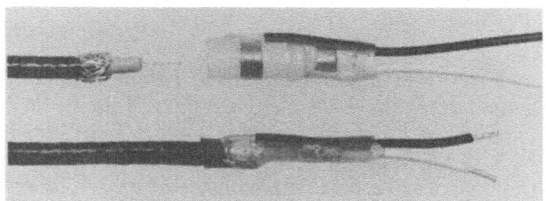

Fig. 8.

A more complex version of this device is the coaxial-cable termination with two leads and two solder preforms which attach to the center conductor and shield, respectively. This provides a rapidly installed, reliable means of attaching high-strength leads to the small, vulnerable coax-center conductor (Fig. 8).

Utilizing the same principles, a fine wire termination has been developed. This device consists of a precise film of flux-coated solder fused to the surface of the terminal pin. Surrounding this terminal is a sleeve of irradiated modified polyolefin. This terminal design is illustrated in Fig. 9.

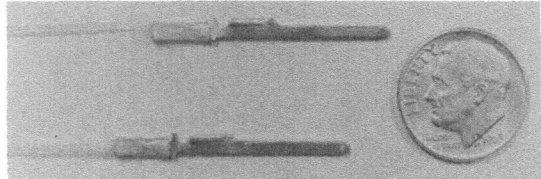

Fig. 9.

Upon heating, the sleeve shrinks, pressing the wire upon the solder. Within a few seconds the solder melts, tightly bonding the wire to the pin. The high-strength sleeve provides strain relief to the joint by gripping the cable insulation tightly. Since the material is clear, the joint may be easily inspected. In one application of this design, the terminal was attached to memory-core wire which had an O.D. of 0.003 in. A crimp type of termination would have flattened and weakened this small wire.

A version of this terminal used to tie together a number of parallel miniature wires in a single operation was designed as shown in Fig. 10.

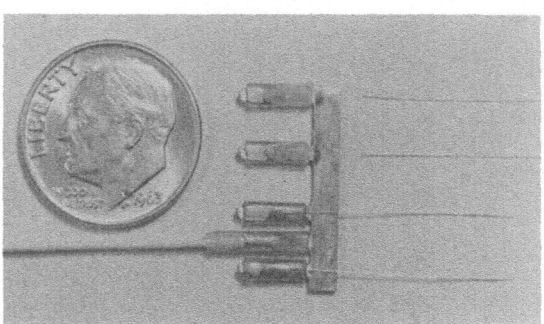

Fig. 10.

CONCLUSION

The most recent cross-linked polymers have high shrinkage ratios and high strength at temperatures above the melting point of solders. In addition, it is possible to produce selectively cross-linked tubing and materials which have very thin walls. These properties have been the basis of a number of new encapsulating and terminating devices which are providing the electronic packaging field with a choice of ways to meet the requirements for lightweight, environmentally resistant, and compact encapsulations and terminations.

Encapsulating with Loose Microballoons

E. C. Neidel

Sandia Corporation
Albuquerque, New Mexico

This paper describes an investigation which led to the successful use of loose microballoons as an encapsulant for electronic equipment. The properties of a loose-microballoon system and the advantages derived from their use are also explained. They include the reduced dynamic response expected; however, the ability to repair and rework the unit is not compromised by the encapsulant.

INTRODUCTION

THE ENCAPSULATION of electronic assemblies designed for missile and airborne applications has become virtually universal. The purpose of the encapsulant is to eliminate environmentally induced failures by offering some degree of protection for the components, their leads and connections, and the interconnecting wires. For the dynamic environments of mechanical shock and vibration this is accomplished in two ways. First, the position of the components becomes fixed and the stress on the leads is thereby reduced. Second, the local thickness and, consequently, the moment of inertia of the assembly is increased; this reduces the vibration response of the unit.

The use of encapsulants, however, introduces other problems and difficulties. The rigid materials and the foam encapsulants require the use of molds to contain the encapsulant until the curing action is completed. During this period, the forces which are applied by the foaming encapsulant are difficult to control and sometimes result in failures. With the more rigid, denser encapsulating materials, the temperature coefficient of expansion is often large enough to cause failure at the temperature extremes. Nevertheless, encapsulation is successful, and the number of shock and vibration failures is being reduced even though the requirements are becoming more stringent.

Unfortunately, while the dynamic characteristics of the unit are being improved with encapsulation, the ease of repair of the unit is being reduced. Polyurethane foam, epoxy, and other materials commonly used as encapsulants virtually prohibit repair and maintenance of the unit. Since chemical solvents which would selectively attack the encapsulant are unobtainable, access to a component or connection for testing, repair, or replacement can only be obtained by digging or cutting. Frequently the circuit boards, components, or leads adjacent to the excavation are inadvertently damaged during the process. If repair is judged to be too difficult, it is not even attempted, and the only benefit which can be derived from the unit is a post mortem report.

Ease of repair and maintenance is frequently relegated to a low priority in the list of design objectives for an electronic unit. Consequently, it is rarely given much consideration. If a unit can be easily repaired, a design and development group can get appreciably more use from prototype units by keeping them functional during an evaluation program. An easily repaired unit also makes possible experimental modifications as well as the evaluation of proposed design changes. In a production facility, an easily repaired unit is more apt to meet its shipping schedule. Customer satisfaction with production units is certain to be enhanced since the cost of repair and the amount of downtime is reduced.

Furthermore, a series of experimental vibration tests performed on a group of assemblies encapsulated in polyurethane foam indicated that the components had a large response ratio. Apparently the foam offers far less damping than had been expected.

For these reasons, an investigation was begun to develop an encapsulation system for electronic units that would make them easier to repair and maintain while reducing the vibration response of the components.

PROPERTIES

An encapsulant used for an electronic unit must obviously be nonconductive. If the unit is to be easy to repair, the material must be easy to remove. This suggests that the material should have low viscosity, remain loose, and not solidify. If it remains loose, it must fill the container to provide any appreciable damping. Therefore, it would be used generously and should be light in weight and low in cost. A hollow plastic sphere, or microballoon, satisfies these requirements.

To determine the damping characteristics of microballoons a set of vibration experiments was conducted using glass and phenolic microballoons. The experiments were made on a functional unit containing components soldered to a printed circuit board (Fig. 1). At this assembly stage, the board of this unit would normally be encapsulated in a polyurethane foam brick about 1 in. thick. For this experiment, however, an unencapsulated board assembly was used. Accelerometers were located as shown.

The results of the experiment are shown in Fig. 2. The unit was first vibrated without microballoons in the axis perpendicular to the circuit board. With a 1 g table input, peak response ratios of 28.0 to 1 at 84 cps, 24.5 to 1 at 217 cps, 7.3 to 1 at 300 cps, and 10.0 to 1 at 400 cps were observed.

The unit was then filled with glass microballoons, and the experiment was resumed. With the 2 g table input, peak response ratios of 12.0 to 1 at 300 cps and 6.85 to 1 at 600 cps were recorded.

The glass microballoons were then replaced with phenolic microballoons, and the experiment was repeated with a 2 g table input. Peak response ratios of 6.2 to 1 at 170 cps, 4.05 to 1 at 430 cps, and 4.25 to 1 at 666 cps were noted.

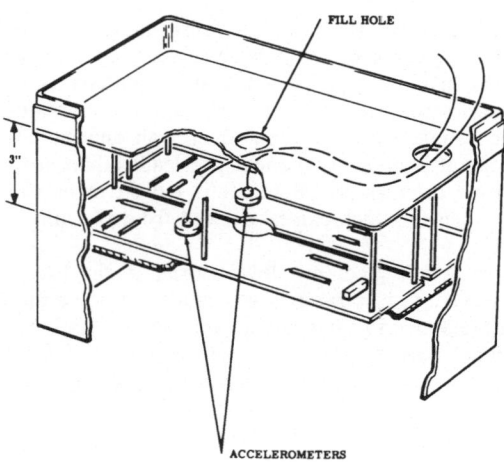

FILL HOLE

3"

ACCELEROMETERS

Fig. 1. Drawing of experimental unit.

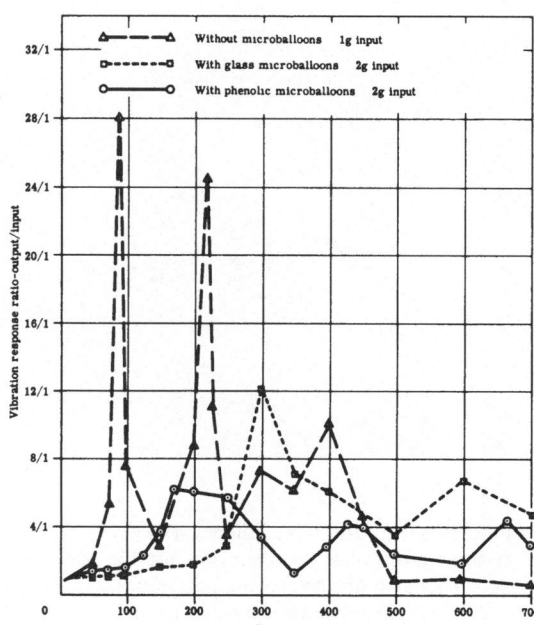

Fig. 2. Results of vibration tests conducted to determine the damping characteristics of microballoons.

The results of this experiment leave no doubts about the damping ability of microballoons. Adding glass microballoons reduced the maximum response ratio from 28 to 1 down to 12 to 1. Using phenolic microballoons further reduced the ratio to 6.2 to 1. It was also noted that the ease with which the glass microballoons were removed and replaced with the phenolic microballoons demonstrated that access to a component or connection could be easily obtained.

Continued evaluation of the two types of microballoons revealed that the glass had a tendency to break up into small pieces. No such tendency was observed with the phenolic. This tendency to break up is an important consideration for three reasons. First, any significant amount of microballoon breakup would result in a considerable decrease in the volume occupied by the material and, hence, a decrease in the damping effect. Second, after breakup the smaller size particles would have a greater tendency to become airborne and might present a health hazard if they were inhaled during subsequent handling. With a size range for the phenolic of 0.0002 to 0.0050 in. in diameter and an average particle diameter of 0.0017 in., no health problems were anticipated with the use of the phenolic packing material by either the Sandia Corporation medical or safety departments. And third, the broken glass microballoons would act as a loose abrasive in a unit. During the above experiment, it was also observed that severe abrasive action had taken place on the surfaces of screw threads which were engaged with glass microballoons.

It was found that the microballoons behaved not unlike sand when exposed to moisture. In a dry atmosphere, the material is readily poured with no detectable difference between the two types. During disassembly, the microballoons fall away like dry sand and can be collected and reused, as shown in Fig. 3; however, with increased humidity, they cake much like wet sand. This appears to be more pronounced with glass microballoons which are usually discarded when they cake. The phenolic material can be broken up quite easily one day after bags of desiccant have been introduced into the storage container.

Consequently, loose phenolic microballoons were chosen as the packing material to fill the free volume of the unit.

Again referring to the above vibration experiments, use of a functional unit made it easy to determine if any deterioration had taken place. Post-vibration electrical tests showed that

Fig. 3. Phenolic microballoons fall away during disassembly like dry sand.

component lead and wire failures had occurred during the experiments. The indications were that some additional local support for the components was needed to reduce the stress on the leads. This also applied to the individual harness wires at their termination points.

A silicone rubber was selected to provide this additional support. The manufacturer states that this material will vulcanize "in sections of unlimited thickness within 24 hr at room temperature" independent of the presence of air or moisture. He further states that "neither thickness of cross section nor degree of confinement has any effect on the uniformity of rate of setup throughout the rubber." Thus far the rubber has lived up to these claims. As expected, the cure time can be accelerated by increasing the ambient temperature. The silicone rubber is a viscous material which is applied by pouring it over all the inside surfaces of the unit and allowing it to run off freely. The result is a cocoon-like, conformal coating about $\frac{1}{16}$ in. thick (Fig. 4). The coating supplies most of the required additional support.

Adding 2% silicone dioxide to the rubber makes it very viscous, and it will not pour. This mixture is applied with a spatula to form large fillets between the bigger and heavier components and the circuit board. The mixture is applied before the conformal coating and provides greater support for the larger components. It sets up rapidly and allows the conformal coating operation to proceed without further delay.

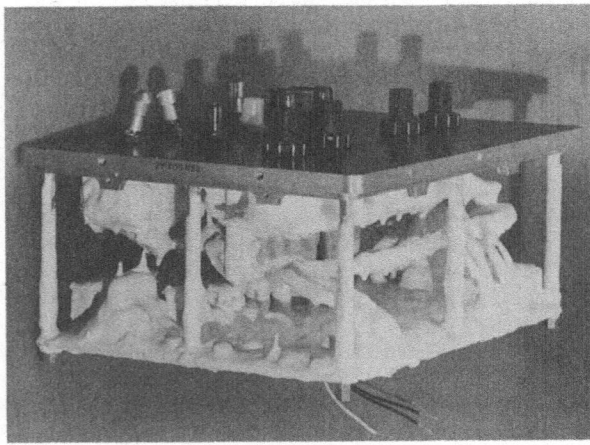

Fig. 4. Cocoon-like conformal coating of silicone rubber.

Fig. 5. Portable computer-type electronic unit.

Both the conformal coating and the thicker fillet mixture are readily peelable after curing. The surface adhesion is low enough to allow easy removal from all surfaces. As a result, the dynamic characteristics of the unit have been improved with the only compromise on the ease of repair being the addition of a readily peelable material. Areas exposed during repair can be recoated with no difficulty.

PROCEDURE

A portable computer-type electronic unit (Fig. 5) was used to test this encapsulating system. The volume encapsulated is about 7 × 10 × 4 in. The unit contains about 200 components in fourteen cordwood modules plus another 200 laydown components. These are soldered to a 7 × 10 in. epoxy glass printed circuit board. The board shown in Fig. 6 is mounted

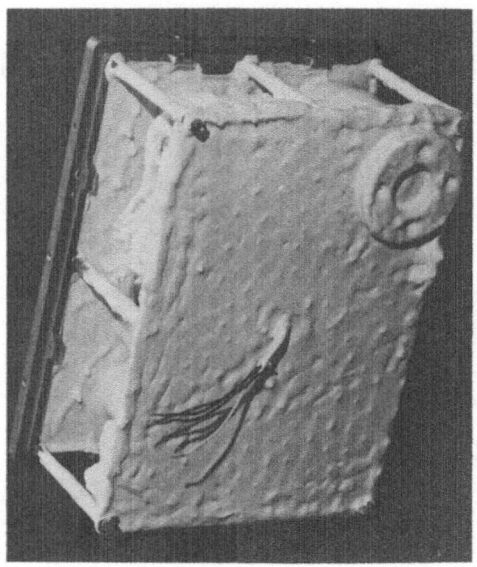

Fig. 6. The 7 × 10 in. printed circuit board mounted and coated.

parallel to a similar size cast-aluminum panel with eight 3-in.-long studs around the periphery of the board.

The four corner studs, together with a flange on the panel, also support a metal shield. The shield is about $\frac{1}{2}$ in. from the circuit board. The panel assembly mounts in the open side of a compression-molded phenolic and glass housing; it is sealed in the housing with a synthetic-rubber sealing compound. The shield clears the inside of the housing by about $\frac{1}{4}$ in.

The encapsulation procedure begins with the application of the silicone dioxide–silicone rubber mixture. Fillets of the mixture are formed with a spatula between the circuit board on all components which exceed $\frac{1}{4}$ in. in diameter. The panel is masked, and the conformal coating is applied by pouring the silicone rubber on all exposed surfaces and allowing it to cure.

The free volume of the unit is then filled with phenolic microballoons through the access hole in the shield which is in a plane parallel to the board. Some rocking action of the unit is used to aid in filling, especially as the filling action nears completion.

The sealing surfaces are coated with synthetic rubber, and the panel assembly is installed in the housing.

The housing contains a partition with two threaded inserts which are located adjacent to access holes in the bottom of the shield. The final filling step is accomplished by introducing additional microballoons through the partition. The principal purpose of this step is to fill the void between the shield and the housing. Since the components are well protected at this stage, this step is accomplished on a vibration machine at 50 cps with a 4 g table input in a plane parallel to the board and partition. Usually, this step requires about 25 min to complete. At this point, threaded plugs are installed, and the unit is ready for its acceptance tests. This final step indicates that the unit has excellent dynamic characteristics since 50 cps vibration at 4 g for 25 min is not considered a destructive test and has caused no discernible deterioration in any units. The silicone rubber–phenolic microballoon combination is now specified as the encapsulant for this unit.

RESULTS

Units constructed in the manner described above successfully pass vibration tests with table inputs of 0.064 in. double amplitude displacement from 10 to 55 cps, 10 g from 55 to 300 cps, and 2 g from 300 to 1000 cps. In addition, the units survive a 30 in. drop test onto concrete.

To determine the damping effect of the phenolic microballoons on a conformally coated board, a group of vibration experiments were conducted on two units. Each unit was surveyed with and without microballoons. Both units had been conformally coated with silicone rubber. The axis of vibration was perpendicular to the printed circuit board. An accelerometer was mounted in the center of the board. The levels of vibration were those specified above. The results are shown in Figs. 7 and 8.

Figure 7 shows that unit #303 had a maximum response ratio of 13.2 to 1 at 105 cps without microballoons, and 5.5 to 1 at 200 cps with microballoons. Figure 8 shows a similar reduction for unit #306; the maximum response ratio of 12.5 to 1 at 95 cps was reduced to 6.0 to 1 at 175 cps. This represents reductions in the response ratio of 42% and 48%, respectively,

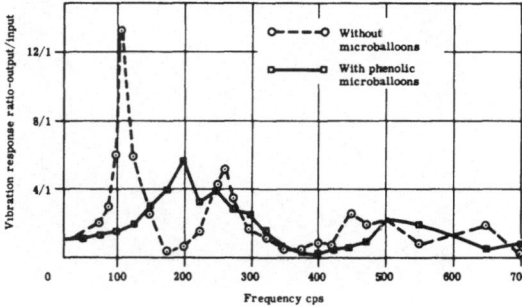

Fig. 7. Results of vibration tests conducted on unit No. 303.

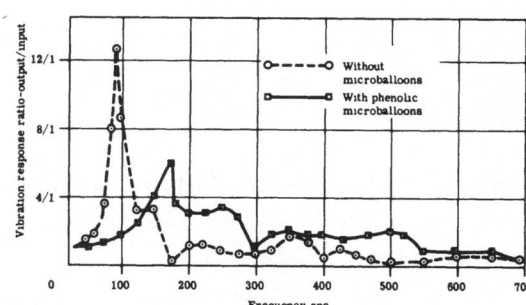

Fig. 8. Results of vibration tests conducted on unit No. 306.

coupled with corresponding increases in the resonant frequency as a result of encapsulation with loose microballoons.

There have been no throw-away units since the introduction of the loose microballoon encapsulating system. Prototype units have been modified and kept functional since depotting is both simple and rapid. Evaluation of proposed design changes at environmental extremes is easily accomplished since repotting is also simple and rapid. The turn around time, i.e., the time required to disassemble, empty, peel, recoat, refill, and reassemble, excluding circuit modification, can be measured in hours if the curing process of the silicone rubber is accelerated by the use of an oven.

LIMITATIONS

An encapsulation system using microballoons is successful as an acceleration damper because of the tremendous number of microballoon collisions taking place while the unit is being dynamically excited. The collisions convert the vibration energy to heat through friction and the elastic deformation of the microballoons. Consequently, whenever the number of collisions is drastically reduced (for example, in a close-fitting, contoured package with little free volume available for microballoon packing), the damping action would also be drastically reduced. Similarly, little microballoon damping action could be expected in a rigidly mounted assembly which already possessed a very low vibration–amplification ratio.

In an encapsulated unit which required the use of a mold, only the components which see the inside of the mold need to be sealed. In a microballoon system, all components must be sealed, including those which penetrate the container. Also, the container itself must be sealed to keep the microballoons in and to keep moisture out.

Finally, this type of system is incompatible with the cleanliness commonly found these days in modern electronic assembly plants. If a separate adjacent location is not available, a transparent enclosure on a bench top together with frequent use of a vacuum cleaner would be necessary.

Packaging Concept for a Miniature Low-Light-Level TV Camera

EDMUND C. DECKER

General Electric Company
Electronics Laboratory, Ithaca, New York

This paper presents a packaging design concept for a low-light-level TV camera that weighs less than 13 lb and is capable of displaying scenes under starlight conditions. The discussion highlights thermal problems, weight considerations, tube mounts, modular packaging of electronic components, structural design, designing for final fabrication without the need for breadboard models, and maintainability features.

INTRODUCTION

THE RECENT DEVELOPMENT of the electrostatic image orthicon (EIO) tube under Contract DA 36-039-SC-88964 sponsored by the U.S. Army Electronics Laboratory has opened up numerous possibilities for applications that require small, lightweight, low-power, low-light-level TV cameras. To further explore these possibilities, the General Electric Company undertook a program to design and build an EIO camera suitable for customer demonstration. The resulting camera will withstand the rigors of handling and shipping and is compatible with present manufacturing techniques and potential environmental requirements. The EIO exhibits thermal and structural integrity, is attractive, and has good maintainability features. It weighs 13 lb, has a volume of 489 in.^{3}, and consumes 22 W. Design to completion was five months.

Present low-light-level TV cameras, whose sensitivity is as low as 300 TV lines at 10^{-6} ft-c, require heavy magnetic coils for focusing and deflection. The EIO accomplishes essentially the same functions with electrostatic techniques. Thus, weight, volume, and power requirements of the miniature low-light-level TV cameras are reduced by $\frac{1}{2}$ to $\frac{1}{3}$ of those utilizing a 3-in. magnetic tube.

HISTORY

The evolution of the low-light-level TV camera has been brief, development having taken place over the last three or four years.* The vidicon (standard commercial TV camera) does well to sense 10^{-2} ft-c (twilight conditions). In contrast, electromagnetic image orthicons are sensitive to light levels as low as 5×10^{-6} ft-c, which is equivalent to starlight conditions. (For comparison, the normal eye is sensitive to approximately 10^{-3} ft-c, full-moon conditions.) The primary disadvantages of the magnetic IO are its size, weight, and power. The EIO tube has now been developed to where its feasibility has been demonstrated, and it has been shown to be a potential substitute for the electromagnetic IO.

* Paper presented 1964.

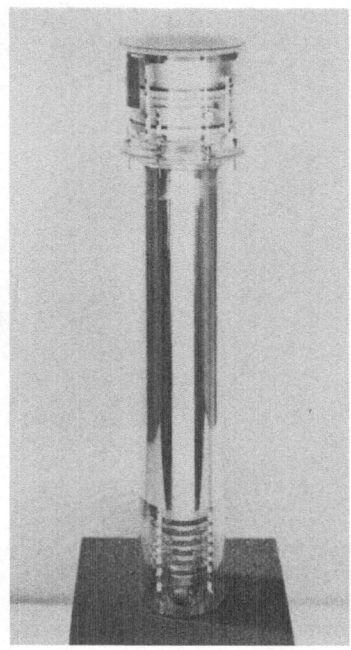

Fig. 1. Typical 3-in. magnetic image orthicon without coil.

TV TUBE PRINCIPLES AND CHARACTERISTICS

Although it is not the intention of this paper to discuss the electronic design of the EIO tube and circuitry in any detail, it may be useful to discuss the television pickup tube generally, and to show the advantages of the new tube. Figure 1 shows the normal electromagnetic image orthicon pickup tube without its magnetic deflection and focus coil. The tube is 15.2 in. long and 3 in. in diameter at the forward end. Figure 2 is a cross-sectional illustration of the interior of this tube.

When an image is focused on the front surface of the tube, a charged pattern of the scene is formed at the target. The target is then scanned by an electron beam which reads out the

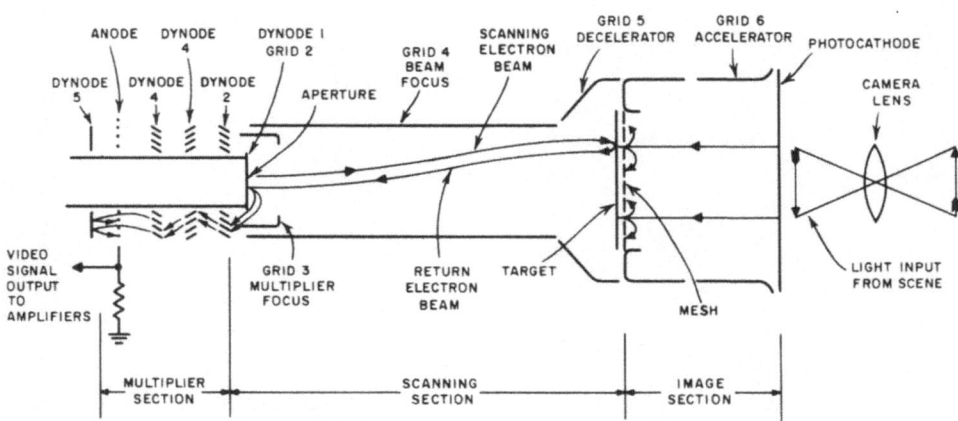

Fig. 2. Cross-sectional illustration of a magnetic image orthicon showing flow of electrons.

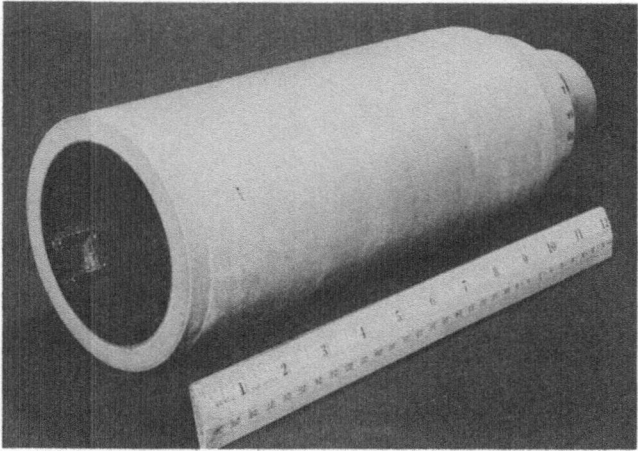

Fig. 3. Magnetic coil for image orthicon.

image in the form of an electronic signal. This signal is later converted back to a visible picture on a TV monitor (not part of this system) where the scene may be viewed. The pickup-tube electron beam is normally scanned and focused by a large (14 lb) magnetic coil assembly, as shown in Fig. 3, which fits over the outside of the glass envelope.

The new electrostatic tube, Fig. 4, utilizes the same type of imaging process, except that now the tube can be scanned and focused by electrically charged plates located inside the glass envelope. No outside coils are required. The tube is approximately 13 in. long and 2 in. in

Fig. 4. Typical electrostatic image or thicon. (Note radial pins across from 9 mm mark on rule.)

diameter over its full length. Electronically, its sensitivity and picture resolution are somewhat less than in the magnetic version, but for many applications the electrostatic camera will be extremely useful.

GENERAL FEATURES

Figure 5 shows the electrostatic image orthicon camera system. The EIO camera features the new General Electric electrostatic image orthicon tube type Z-7804. This tube is in the advanced development stage at the Power Tube Operation, Syracuse, New York, and sample tubes are presently being evaluated in this camera.

The advantages of the EIO camera over a conventional magnetic version, as mentioned previously, are primarily its size, weight, and power. The following tabulation compares an earlier 3-in. transistorized magnetic image orthicon camera to the EIO.

	EIO	Magnetic IO
Volume (in.^{3})	489	1345
Weight (lb)	13	50
Power (W)	22	100

In addition to these reductions, the EIO offers essentially the same sensitivity potential and approximately 70% of the resolution potential of the magnetic camera. Industrial design objectives were to keep all external surface breaks to a minimum, to proportion the control box to suggest stability, and to keep all external lines long and straight.

CAMERA HEAD

The camera head assembly has a diameter of 3 in., is 17 in. long, and weighs approximately 3 lb (excluding lens, lens flattener, handle, aluminum fin, base, and camera head cable). Material selection, the weight-saving design features, and substitution of the aluminum fin for the blower* resulted in a weight reduction in the camera head of $1\frac{1}{2}$–2 lb. It should be noted at this point

* See Thermal Analysis.

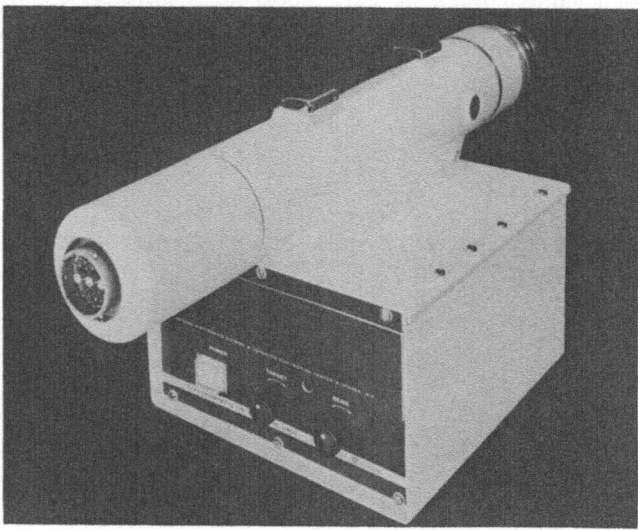

Fig. 5. Complete electrostatic image orthicon camera system.

that the lens, lens flattener, handle, base, aluminum fin, and cable are not included in the weight for the following reasons:

1. Lens weight would depend on system requirements—wide angle, zoom, telephoto, etc.
2. The lens flattener would not be required in the final tube configuration with the incorporation of the flat fiber optics faceplate.
3. Handle and base are arbitrary, and were selected for camera demonstration purposes.
4. Aluminum fin—although present design dictates a maximum temperature differential of 5°C across the tube envelope, some systems requirements would preclude this requirement as on short-lived missions. In addition, future tube design is expected to advance to the point where the cathode-heater output is reduced, and the cesium migration phenomenon mimimized.
5. The camera head cable is also excluded since it again would depend on the system requirements.

The camera head contains the following elements: (1) EIO tube; (2) preamplifier; (3) field flattener, manual iris control, and lens; (4) focusing magnet (not to be confused with the electromagnetic image orthicon focus coil); (5) magnetic shield; (6) heat sink-aluminum fin; (7) housing; (8) mounting base and handle; (9) connector; (10) tube mounts. These elements are combined into an attractive magnesium case which is compact, lightweight, lighttight, and moistureproof.

Figure 6 is a cross-sectional view of the camera head assembly. The tube and preamplifier are plugged together and suspended from the housing by silicone rubber mounting rings. The head is assembled by attaching the field flattener and holder to the housing, and then inserting the EIO tube and preamplifier into the heat sink and magnetic shield, clamping the focusing magnet to the glass face of the tube, and installing this assembly into the housing. Finally, the housing cap is placed over the preamplifier and hand-tightened into place. As the cap is tightened, the rear silicone mount will push against the preamplifier, the preamplifier against the tube, and the tube against the forward O ring. No tools are required during assembly operations. The housing and housing cap are fabricated from standard magnesium tubing in order to take advantage of its lighter weight. This results in a weight saving of 3 oz. A fine thread was incorporated in the housing in order to keep the wall thickness to a safe minimum. The results were satisfactory except that continued disassembly scored the housing cap threads. On future units a coarser thread will be used. To preclude the possibility of the threads binding or seizing, an aluminum lubricant was found to be satisfactory. Tube and preamplifier annulus-shaped silicone rubber mounts were selected because their simplicity and availability lent

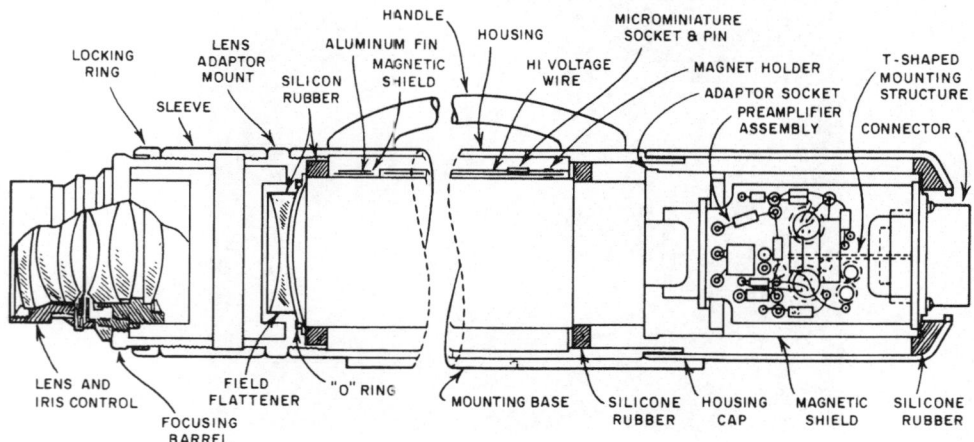

FIG. 6. Cross-sectional view of camera head showing parts assembled.

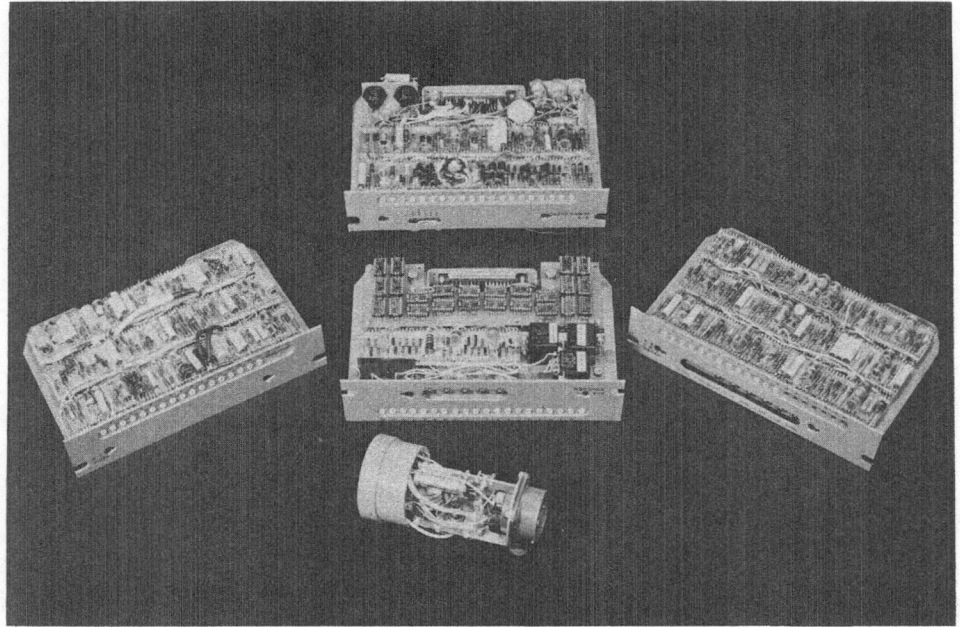

Fig. 7. The four modular slide-in cards and preamplifier.

themselves well to a lightweight design and for ease of manufacture. Also they do not require breaks or mounting screws in the external surfaces, resulting in a neater appearance. Vibration of a prototype glass tube envelope and housing incorporating the silicone rubber mounts showed mechanical resonance in the 250- to 350-cps range with a transmissibility factor of eight.

The preamplifier components are assembled on a T-shaped supporting structure which is the load-bearing member for the compressive force of the housing cap (see Fig. 7). Radial pins, located toward the front of the EIO-tube envelope, for target, grid, image focusing, and the photocathode, are connected to the preamplifier by individual microminiature sockets and pins. This feature, along with the removable cap and preamplifier socket, permits quick and convenient tube replacement.

The imaging optics and manual iris control assembly are mounted on the front face of the housing. Focusing is accomplished by means of a threaded focusing barrel, sleeve, and locking ring. The field flattener is cemented to the lens adapter mount.

The camera head can be mounted on the control unit or removed and remoted as shown in Figs. 5 and 8. A quick-disconnect type of coupling for removing the head from the control unit allows ease of handling both as a system and for remoting the head from the unit. The camera head may be remoted over 200 ft without additional special electronic circuitry.

CONTROL UNIT

Camera-control circuitry and adjustments are contained in the control unit (see Fig. 9). The unit, which weighs 10 lb and is $9\frac{3}{4}$ in. long $\times$ 8 in. wide $\times$ $5\frac{1}{4}$ in. high, is of dip-brazed aluminum construction. Dip-brazed aluminum construction was chosen because of its good strength-to-weight ratio and it lends itself to simple, attractive design, since a minimum number of external surface breaks are necessary, and materials and processes are readily available. Thicknesses of the wall of the box (0.040 in.) and of the module cards (0.032 in.) and panels (0.060 in.) were kept as thin as possible. Stiffeners were added to the sides and bottom of

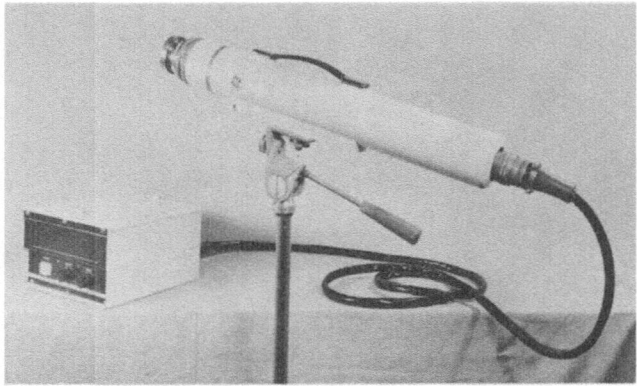

Fig. 8. The camera head shown remote from the control unit.

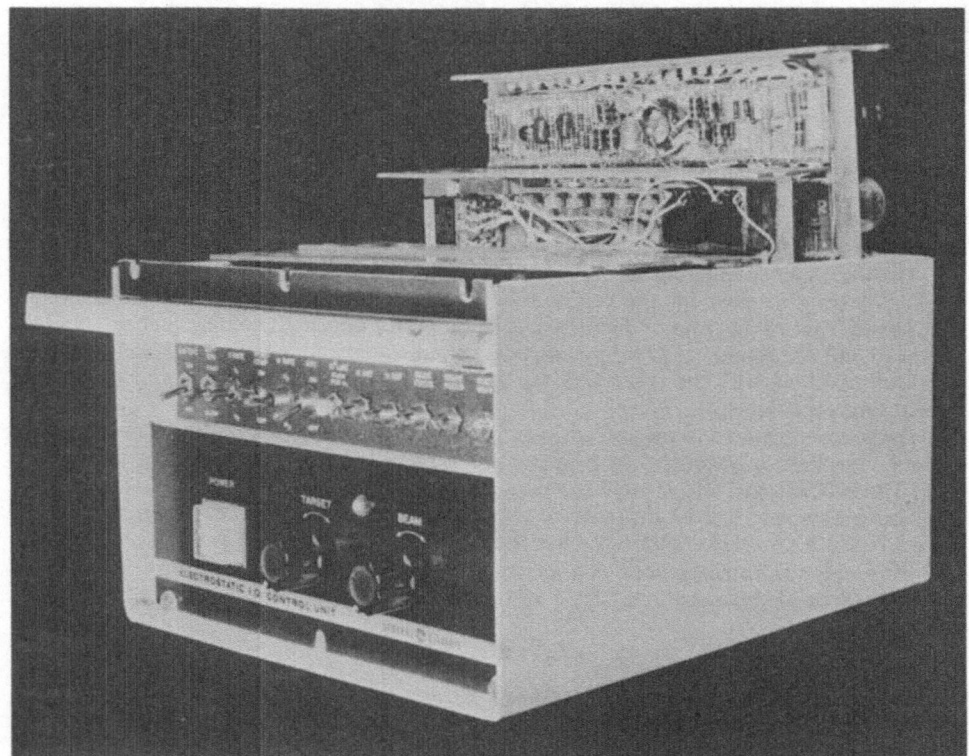

Fig. 9. Modular slide-in cards are featured in the EIO camera. Shown partially elevated are the power supply (background) and synchronous generator (foreground). The spring-loaded door is opened to reveal the recessed alignment controls. The power On-Off switch, and the Target and Beam controls are also visible.

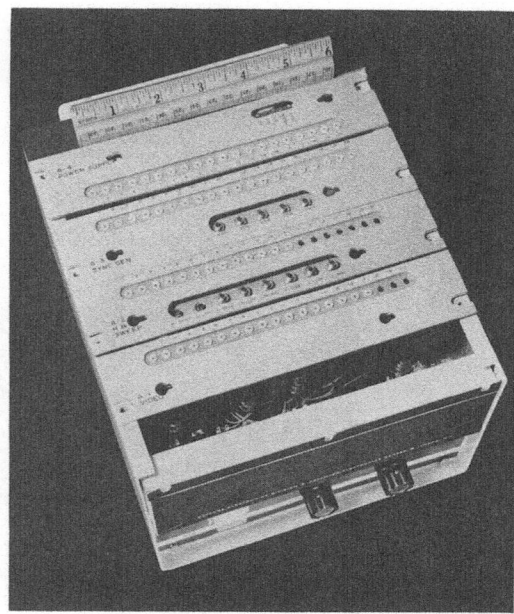

Fig. 10. Top view of control unit showing four modular slide-in cards with their adjustment controls and test points.

the box to increase strength and rigidity and to reduce the possibility of warping in the dip-brazing process. The cover was made of 0.060-in.-thick material and stiffeners added since the camera head mounts directly to it when in the carrying mode. The control unit as well as the camera head were finished in a white epoxy paint and trimmed with black and brushed aluminum-colored façades.

Featured are four modular slide-in cards: video, horizontal and vertical sweep, synchronous generator, and power supply (see Fig. 7). If desired, any card can be removed and adapted for "desk top" maintenance with an interconnecting cable. Test points and adjustments are easily accessible from the top of the card (see Fig. 10). The cards are thermally connected to the control box by fastening the removable cover in place so that it pushes the modular cards against the top of the control box. Cooling is by radiation and free convection with the outside environment.

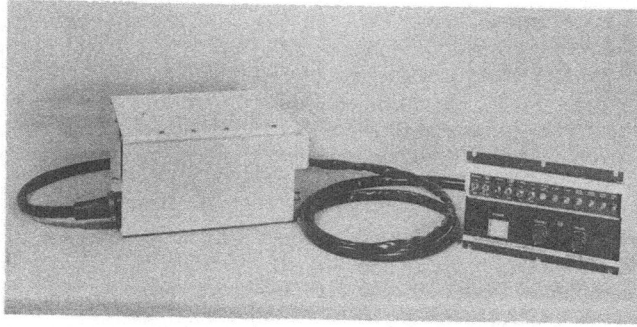

Fig. 11. Control panel shown remote from control unit. The panel can be remoted a long distance from the control unit.

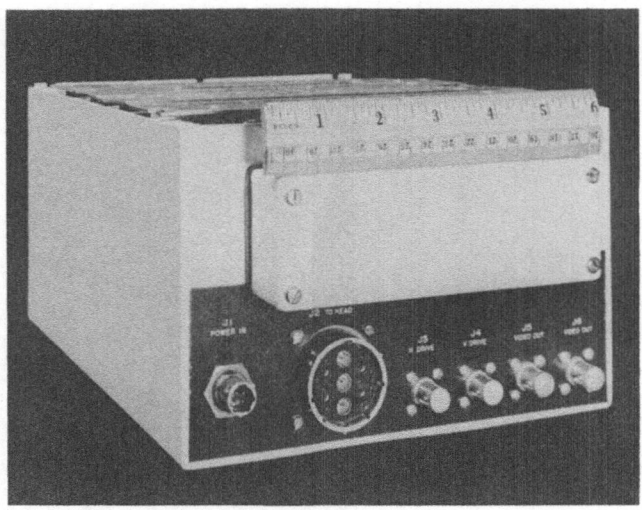

Fig. 12. View of back panel showing Power In, Head, H (horizontal)
Drive, V (vertical) Drive, and two Video Out connectors.

The alignment controls are recessed into the front panel and covered by a two-position spring-loaded door for ease of access and maintainability. These adjustments are usually required when the EIO tube is replaced. The power On–Off switch and Beam and Target controls are also available on the front panel. The control panel may be removed and remoted up to 100 ft (see Fig. 11), providing flexibility of operation, as may be required, for example, when available space around the operator is at a premium, or for meeting difficult mission requirements. Provisions are available for installing weatherproofing and RFI sealing around the front panel and the top access cover. The rear panel (see Fig. 12) contains connectors for power in, camera head cable, horizontal drive out, vertical drive out, and two video outputs.

The electronics are assembled using simulated welded-wire matrices (WWM) which permit the design and assembly of the engineering model without going through the usual breadboard stage. This flexibility in design allows interconnecting wiring to be rerouted and reworked as required (see Fig. 13); at the same time the system can be designed so that it is easily converted to the standard General Electric WWM as soon as the electrical circuitry is optimized. The only additional efforts required then would be to update the original drawings and release them for production.

The basic General Electric Welded-Wire Matrix consists of an assemblage of wires placed side by side and spaced at standard grid increments of either 50 or 100 mil. A layer of insulation is placed over these wires, and a second set of wires, at right angles to the first, is similarly placed over the insulation (see Fig. 14a). Necessary connections between the wires in the two planes are made by welding through the insulation. The end result is a grid upon which components can be secured by either welding or soldering to the slotted terminals which are welded to the transverse wires of the matrix grid (see Fig. 14b).

The production version of the WWM utilizes a computer which assumes the task of circuit layout by processing information coded from the schematic diagram. In determining matrix layout the computer allows for component sizes and location and the required electrical connections. The computer controls the automatic weld machine. An automatic circuit test is conducted for quality control purposes.

OPTICS

The camera is currently equipped with an advanced-development EIO tube having a curved faceplate. Eventually a flat fiber-optics faceplate will be incorporated into the tube and a final lens design will then be established.

The choice of the objective lens to be used with the curved faceplate EIO tube provides two challenges: (1) optimizing performance of the lens because of the sharply curved faceplate; and (2) reducing the cost of the lens for this application since it is a "one of a kind" item.

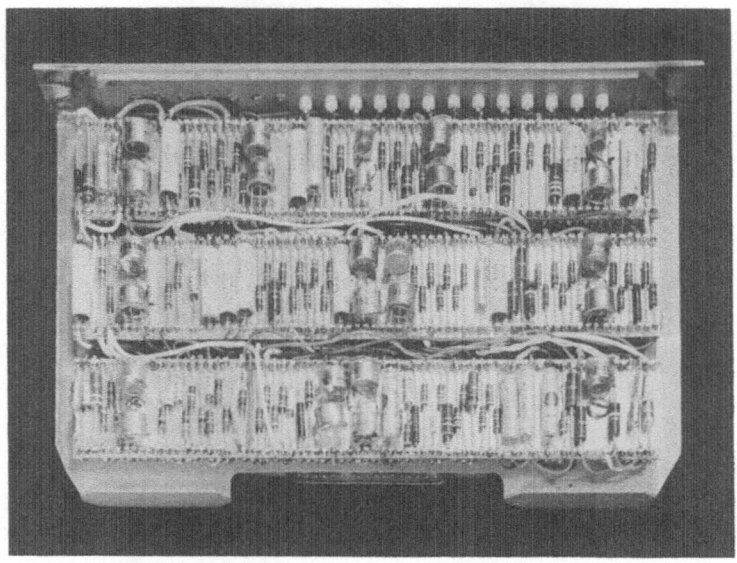

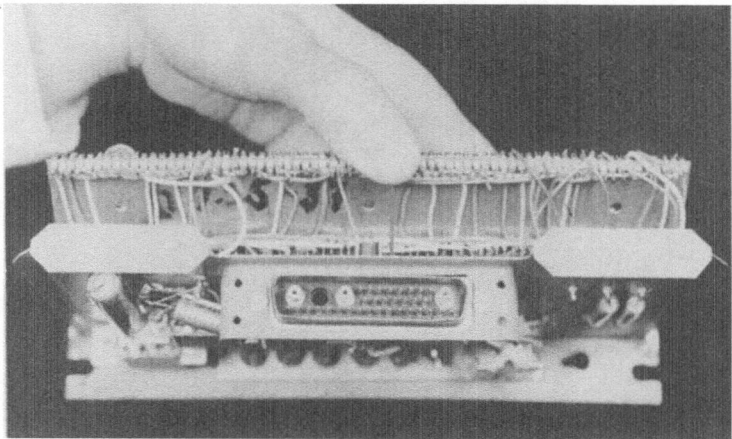

Fig. 13. (a) H and V sweep card, showing three simulated welded-wire matrices with components; (b) bottom view of card, showing underside of simulated welded-wire matrix.

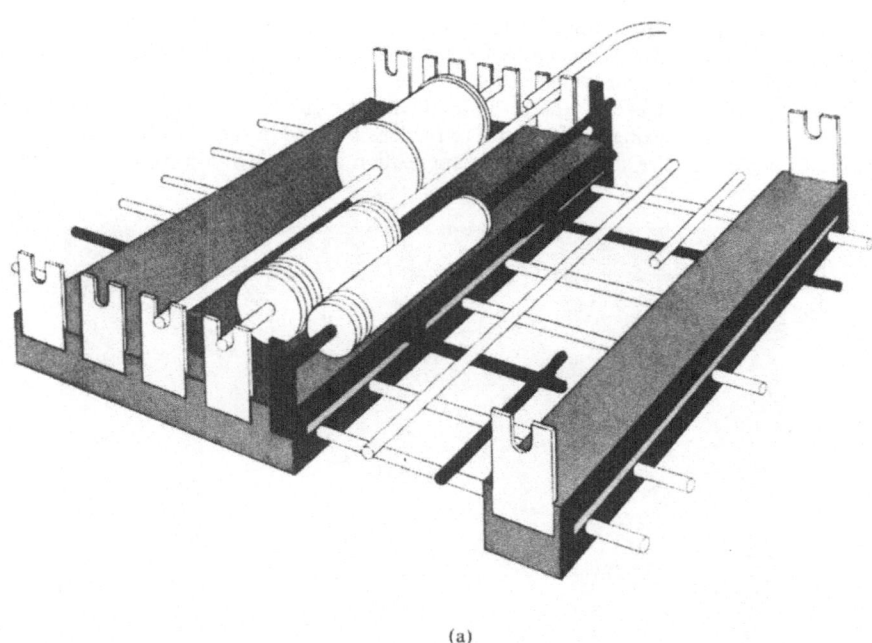

(a)

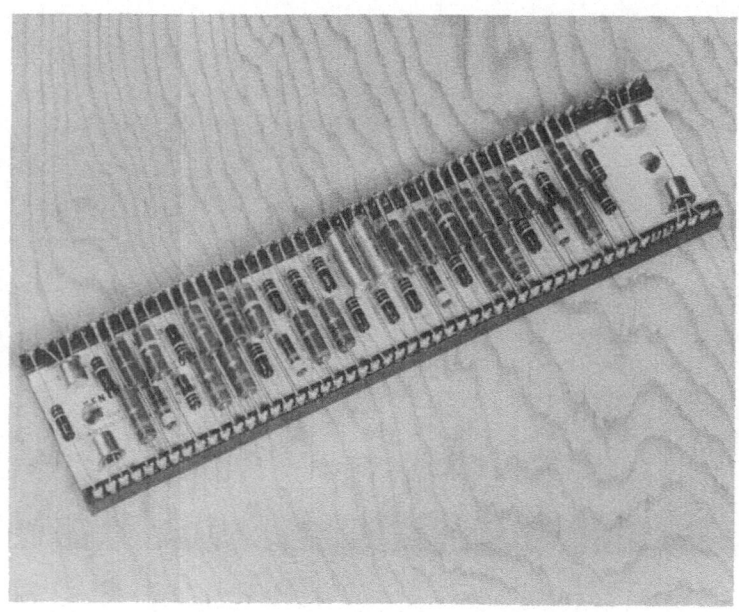

(b)

Fig. 14. (a) Partial illustration of standard welded-wire matrix; (b) standard welded-wire matrix.

The tube faceplate, which has a radius of 57.3 mm, is a rather sharp curve. The normal field curvature of a lens is toward the lens, which means that a strongly negative lens has to be used against the end of the tube to bend the focal plane of the lens to match the sharp curve of the tube. A standard 3-in. $f/1.9$ commercially available objective is used with a standard negative lens as a field flattener. These elements were combined and mounted on the forward end of the camera tube. The $f/1.9$ lens is held in the focusing barrel and the field flattener is cemented in place in the housing so that its position is fixed to the camera tube.

Resolution tests were made on the lens with the field flattener in place. Microfilm emulsion was used at the center of the system and showed the following results:

Position	(Line pairs/mm)
Best focus	151
−0.06 mm	76
+0.06 mm	107
+0.12 mm	60

Obviously, from the above test, the focus of the objective is critical and means for accurately adjusting the focus and locking it were required and have been provided by the focusing barrel and locking ring (see Fig. 6).

EIO TUBE THERMAL ANALYSIS OF HEAT SINK

One of the requirements often quoted for image orthicon tubes is that the temperature difference between the target section and any part of the bulb does not exceed 5°C to preclude or minimize the possibility of cesium migration to the target. The primary source for a temperature differential along the bulb is the concentrated heat at the cathode heater, and the maximum allowable temperature differential along the tube is often controlled by a blower. However, a blower imposes undesirable power and weight penalties on a system in addition to presenting vibration and electrical noise problems. Replacement of the blower by a cylindrical fin mounted around the tube was therefore considered. The fin proposed consisted of an aluminum cylinder 11 in. long and assumed to be closely fitted around the tube envelope for good thermal contact (see Fig. 15). Aluminum was selected because of its high thermal-conductivity-to-weight ratio.

If we assume: (1) Perfect thermal contact between the cylinder and the bulb (this can be approached by using silicone grease between the cylinder and tube); (2) negligible heat conduction along the glass bulb; (3) negligible heat conduction through the leads and connector; (4) the heat transfer coefficient along the surface of the cylinder to the environment constant; then, the maximum temperature differential along the surface of the bulb may be computed from the following equation:*

$$\Delta T = \frac{q}{\pi(D + 2\delta)L_0 U}\left[1 - \frac{\sinh \lambda L_0 + \sinh \lambda L}{\sinh \lambda(L_0 + L)}\right]$$

where

ΔT = maximum temperature difference on the bulb surface
q = heat dissipated by cathode heater
L_0 = effective heater length
L = total fin length minus L_0
D = inside diameter of cylindrical fin
δ = fin thickness
U = heat-transfer coefficient to environment
$\lambda = \sqrt{U/k\delta}$
k = thermal conductivity

* See Reference 1 for a more detailed derivation.

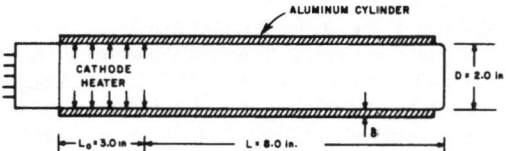

Fig. 15. Illustration of aluminum fin over EIO tube.

Temperature difference as a function of fin thickness calculated from the above equation is shown in Fig. 16 for the following conditions: Cathode heater power, 3.6 W; total length of fin, 11.0 in.; effective heater length, 3.0 in.; inside diameter of cylindrical fin, 2.0 in.; thermal conductivity (aluminum), 100 Btu/hr · ft · °F.

Actual temperature measurements were obtained from a dummy magnetic image orthicon tube using the heater to check the values obtained analytically. The measured values shown by circles and the analytical curve are presented in Fig. 16 for comparison. Note that the measured values are generally lower than those computed. This discrepancy is primarily due to heat conduction through the leads, end connector, and glass, which was neglected in the computations.

Measurements were performed on circular aluminum fins 0.032 and 0.096 in. thick, and with circular magnetic-shielding material 0.008 in. thick. Silicone grease was applied to the bulb to insure good thermal contact. The image orthicon tube was operated in the horizontal position at 20°C with 3.6 W of power dissipated by the cathode heater. Heat transfer was accomplished by natural convection and radiation. The temperature distribution measurements obtained are shown in Fig. 17. (Note also the significant reduction in temperature differential rise on the 0.032-in. fin by coating its outside surface with flat black paint.)

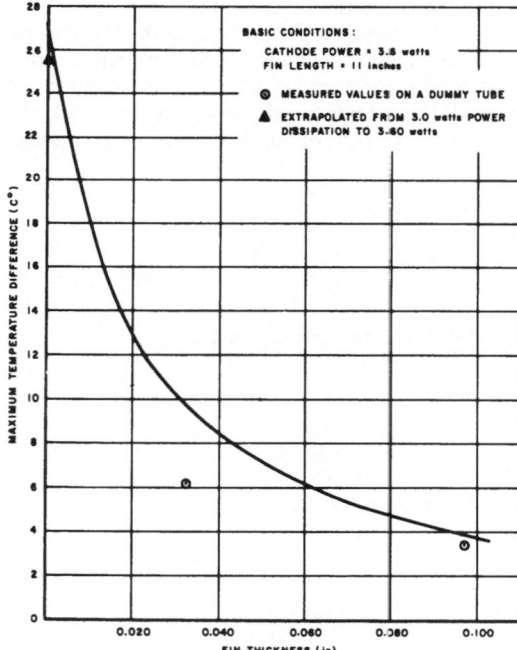

Fig. 16. EIO maximum temperature difference along bulb as a function of fin thickness, illustrating the effect of reducing the temperature differential by increasing the fin thickness.

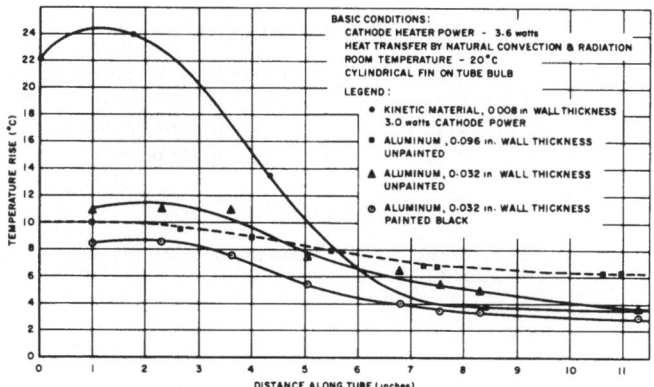

Fig. 17. Temperature rise *vs.* distance along image orthicon tube.

ENVIRONMENTAL CONSIDERATIONS

The operating camera control box was environmentally tested to the following specifications without the benefit of conformal coating or potting: High temperature operating, 120°F for 3 hr; humidity, 95% RH, 80°F for 3 hr; altitude, 30,000 ft, 80°F, 50% RH for 3 hr; cold, −30°F for 3 hr; vibration,* 20–2000 cps ± 1 g, resonance search (transmissibility was less than eight on all major components except the H and V sweep, which went as high as eighteen in an area of about the center of the card).

At the time this report was written, a ruggedized version of the EIO tube was not available and consequently no environmental testing was performed. Image orthicon tubes by virtue of their extremely thin targets and photocathode are limited, however, to low-amplitude vibrational imputs and a maximum operating temperature of 150°F. It is expected that the vibration limitation can be significantly improved with the ruggedized tubes expected in the latter part of 1964, and by designing the system so that the structure is free of mechanical resonances in regions which destroy or damage the target. The optics, preamplifier, and housing are expected to meet the environmental requirements. With a ruggedized tube, the present camera head is expected to meet the following environmental tests: High temperature operating, 160°F; humidity, 95% RH at 80°F; altitude, 30,000 ft; cold, −30 F;† vibration, 5 g's, 20–2000 cps; shock, 10 g's for 11 msec.

SUMMARY

Initially the program required that a low-light-level TV camera and control unit be built incorporating the EIO tube. It would have to weigh less than 20 lb, require less than 30 W, be no larger than 500 in.³, and be suitable for customer demonstration. Since then the results have been significant. The weight was reduced by 35% to 13 lb; the power required was reduced by 27% to 22 W; the volume reduced 2% to 489 in.³; and the equipment successfully withstood the rigors of handling and demonstration.

The completion of the engineering model of the EIO camera represents a firm step forward toward a camera for military usage. However, packaging design should be further improved when the ruggedized EIO tube becomes available in the latter part of this year.

---

* Based on experience with equipment of similar design, it is expected that with proper encapsulation the control unit will withstand a minimum sinusoidal vibration of 20 g's and 11 msec shock pulses of 100 g's.
† NOTE: To meet the cold test, laminated silicone rubber heaters and a thermostat may be required. The weight penalty would be about 3 oz, and the power consumption will depend on the system requirements.

ACKNOWLEDGMENTS

The author wishes to acknowledge the assistance of those whose cooperation has been an asset in the preparation of this paper. In particular, I would like to thank Loren Ford, project manager, for his assistance, Ernest Elovic for his contributions to the thermal analysis, and Kennard Harper who analyzed and specified the optics requirements.

REFERENCE

1. L. Ford, E. C. Decker, and R. P. Warner, "Electrostatic Image Orthicon," General Electric Company, Ithaca, New York, Report No. R63ELC25, November 1963.

The Heat-Sink Module

R. E. KLEIN* AND J. GAMMON†

ITT Kellogg Communications Systems
Chicago, Illinois

This paper explains the bases for the design of a heat-sink module, a module desirable because of its great flexibility in component arrangement and adaptability to new usages. The methods by which the module would meet major design problems, such as heat management, standardization and internal flexibility, and component interconnection, are outlined, and manufacturing and maintenance considerations are discussed.

INTRODUCTION

BECAUSE OF THE GROWING NEED for a reliable, standardized, and flexible system of electronic modules for use in the development of military and "space" electronic equipment, ITT Kellogg has designed a module that has the unique flexibility of accommodating integrated circuits, thin films, discrete components, and small variable elements in one standard miniature assembly size. The new module can operate in most alien environments. This paper is a discussion of the design and assembly techniques which evolved from a study program of the requirements for such a module.

The basic design of the module is shown in Fig. 1. The main constructional features are a central longitudinal heat sink, with notched printed circuit boards on the top and bottom. Components are mounted on both sides of the heat sink, with their leads in the notches of the boards. Because of the internal construction of our module, we refer to it in this paper as the heat-sink module.

FACTORS CONSIDERED IN THE HEAT-SINK MODULE DESIGN

At present, our standard line of digital welded cordwood modules, an example of which is shown in Fig. 2, has been used with a high degree of success on several military contracts. This cordwood module design has been found satisfactory in most applications where minimizing size and weight were important considerations. However, its use has been limited primarily because of two factors: Heat dissipation is a major problem when maximum density and volume efficiency is required, and the module is not sufficiently flexible to accommodate the variety of applications and component arrangements now required. It was therefore apparent to us that the existing cordwood type of design would not adequately meet our requirements. A new concept was required to satisfy all of our requirements.

The following list points out the problems that we considered in elaborating the design of our heat-sink module. The major problems are discussed in the next section in the same order as listed, and the secondary problems are considered in the section after.

---

* Present address: Symington–Wayne Co., Ft. Wayne, Indiana.
† Present address: The General Electric Co., Cocoa Beach, Florida.

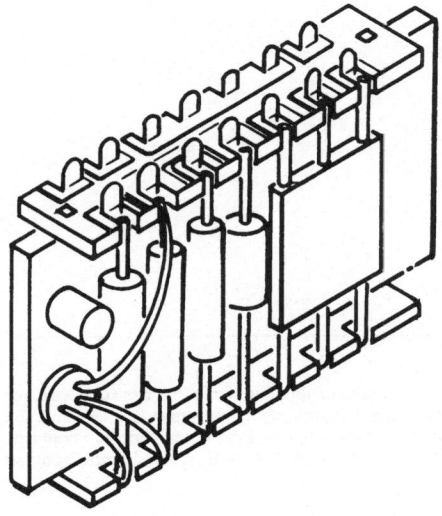

Fig. 1. Basic design for heat-sink module.

Major Problems
1. Heat Management. Designing the module for the adequate dissipation of internally generated heat proved to be a major consideration.
2. Standardization and Internal Flexibility. Standardization of size and construction is inherent in the modular concept itself; this had to be combined with a design that would provide the internal flexibility required for adaptation to many types of circuit elements and functions, both analog and digital.
3. Component Interconnections. This involves the method by which the components within the module are connected with one another.

Secondary Problems
1. Design for System Usage. How the module should be designed for system integration—external connections and mounting.
2. Manufacturing Considerations. The use of simple parts, interchangeability and ease of assembly.
3. Maintenance Considerations. The ease of testing and replacement of modules.

In addition, there were also the obvious considerations of small size, light weight, reliability, and cost.

HOW THE HEAT-SINK MODULE COPES WITH MAJOR DESIGN PROBLEMS

Heat Management
The term "heat management" refers to the many and complex considerations, both electrical and mechanical, involved in providing the required heat dissipation, consistent with other design requirements.

Circuit Factors Related to Heat Management. Since transistor circuits are widely used in miniature modules, the factors which affect the generation of heat in a transistor junction are of primary importance in heat management. In general, heat generation is a function of the mode of transistor operation (switching, amplification, etc.), the level of bias, and signal applied to the transistor.

Operation of a transistor as a saturation device, or as a Class B amplifier, results in less junction heating than results from operation as a Class A amplifier. When used as a saturation

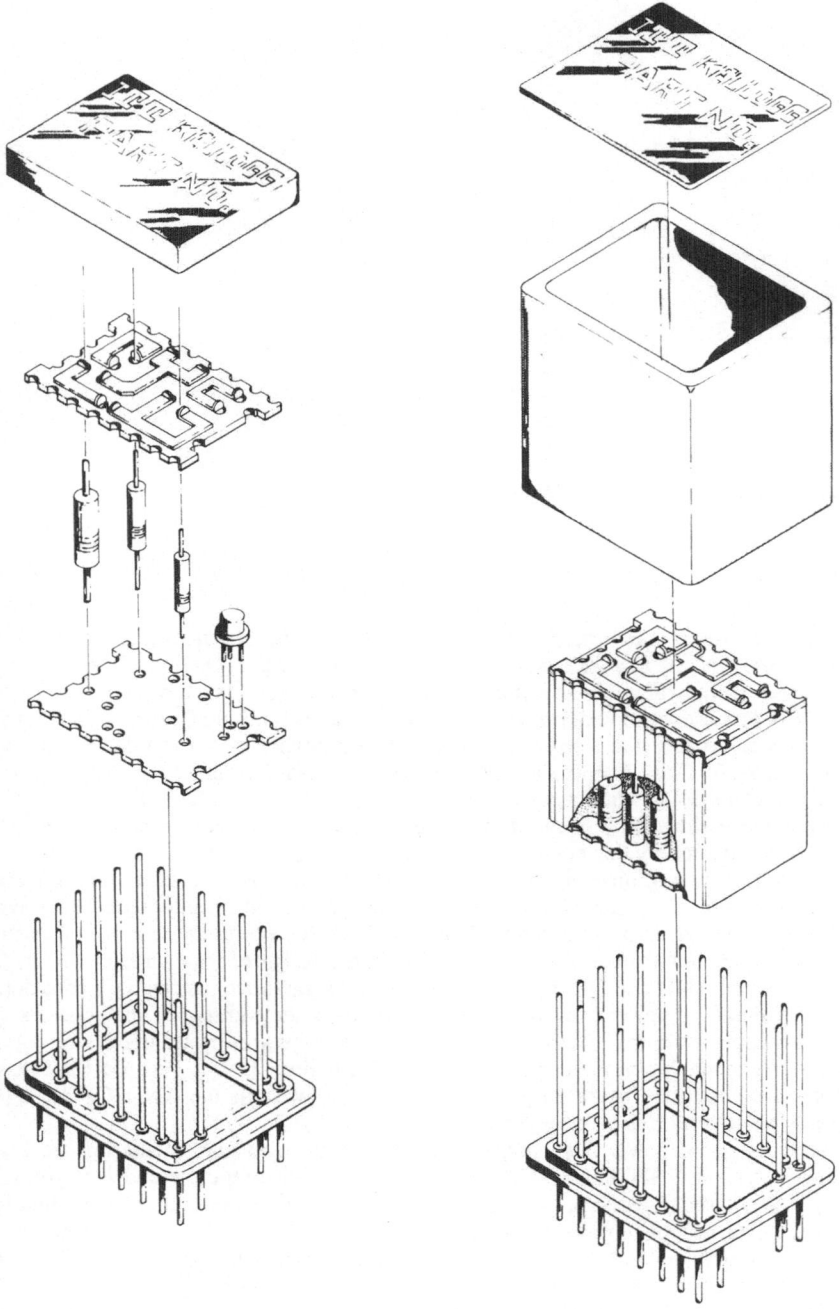

Fig. 2. Standard welded cordwood module.

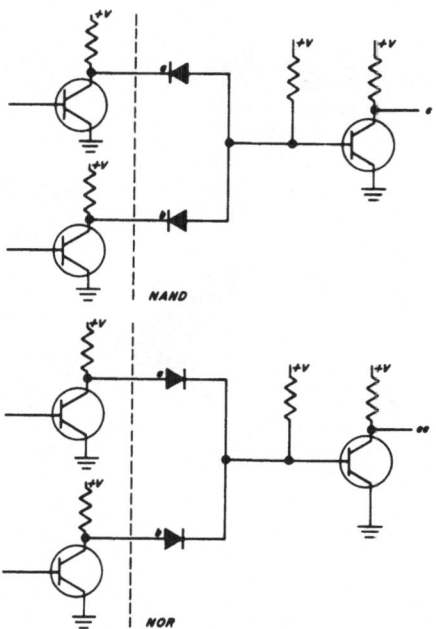

Fig. 3. Typical NAND and NOR gates.

device, the transistor is alternately cut off and saturated. When cut off, very little heat is generated, because of the low leakage current flowing through the junction. When saturated, the resistance of the junction is so low that practically no voltage is developed across it to cause a power loss. Class A operation of transistors results in the dissipation of power in the transistor junction. This is because the transistor is operating essentially as a variable resistor rather than as a switch. In Class A operation, the transistor is biased to maintain operation in the linear portion of its I_C–I_B characteristic curve. Thus, there is a no-signal collector current flowing which is equal to about one-half of the peak value of the collector current. The power dissipated in the junction by this no-signal current is in the form of heat.

For good heat management, it would be desirable for all power to be turned off when a system is in the quiescent state. Since this is usually impossible or impractical, the system designer must select types of circuits which provide both the optimum heat management as well as the required circuit functions. For example, in the design of logic circuitry the method of implementing the AND and OR (or NAND and NOR) logic functions can have considerable effect on the amount of power consumed and dissipated by the transistors and resistors. The power consumption is approximately the same, but the dissipation by transistors in NAND–NOR circuits is much lower than that in AND–OR circuits. This is because the NAND–NOR usage involves either saturation or cut-off operation, whereas the AND–OR logic (e.g., emitter followers) operates in a linear mode.

For an example of power heat management by system logic implementation see Fig. 3, which shows a typical NAND and NOR gate. In order to have inputs a and b at ground or zero voltage, the driver transistors, which may be considered as the outputs of preceding logic blocks, must be turned on, drawing maximum current. If the gates have many inputs, the current drain through the driver transistors may be considerable. This can be avoided by inverting the circuit functions while still providing the same signal at output terminal C or CC by using NAND's for NOR operations or NOR's for NAND operations.* In this manner, the

---

* The interchange of NAND's for NOR's can be justified by the Boolean equation:

$$\overline{a + b + c} = \bar{a} \cdot \bar{b} \cdot \bar{c}.$$

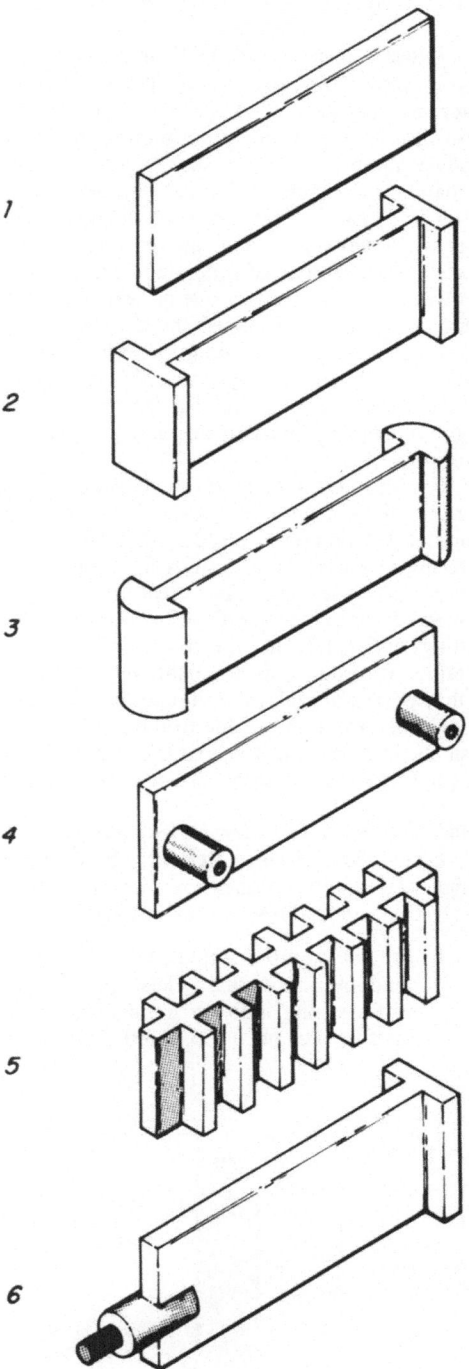

Fig. 4. Different heat-sink diagrams.

quiescent state of the inputs will be at a high voltage, requiring the transistors of the prior stage to be turned off, thus drawing negligible current.

Other Factors Related to Heat Management. After the system designer has determined the circuit types to use, the packaging engineer must consider the possible heat sources and must place the heat generative elements as close as possible to a low-resistance thermal path. He must also provide a minimum number of thermal junctions between the heat source and the module mounting. Where such junctions exist, they should have a mechanical load applied to maintain a minimum thermal resistance across the junction.

The physical arrangement of the components, shape of the heat sink, heat-sink material selected, adhesive, coating and potting compounds used, and heat-sink orientation, are all related to the control of generated heat.

We have investigated several heat-sink configurations, some of which are shown in Fig. 4. The investigation considered the thermal resistance of junctions with large surface areas but with small mechanical loading forces versus small surface areas with large loading forces. The loading is, of course, physically limited by the strength of the assembly frame. We are still studying this problem; but for the present, the flat plate design is preferred because of the possibility of improved shields and ground in conjunction with the high mechanical loading possible.

For the heat-sink module, we selected type 6061 aluminum for the heat-sink material, because of its high heat conductivity, its good structural properties, and its light weight. During the evolution of our design, it was found that heat dissipation and component retention were improved by using a silver-filled plibond adhesive to retain the components in intimate contact with the heat sink. Heat transfer was somewhat further enhanced by coating the assembled components with a modified RTV compound and potting the entire module. For potting, either a foam-type compound or a glass-filled or aluminum-filled epoxy was found suitable. In order to save weight, we decided to use the foam-type potting compound instead of the epoxy material, even though the heat dissipation of the latter is slight better. (The effect on module heat transfer due to the difference in heat conduction properties of these materials is small, since most of the heat is dissipated through the component body and is conducted by the plibond to the adjacent surface of the heat sink in either case, as shown in Fig. 5.)

Standardization and Internal Flexibility

Dimensional Elements Related to Standardization. Our method of standardizing the module design started with a dimensional analysis of the component types to be packaged. The steps involved in the analysis were as follows:

1. The selection of components of standard size, where possible. The dimensions of semi-conductor networks, pellets, and thin-film components are being standardized. There is also a strong industry trend toward standardizing discrete component diameters and lengths, such as 0.050, 0.100, 0.250 in. diameters and 0.100, 0.250, 0.50 in. lengths.

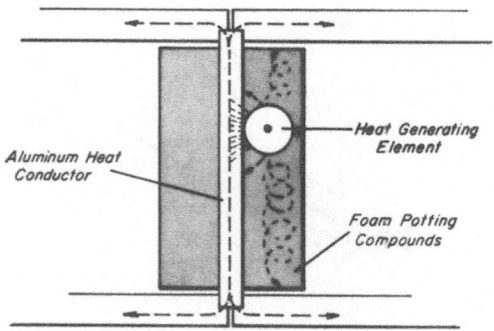

Fig. 5. Heat flow in the heat-sink module.

2. To minimize the total number of different dimensions of components. Higher priced components can often be obtained in sizes smaller than standard off-the-shelf components. In those cases where necessary, such components should be designed into low-usage circuits.

Thus, standardization can be achieved by considering the major usage circuits to establish the module dimensions. At a slightly increased price, special circuit modules can then be designed to these dimensions.

Optimum module dimensional increments were determined by referring to several published analyses and development techniques. The standard dimensional elements chosen are 0.050, 0.100, and 0.250 in. The dimensions of the completed module were determined by an evaluation of the standard circuits to be packaged and by the use of these dimensional elements.

Circuit Factors Related to Standardization. The determination of standard module dimensions was influenced by the number of integrated circuits required per module, the size of thin-film circuit elements, and the number of standard discrete components needed for the various circuits. For the purpose of the size analysis, the following circuits were used:

1. Six integrated, five-input NOR gates, in "flat pack" configuration.
2. Four thin-film NOR gates with standard semiconductors (see Fig. 6, which shows one of the gates).
3. A dual four-input NOR gate with standard components (see Fig. 7).
4. A mixer module with tunable transformers (see Fig. 8).

The simplest module design was one using the integrated circuits. Three NOR gates were placed on either side of the heat sink. The length of the module was 0.85 in., with a 0.050-in. clearance between the flat packs. A 0.050-in.-thick aluminum plate was chosen as the heat sink. This module had the smallest size requirements of any investigated: It would require a volume of 0.85 by 0.25 by 0.21 in. These tentative dimensions were replaced, however, because of the requirements of the other module circuits, as described below.

Figure 6 shows a NOR gate which comprises one-fourth of the next module considered. These thin-film gates use standard TO-18 transistors and 0.85-in.-diameter diodes. The four transistors required a heat-sink area of 0.25 by 0.9 in. With the module length defined as 1 in. and the height set at 0.50 in., the transistors could be placed in two pairs at the ends of the heat sink. There was thus an area of 0.55 by 0.50 in. on each side of the heat sink for two

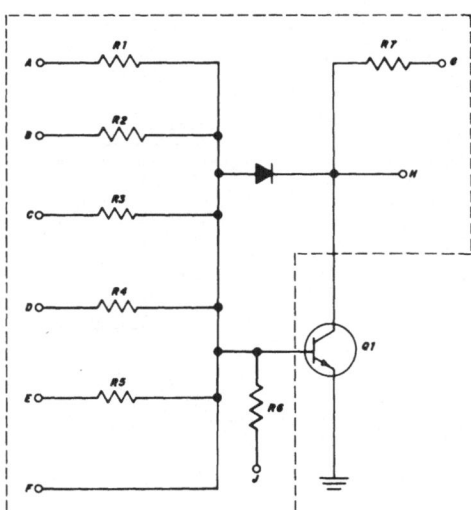

Fig. 6. Thin film NOR gate with standard diode and transistor.

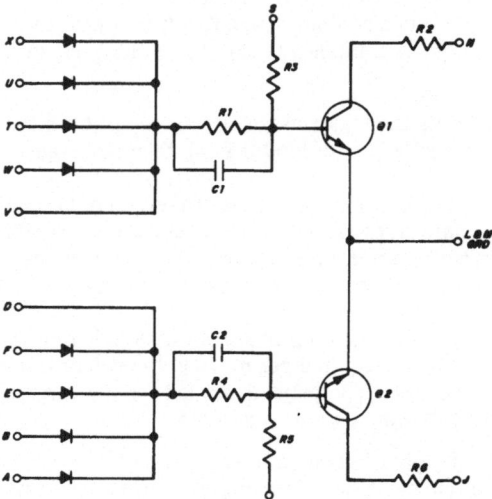

Fig. 7. Dual four-input NOR gate with standard components.

thin-film circuits, and also space for the four diodes. On this basis the module size was re-established as 1 in. long by 0.50 in. high by 0.25 in. wide.

The dual four-input NOR gate made of standard components was the next most complex circuit considered. (See Fig. 7.) Both transistors were placed at one end of the heat sink and required a length of 0.21 in. The resistors used 0.30 in. and the diodes required 0.40 in. All of the components fitted within the same volume as did the thin film NOR gate components: 1 in. by 0.50 in. by 0.25 in.

The mixer module with tunable transformers is shown in Fig. 8. For this circuit the component quantities and sizes are as follows:

Tunable transformers, two required, $\frac{1}{4}$ by $\frac{1}{4}$ by $\frac{1}{4}$.
Capacitors, five required, 0.10 by 0.25 in.
Resistors, four required, 0.070 by 0.0187 in.
Transistor, one required, standard TO-18.

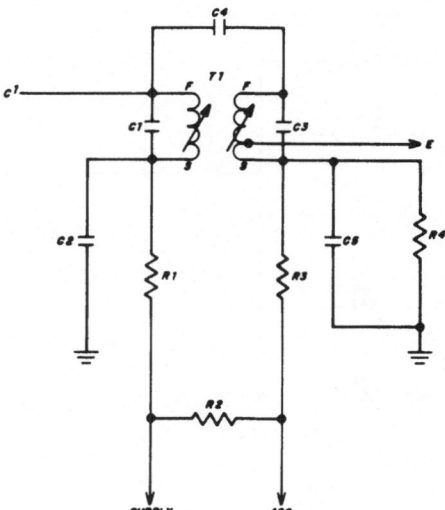

Fig. 8. Mixer module with tunable transformers.

The transformers were placed through the heat sink at opposite ends of the one-inch dimension, and the transistor was located next to one transformer. Elements C1, C2, C5, and R1 were placed on one side of the heat sink, and C3, C4, R3, and R4 were located on the other side. The circuit assembly for this module also fits the proposed standard dimensions of 1 by 0.50 by 0.25 in.

Component Interconnections

The choice of printed circuit boards with notches to accept component leads solves several interconnection problems within the module. This method requires no preforming of the component leads, and it eliminates most preassembly (prior to welding) operations. It also allows considerable circuit flexibility, the use of different printed circuit boards, in conjunction with the required components, to accomplish different functions. If necessary, wires can be connected from certain contacts of the top board to those of the lower one, so that all external connections could be made from the bottom of the module.

Because of the compactness of the component assemblage, welded or soldered component connections are required. We selected the welding method because of better reliability, lighter weight, and because of degradation of soldered contacts in a high-vacuum environment such as space vehicles would encounter.

Figure 9 is a sketch of the type of printed circuit board used in the heat-sink module. It is a multilayered board, notched, etched, and tabbed. Several companies, such as ITT Kellogg, Borg-Amphenol, Sanders Associates, and Methode, have the capability of manufacturing these items.

If connections within the module become so complex that the multilayered boards in themselves are insufficient, the problem could be resolved by placing an interconnection tape from the top board to the bottom board outside of the components after they are installed. This would make manufacturing and assembly somewhat more difficult. Therefore, any scheme of this nature should be carefully considered. We have reviewed several published concepts very similar to this except that the interconnection tape is placed between the components and the heat sink. In these cases, the manufacturability problem is solved, but good heat management is more difficult.

THE HEAT-SINK MODULE

The heat-sink module is, we believe, a successful attempt to cope with the problems encountered in designing a new versatile module system. While retaining all of the advantages of the cordwood module, the heat-sink module eliminates most of the undesirable characteristics of the former and presents several distinctive system characteristics.

Design for System Usage

The interconnection and mounting of modules to form a system is, of course, integral to the design of the module itself. Not only must positive contacts be made for electrical and thermal connections, but the technique of assembly must allow for relative ease of module replacement.

Figure 10 illustrates a solution to the problem of obtaining solid, positive connections for both electrical and heat transfer junctions. If the method of mounting shown in the upper portion of Fig. 10 were used, with a single force F1 accomplishing both the heat-sink connection at A–B and the electrical connections at C–D, there would be no means of controlling these two independently. We have preferred, therefore, an arrangement like that shown in the lower portion of the figure. Here the thermal junction surfaces are held together by the force F2, while the electrical connections are made, independently, by the force F3.

Figures 11 and 12 show a method of system assembly using modules with flat heat sinks and flexible printed wiring contacts on both sides. Silicone rubber inserts are installed over ribs on the module mounting rail. The multilayer flexprint is then looped down between the inserts. Each module is pressed down between two of the adjacent rubber inserts. This provides the electrical interconnection force F3 shown in Fig. 10. Then the loading rail is attached to the

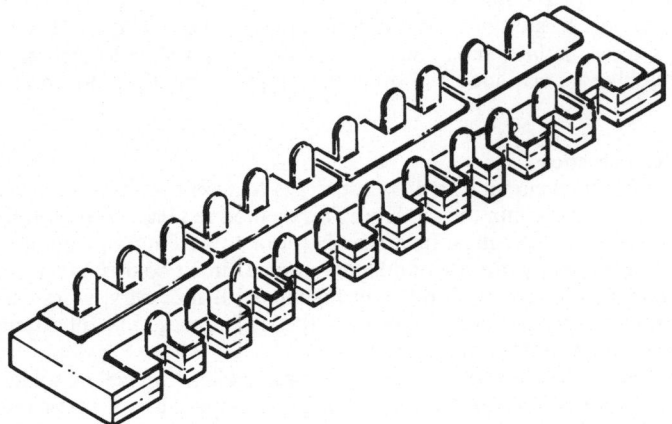

Fig. 9. Type of printed circuit board used in heat-sink module.

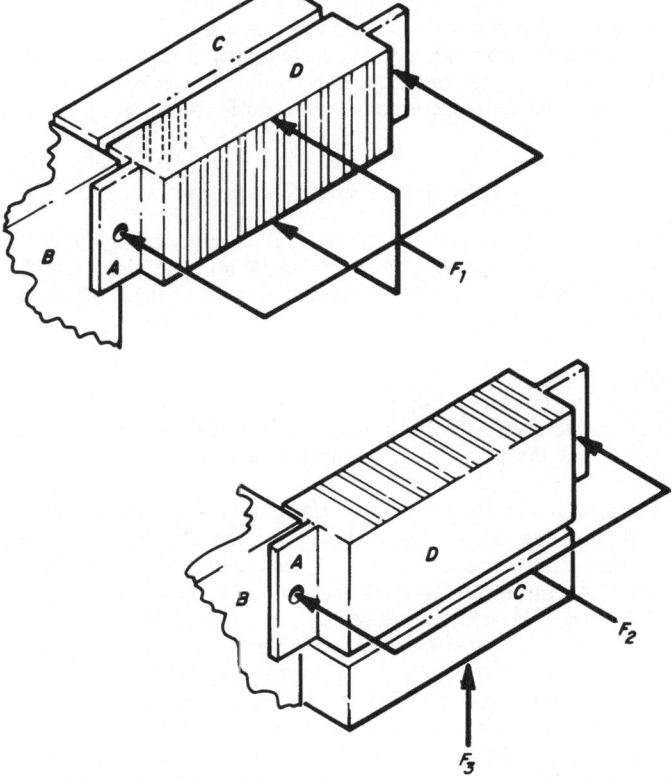

Fig. 10. Force diagram of electrical and mechanical loading.

module mounting rail by means of screws. This accomplishes the heat transfer connection indicated as force F2; it also provides a very solid and rigid method of mounting. The top half of Fig. 12 illustrates the front and rear of such an assembly; the bottom figure shows the same assembly with the two loading rails detached and two modules removed.

Manufacturing Considerations

The heat-sink module uses a minimum number of parts, and these are relatively simple. The flat heat sink is, of course, the ultimate in simplicity, as well as being applicable to many different kinds of module circuit elements. The printed circuit boards are single, vendor-obtained items. These are the basic parts of the module—the heat sink and the two boards—in addition to the components themselves and the interconnecting medium. There is nothing intricate or complex about any of the module constituents.

In the area of interchangeability, the same basic, flat heat sink can be used for all of the different kinds of circuit elements required. Also, the physical dimensions of the printed circuit boards would be the same for many different kinds of modules, possibly for all of them. Since the only differences from one module to another would be in the printed wiring, this would simplify manufacture for the printed circuit board vendor.

The module is quite easy to assemble since there is no need for preforming leads and no requirement for any preassembly processes. There are only a few simple steps needed to assemble the entire module—staking the boards on the heat sink, welding the components into place, and providing the external connections. Also, we believe that with automatic loading and welding equipment used the heat-sink module will be within the manufacturing price range for commercial applications.

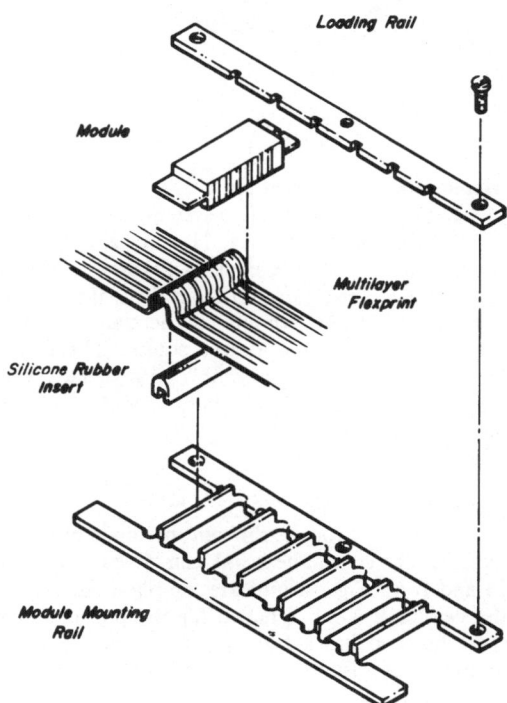

Fig. 11. Example of module assembly technique.

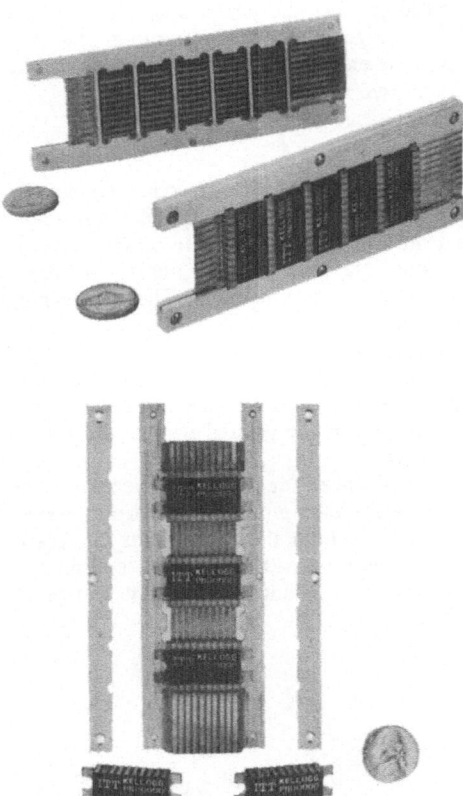

Fig. 12. Example of module assembly.

Maintenance Considerations

Replacing a module in a system can be accomplished in a minimum of time. In the example of assembly shown in Figs. 11 and 12, it is only necessary to remove the two loading rails, which bear against the ends of the heat sinks, to make the modules accessible. Any of the modules can then be pushed out from the rear, without disturbing the others. Also, with the flexible wiring used for interconnection, it is very simple to make all of the connections to the modules available as test points. The flexible wiring could be extended on the top or bottom of the module assembly to a surface where all of the leads could be exposed. Thus, with the heat-sink module, both module replacement and input–output testing become quite simple.

Variations in Design

To provide for the different methods of mounting and different degrees of heat dissipation required, the central heat sink can have any of several shapes. The modules shown as B on Fig. 13 and B on Fig. 14 indicate the plain flat heat sink also shown in Fig. 15. In A of Fig. 13, the heat sink is "T" shaped for the purpose of interconnection with a larger heat sink, and the other end has a bolt-type provision for mounting. Item C of Fig. 13 shows sleeve-type mounting provisions on both ends of the heat sink for the purpose of direct chassis mounting. Items A, C, and D of Fig. 14 show heat sinks with both ends enlarged for the purpose of

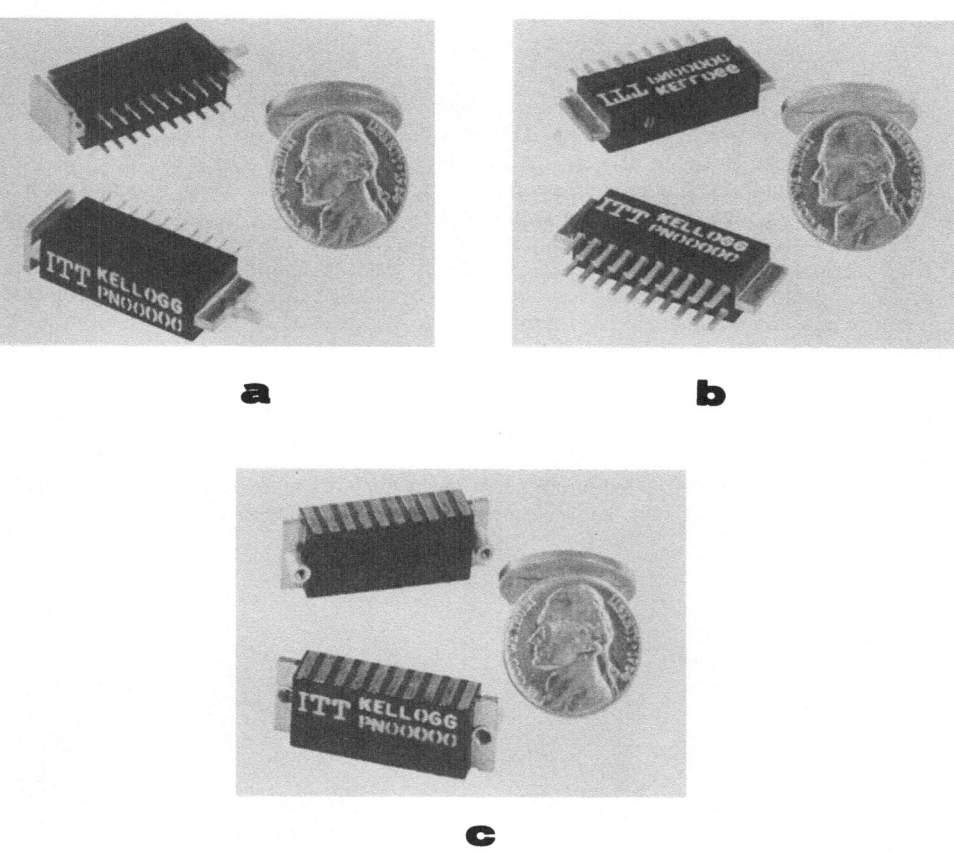

Fig. 13. Heat-sink modules.

increasing the mass of the heat sink at the point of thermal interconnection. This added mass minimizes the possibility of damage to components in the module from external transient thermal pulses.

The possibility of controlled dimension variations is also considered in the design of the new module. Although the length dimension is held constant at 1 in., either or both of the other dimensions can vary according to certain definite increments, to accommodate different component arrangements.

The mounting of components in relation to the heat sink can also vary, according to the shape and heat generation of the component. Transistors, for example, can be mounted in holes in the heat sink or inserted within a collar which would be prewelded to the heat sink or stamped from the heat-sink material.

There are several variations possible in accommodating the external design connection to the module. Flexible wiring can be wrapped around the module and soldered to the module terminals on the top and bottom of the module, or on only one of these surfaces. Most of the modules depicted on Figs. 13 and 14 show this type of external connection capability. The terminals can be brought out as plug terminals on one surface of the module (see A of Fig. 13) or the module terminals can be brought out for wire-wrap connections, as shown in B of Fig. 13.

Derived Features

The heat-sink module has heat management as its primary design consideration; however, many useful features are inherent in a low-thermal-resistance system. Shielding, self-contained within the module, is just one of these features. Such shielding could be augmented by inserting small metal plates between the modules. This provides isolation between modules in addition to the self-contained shielding between the two sides of the module. By using the heat sink as a common ground point, a good, repeatable ground path can be achieved, thus making it a likely candidate for high-speed logic or high-gain analog circuits.

The hot spots normally experienced in modular packaging are nonexistent in our module. The low thermal resistance between the internal circuit elements and the heat conductor causes a smaller temperature differential between the various circuit elements, thus simplifying temperature compensation. The thermal lag from turn-on to stabilization is much less than in most potted modules.

SUMMARY

We have, of course, considered future developments in working out the design of the heat-sink module. As Fig. 15 shows, and as other portions of this paper have brought out, the

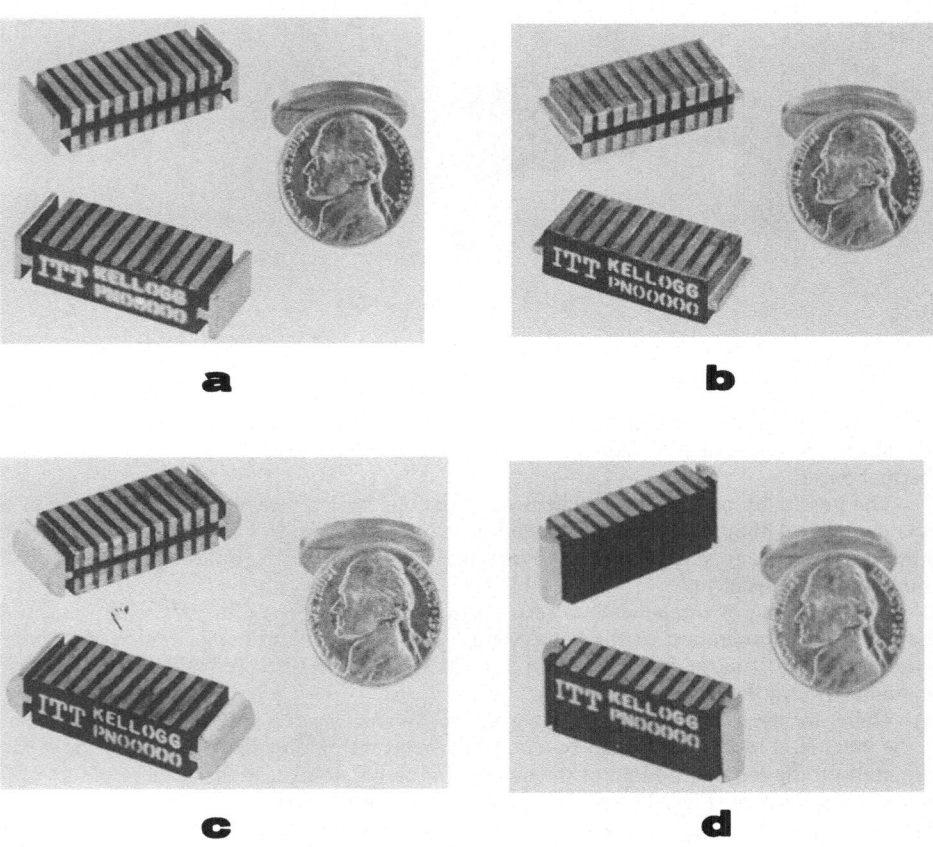

Fig. 14. Heat-sink module variations.

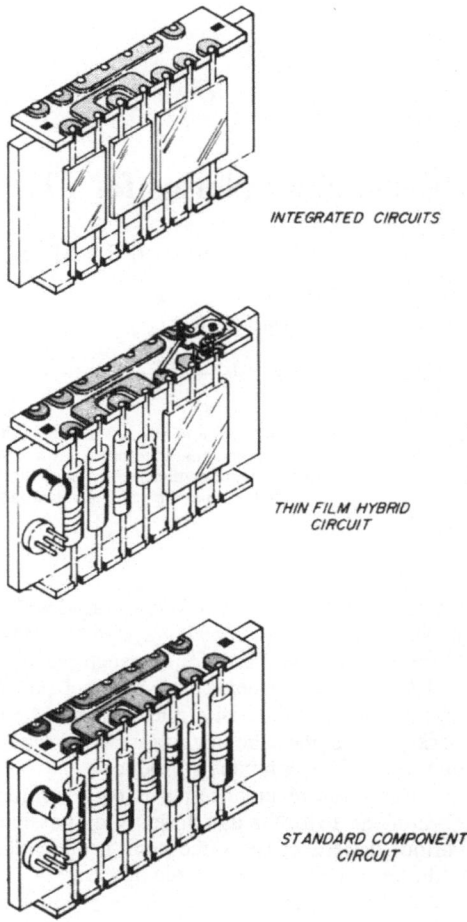

INTEGRATED CIRCUITS

THIN FILM HYBRID
CIRCUIT

STANDARD COMPONENT
CIRCUIT

Fig. 15. Internal module flexibility.

module allows for great internal flexibility of component arrangement. Because of the diversity of possible applications, because of its adaptability to new usages within controlled modifications of the dimension standards incorporated, and because of the solution which the module offers to numerous and conflicting design problems, we believe that the heat-sink module definitely has a future.

Producibility Norms for Electronic System Packaging

E. I. MOORE AND L. M. SCHNEIDER

The Martin Company
Orlando, Florida

Early evaluation of the producibility of a systems packaging concept can result in considerable cost savings to manufacturers. It is essential that a team effort be conducted among design and manufacturing engineers to assure that the end product can be produced using techniques, processes, and equipment which are available to manufacturing.

To REMAIN COMPETITIVE in today's rapidly changing electronics industry, a company's products must, in addition to meeting system functional requirements, be capable of being produced at minimum costs. Electronic system packaging, the operation of translating circuit diagrams and system requirements into physical hardware, has the largest single responsibility in the determination of the cost of the finished system to the customer. It is therefore essential that the packaging engineer be cognizant of the many facets of producibility which influence the end-item cost of his company's products.

Furthermore, engineering is by definition the applied science of causing desired physical phenomena to occur as efficiently as possible. Using this yardstick as a tool for evaluating various designs of the same item, producibility becomes important as a professional ideal, in addition to being an economic necessity.

As implied above, a requisite of producibility is close coordination early in the design phase of the project between packaging and manufacturing engineering personnel. Manufacturing has the responsibility of keeping packaging abreast of the facilities, tooling, and operator skills available for production, as well as the relative costs of the several methods for fabricating the required hardware. Packaging has the responsibility for notifying manufacturing, as early in the project as possible, of the skills that will be needed in the production phase. This will allow them to be fully developed when needed, minimizing the learning costs, costs of redundant operation and effort, and the scrap and rework costs.

The additional benefits of this approach are many: (1) Early screening of the several approaches to packaging will enable a more comprehensive qualification program to be run, including perhaps alternate concepts, with the final decision to be made on the basis of performance and cost rather than on performance alone. This may also eliminate the need to requalify the circuit as a result of a cost-reducing design change during the pilot production phase. (2) Early coordination between packaging and manufacturing will allow equipment procurement, tool fabrication, and process–plan generation to occur in parallel with design detailing, shortening the overall project span time. (3) Manufacturing follow-up during the detailing phase of the project should minimize the rash of engineering changes which usually result when a fresh design hits the production floor. By making these changes prior to documentation release the high costs of formal documentation change are eliminated. The decisions required early in the project to assure a producible end item are graphically depicted in Figs. 1 and 2.

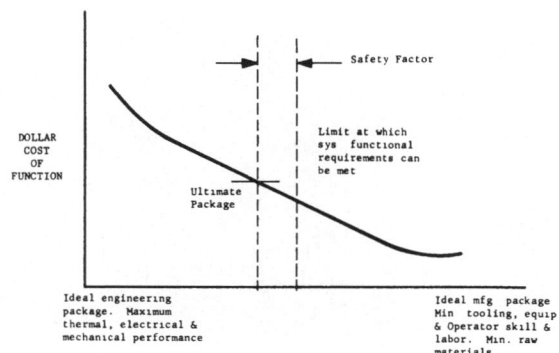

Fig. 1. Determination of packaging philosophy and configuration.

The horizontal axis of Fig. 1 is a continuum of packaging concepts for a given system having, on the one hand, an ideal engineering package with maximum heat transfer, mechanical strength, and electrical circuit properties. This package may be impossible to fabricate. On the other hand, it is the ideal manufacturing package which can be assembled by untrained operators using no tooling or equipment and minimum cost raw materials. No inspection steps or prepared processes are required with this package. Obviously it will not meet the system requirements.

Engineering judgment and experience on the part of both the packaging and manufacturing engineer are required to determine the point along this continuum at which a functioning system can be manufactured at minimum cost. A reasonable safety factor, also chosen by experience, will lead to the ultimate package and to the manufacturing and quality procedures to be used for its production. A value-analysis approach would refer to the cost of the ultimate package as the value of the function it is to perform, all costs above this value being superfluous.

The expected production quantity of the item must be factored into the determination of the ultimate package as it will affect the operator-to-tooling interrelation used in production.

Figure 2 indicates the extent of tooling required to produce a given quantity of a product at minimum cost. The use of less tooling and equipment than that indicated by the curve leads to higher labor, inspection, and rework costs per module. The use of more tooling and equipment than that indicated by the minimum point on the curve leads to a higher equipment amortization cost per module, which more than offsets the lower labor, inspection, and rework figures. Obviously, the greater the number of units to be produced the lower the equipment cost per unit. In most cases this will justify the use of more equipment and mechanization. As the feasibility and cost of mechanizing the manufacture of various packaging concepts differs, the quantity to be produced will influence the choice of package. The manufacturing engineer must supply this information to the packaging team.

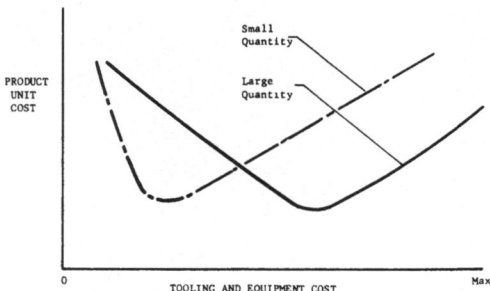

Fig. 2. Product cost as a function of production quantity and equipment used.

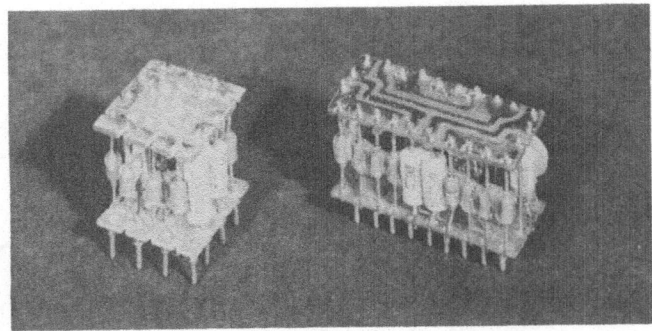

Fig. 3. Four-input gate circuits. Module "A" (left) random load.
Module "B" (right) edge load.

The manufacturing engineering representative, or producibility engineer, on the conceptual packaging team must provide factual cost data to allow intelligent decision making, guide the design and drafting details away from high-cost pitfalls, and educate engineering personnel in the manufacturing capabilities at their disposal. It should be pointed out that friction which develops between manufacturing and engineering can be healthy, for if manufacturing is weak, the product will tend to be overdesigned for the function it is to perform, and thus will be overpriced. If engineering is weak, the reliability of the product may be endangered by numerous concessions to lower cost. The "battle" must be fought, not dodged, and constructive trade-offs made in order that the final design be functional yet producible at minimum cost.

To illustrate in more detail our philosophies on producibility let us consider a hypothetical case study. A comparison between two alternate techniques for packaging the same circuit is shown in Figs. 3 and 4. Both configurations are normally soldered or plugged into a mother board for interconnection with other modules of the system. The techniques for fabricating the modules are different, however, and this is what we are considering in our conceptual packaging team's interchange of ideas. It is this close coordination between manufacturing and packaging which determines the necessary trade-offs for producing the most economical end product which will be consistent with functional requirements. Module "A" in both cases is a combination of random and edge-loaded components, while Module "B" is strictly edge loaded. A general critique of the trade-off inputs for producing the modules would be as follows.

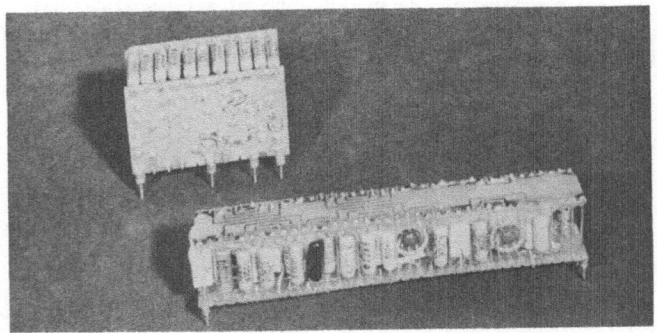

Fig. 4. DC amplifier circuits. Module "A" (left) random load. Module
"B" (right) edge load.

Module Size

The size of the module must be consistent with functional requirements. It must be determined if one circuit or several circuits will be packaged in one module. If one circuit per module, the throw-away cost will be less, but this will result in more interconnections on the interconnection board, which in effect decreases reliability. One circuit per module may increase the number of module types which in turn increases tooling and spares, and affects logistics in general. On the other hand, the system may require electrically that several circuits be closely integrated into one minimum size module. In this case there would be fewer interconnects on the mother board because of fewer pin-out requirements per functional group, many interconnects being made inside the module itself. Another consideration is the form factor of the module, i.e., square, long, flat, etc. The square as opposed to a long or a flat may result in less tolerance buildup in pin-out locations, thereby lessening the need for close-toleranced tooling for interconnection board fabrication.

Module Repairability

The need for module repair or rework is an important consideration in determining the technique to be used. Application of the system, ground or air, may have a bearing on this. Test philosophies associated with the program determine to a great extent the repairability needs of the modules. The modules may have to be tested before and after encapsulation, or in the case of high production, low throw-away costs, only one test may be used, either before or after, since the costs of repair may outweigh the initial module assembly and component costs. However, in normal programs there is a need for rework and repair usually as an in-line function. Techniques for repair must be considered along with module assembly techniques. It is obvious that repairs or rework should be kept to a minimum by the proper application of components and good assembly practices. Also, the type of encapsulation used must be consistent with the repair philosophies. Throwing away a repairable module because of lack of forethought is expensive, and much more important, foolish. Repairability requirements must be scrutinized carefully and then followed.

TECHNIQUE

Random Load (Module A) vs. Edge Load (Module B)

The choice of assembly techniques is perhaps the major contributor to electronic circuit module producibility. Here the producibility engineer will have to determine where, along the manufacturing-cost continuum, the alternate techniques will fall. This is necessary so that intelligent trade-off decisions can be made for the system to fulfill functional requirements at minimum costs. Let us consider some of the requirements of the hypothetical case techniques. (See the chart in Table I.)

It can be concluded from the chart that an edge-loaded module is much more producible than a random-loaded module, when considering manufacturing costs alone. However, it will be up to the conceptual packaging team to determine the trade-offs which are necessary to best fit their program and company situations.

TABLE I

Requirements	(A) Random	(B) Edge
1. Edge-load tooling	Yes	Yes
2. Center-load tooling	Yes	No
3. Intensified operator training	Yes	No
4. More detailed assembly write-ups than normal	Yes	No
5. Increased time for rework and repair (center-loaded repairs)	Yes	No
6. Readily adapted to mechanized production	No	Yes

Producibility

Attainment of highly producible end products is a task requiring full cooperation between a company's engineering and manufacturing departments. Experience with a particular technique may be the key factor in the determination of which way to go. Contractual requirements is another; that is, will the program be relatively short-run or long-term high production. For high production runs it might be well to consider packaging techniques which are adaptable to some higher degree of mechanization. A knowledge of in-house facilities is essential in this case to determine what additional equipment costs must be considered. In the case of the hypothetical illustration, the additional tooling, assembly, and repair costs involved in the random-loaded technique would have to be more than justified by system requirements (module size, electrical characteristics, pin-out arrangements, etc.) before an intelligent decision to use that technique could be made. Knowledgeable engineers—electronic, packaging, and manufacturing—working in harmony from the early phases of a project, can contribute more than any other one factor toward the enhancement of a company's competitive ability through the influence of good producibility practices.

Development of Packaging Techniques for a 960-Bit Plated-Wire Memory

GILBERT R. REID

Electronic Packaging and Connections Unit
Univac, Philadelphia, Pennsylvania

On the basis of data gained from development and testing of packaging techniques used in the 960-bit memory it has been proposed that a complete buffer memory system be constructed by use of the same techniques. Some of the characteristics of the proposed system are as follows: (1) 100,000-bit storage capacity. (2) 100-kilocycle serial-bit information rate. (3) Nondestructive readout. (4) 0.3 W maximum power consumption. (5) Approximate volume of 145 in.³ (6) Approximate weight of 3.5–4 lb. (7) Operating temperature range of $-20°C-+100°C$.

INTRODUCTION

THIS PAPER DESCRIBES the approach used in the packaging of a 960-bit plated-wire memory. The presentation is arranged so that each of the three main components of the system is discussed separately; these components are the circuit modules, ground plane, and backboards. An overall picture of the assembled components is shown in Figs. 1 and 2.

Attempts were made to achieve higher-than-normal component density in packaging the circuit modules; however, the density was limited by the use of standard components. The use of TO-5 cans was particularly nonconducive to high-density packaging, but we were unable to obtain proper transistor parameters in a smaller package; as a consequence, many of the modules have very low component density.

An appendix is provided which described a procedure to be followed for lamination of a multilayered backboard.

CIRCUIT MODULES

The memory consists of a total of 67 circuit modules. Consideration was given to various means of module construction before the selection of a final construction technique. The technique finally used for the module construction was determined by the ability of the module to satisfy the following requirements:

1. Be able to contain all the components in the smallest practicable volume.
2. Be able to be freely removed from and inserted into the backboard.
3. Be repairable at component level before potting or encapsulation.
4. Be rugged enough to withstand vigorous handling both before and after encapsulation.
5. Be convertible to modules which could withstand both the rigors of launching into outer space and continued operation there.

The first requirement eliminated the use of conventional mounting of components on printed-wiring boards. A choice had to be made between the use of vertically mounted components in cordwood construction, and horizontally mounted components in cordwood

63

construction. Because of the advantages in system packaging, horizontal cordwood packaging was selected.

The requirement of freely inserting and removing modules restricted the design to a limited number of choices in commercially available hardware. Available two-piece headers and inter-mating connectors added unnecessary volume to the system and did not provide convertibility to a flight model without requiring significant engineering changes.

The final selection depended upon the use of a wire-formed male contact and a metal female sleeve. The sleeve was soldered into the backboard to allow insertion and removal of the modules. The wire-formed male contact was molded directly into the module header. The female sleeve was later discarded in favor of direct insertion into a multilayered board. Figure 3 shows female sleeves soldered into a multilayered board.

A soldering operation was selected which uses redundant printed wiring rather than welding for interconnection of components within the modules. This decision was based upon our broader experience with printed-wiring construction, the ease of repair at the component level, and the high confidence level achieved in soldered redundant printed wiring used in earlier UNIVAC equipment. Figure 4 is a photograph of the module and header construction.

The ruggedness of the soldered assemblies was adequately demonstrated during design and testing of the system. Before final testing many of the modules were potted with epoxy resin filled with hollow glass spheres; this was done to demonstrate the resistance of the modules to environmental temperature, shock, and vibration. This information will be used during

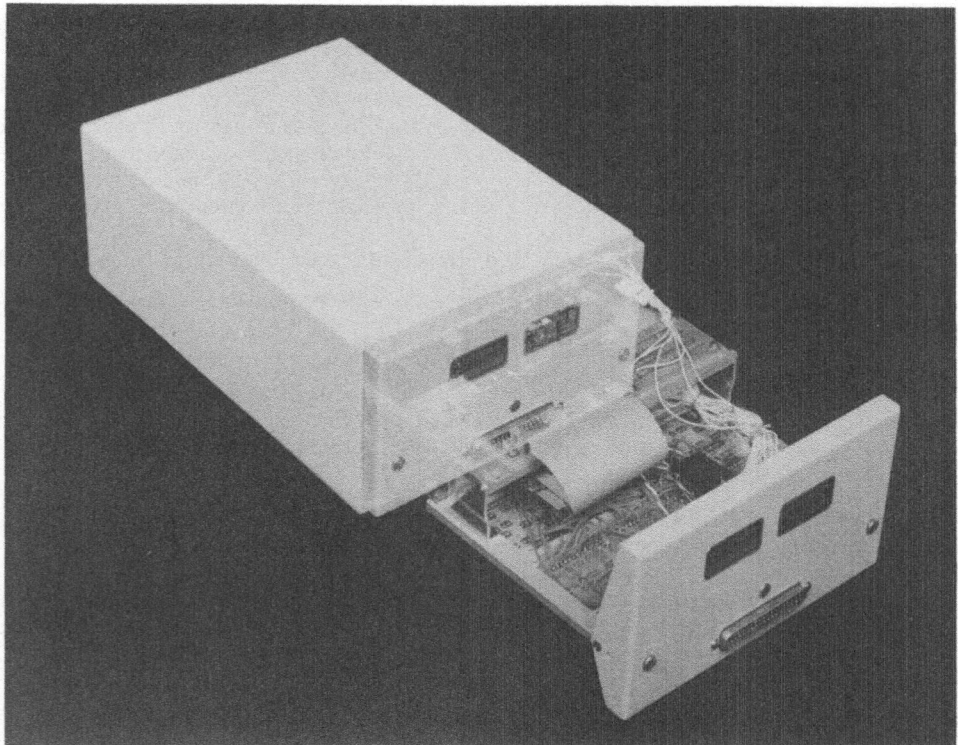

Fig. 1. Plated-wire memory (cased view).

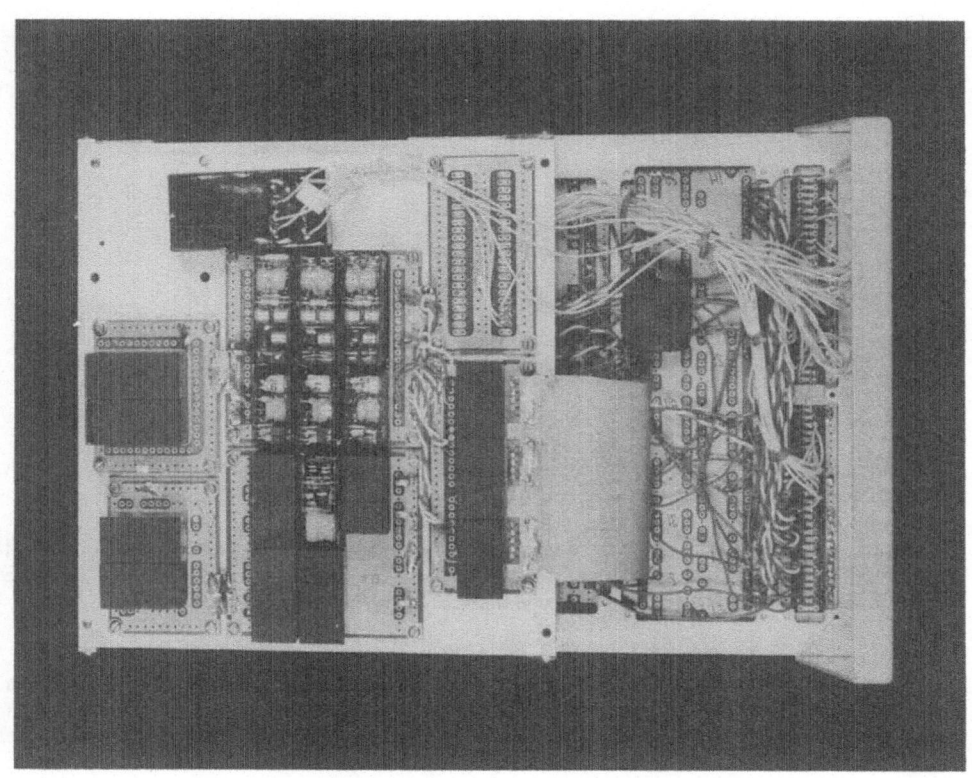

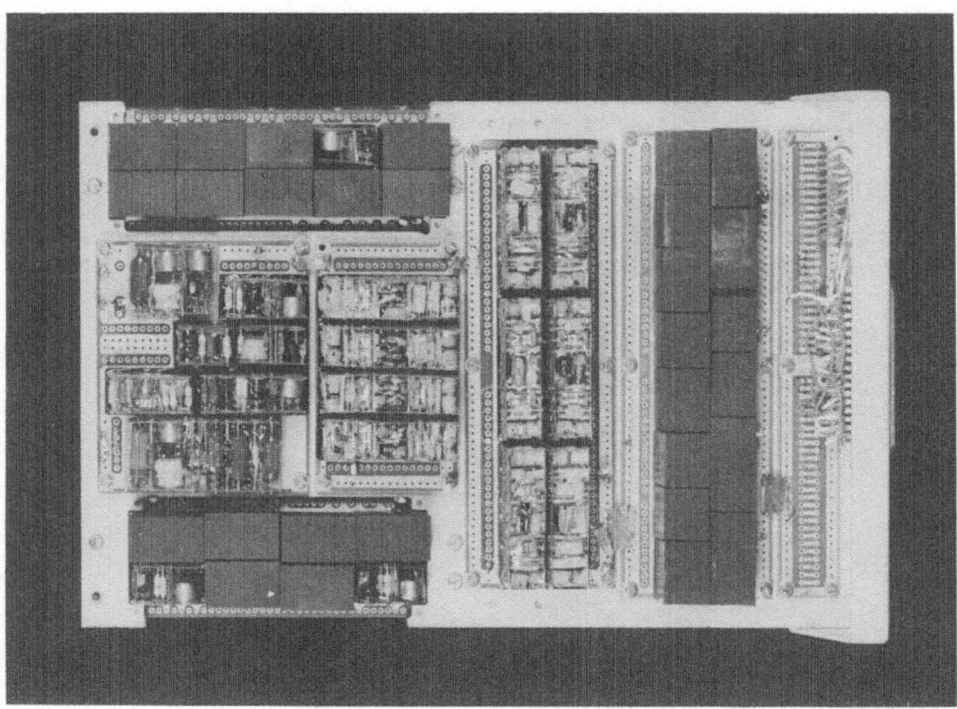

Fig. 2. Plated-wire memory. Upper figure, top view; lower figure, bottom view.

Fig. 3. Backboard construction with sleeves soldered in place.

future flight packaging. The remaining modules were dip coated with epoxy–polyamide resin for protection against changes in temperature and humidity; protection against severe mechanical shock and vibration was not required.

Results of more than 100 temperature cycles and of tests performed on the individual modules which were potted with epoxy resin show that the modules will withstand the severe environmental stresses encountered in space flight.

Tooling for Circuit Module Headers and Final Potting

In order to achieve maximum density in circuit modules which contain varying numbers of components, it was decided to use modules of many different sizes rather than those of one standard size. To minimize the tooling cost for the various headers, a master pattern of the largest header was machined from aluminum (Fig. 5). Many silicone rubber molds were made from this pattern, and length adjustment for the various modules was accomplished by placing paper-base epoxy inserts into the silicone rubber mold. Pin displacement in the cast epoxy headers was controlled to less than one percent.

Potting of the modules was accomplished by the use of vinyl plastisol molds (Fig. 5). The vinyl plastisol, which has a shrinkage of slightly over one percent, produced finished modules which were separated from other modules to such an extent that binding caused by differential expansion during temperature cycling was prevented.

COPPER GROUND PLANE AND ASSOCIATED PRINTED CIRCUITRY

The purpose of the ground plane is to provide rigid, undistorted support for the wire memory elements and to provide proper ground return paths. The requirements for the copper

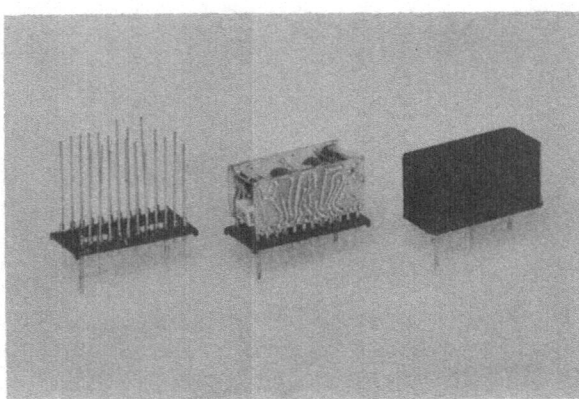

Fig. 4. Module and header construction.

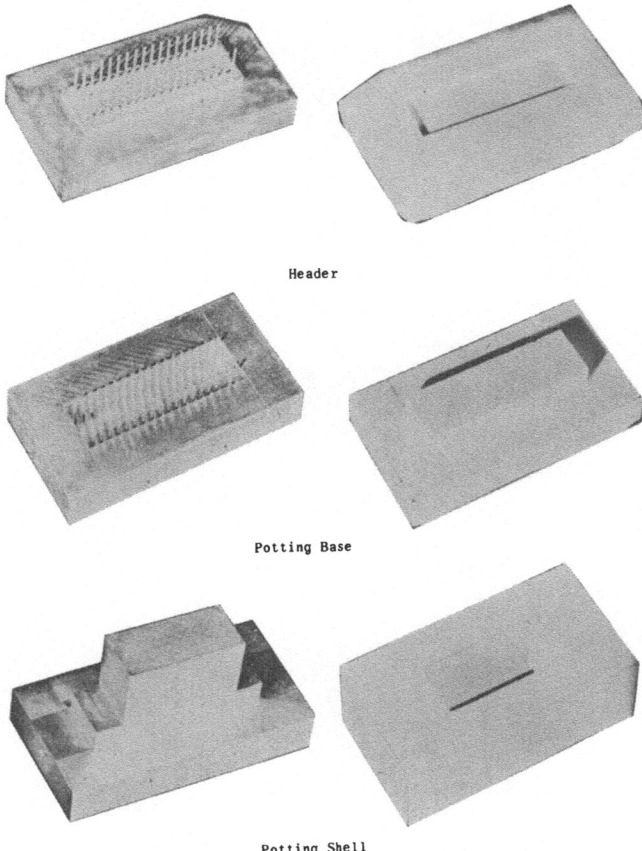

Header

Potting Base

Potting Shell

Fig. 5. Master patterns and molds.

ground plane restricted the choice of construction methods and materials; these requirements were as follows:

1. The overall structure had to have substantially the same coefficient of expansion as the beryllium–copper base wire on which the active memory elements were plated.
2. The structure had to weigh as little as practicable.
3. The structure had to be rigid and flat within $\frac{1}{32}$ in./ft in length or within $\frac{1}{64}$ in. for this particular plane.
4. Accurately sized and spaced grooves had to be formed into the surface of the plane.

The first requirement could have been met by the use of solid copper, but solid copper of sufficient rigidity would have been too heavy. Aluminum and magnesium were suggested but were eliminated because of the great difference in thermal expansion. The structure finally decided upon was made of a thin copper shell reinforced with aluminum honeycomb.

The formation of grooves 8 mils wide by 8 mils deep in the copper surface presented a significant problem. Deep etching was attempted, both in the laboratory and elsewhere, but the grooves were not sufficiently smooth and uniform to support the memory elements properly. Next, electroforming on a stainless steel master was tried, but this met with limited success. The method which was most successful involved the use of a polished brass master for

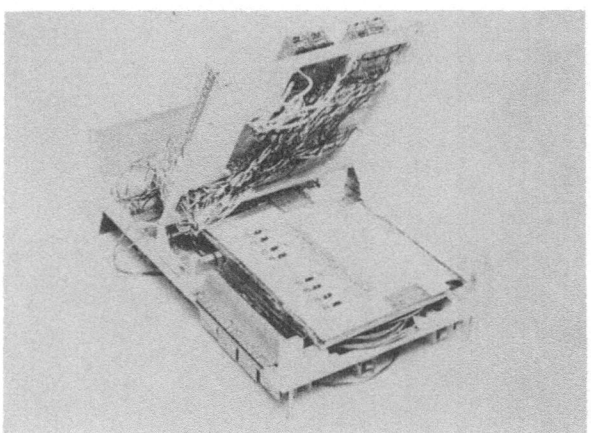

Fig. 6. Complete memory system and electroformed ground plane.

heat-forming a sheet of Plexiglas into a negative of the required plane surface. The Plexiglas form was made electrically conductive by silvering and was then copperplated to a thickness of 10 mils. Two such electroforms were reinforced with aluminum honeycomb and bonded together with epoxy adhesive. The edges of the electroforms were connected with a bead of solder. Figure 6 is a photograph of the complete memory system; also shown is the copper ground plane.

The word-line overlay and the printed wiring interconnecting the memory plane with its associated circuit modules were fabricated from copper-clad glass laminate. Bonding of the word-line overlay was done with polyurethane adhesive. All the electrical connections on the memory plane were soldered.

MULTILAYERED BACKBOARD WIRING

The use of multilayered wiring was decided upon because of its compactness, its adaptability to controlled impedance interconnection, its ability to function as a mechanical support for the modules, and its adaptability for use in more condensed flight-model memories.

Several small multilayered backboards were used, and each of these backboards served a discrete functional part of the memory circuits. Use of small subbackboards, rather than one large board, provided for ease in making changes during design stages and for ease in fabrication involving the use of laboratory scale equipment. The multilayered boards are built up of

Chemically Treated Before Final Plating

Fig. 7. Cross section of holes.

Finished Plated-Through Holes

copper-clad glass epoxy laminate bonded together with B-stage preimpregnated glass epoxy fabric. The temperatures and pressures used are included in the appendix of this paper.

The laminated thickness of all the subbackboards is approximately 0.150 in. Holes of 0.035 in. were drilled at all required hole locations. Although solid carbide drills were used and care was exercised in selecting the proper speed and feed in drilling, there was evidence of resin contamination along the edges of the exposed copper on the internal layers. The resin was removed and all holes thoroughly cleaned by immersing and agitating in concentrated sulfuric acid for from 15 to 45 sec. A water rinse and a one-to-two-minute dip in concentrated hydro-fluoric acid were used to prepare the holes for electroless plating. As can be seen in Fig. 7, the etching technique has not only cleaned away the unwanted epoxy drilling smear but has also undercut the glass epoxy at the internal copper interfaces so that a greater area for plated connection to the circuit layers is provided.

After electroless plating, the final electroplating is accomplished by the use of a pyro-phosphate copperplating bath. Use of the pyrophosphate bath has resulted in plated-through holes which meet the requirements of an internal diameter variation of from 0.0295 to 0.037 in. without the need for extra reaming operations.

The multilayered backboard was designed to allow the use of Mylar artwork and to allow for circuit drilling without the need for precision-machined drill jigs. This was accomplished by using 75-mil pads on 100-mil centers and by allowing no interconnecting circuitry to pass between any two adjacent pads. In all cases, crossovers in layout were made on layers where the pad could be removed. This arrangement allowed for noncritical registration between layers. The use of 75-mil pads with 30-mil plated-through holes assured positive interconnection between layers even though the layers were slightly misregistered.

Registration of artwork was acomplished by printing a master drilling pad pattern in register with a thin sheet of steel which had previously been drilled with a two-hole drill plate. Each successive layer of artwork was registered by alignment with the printed master drill pattern, by taping in place, and by drilling through the holes in the drill plate.

A drill plate was made for each multilayered circuit board by printing and etching a pattern on $\frac{1}{16}$-in. copper-clad glass epoxy. Each pattern was drilled by eye by using the etched-out center of the pads as locating means. With a little extra care in drilling at this stage it was found that sufficient accuracy could be obtained without the need for metal drill jigs.

APPENDIX

Multilayer Backboard Laminating Procedure

1. Select proper thickness of material for layers and cut the material to predetermined size (pre-preg, kraft paper, foil, and copper-clad).
2. Drill registration holes in materials (use drill fixture and $\frac{1}{4}$-in. drill).
3. Place outer (film) image in drill fixture, punch registration holes by use of $\frac{1}{16}$-in. steel plate, and sensitize and print image. Align each film (one at a time) with the printed image plate. Place in drill fixture and punch registration holes.
4. By use of negatives, fabricate straight-etched circuits for inner layers.
5. Clean resist off circuits (use photo resist stripper) and clean with soap and water.
6. Degrease filler boards (MEK wipe).
7. Oven-bake all layers at 200°F for 10 min.
8. Coat jig and related parts with mold release.

CAUTION
Do not coat multilaminate materials.

9. Assemble layers on fixture (use $\frac{1}{4}$-in. dowel pins for alignment). A minimum of five layers of 0.002-in. pre-preg cloth must be used between each layer. Ten layers of kraft paper are used as cushions between each side of stack and laminating plates.
10. Place fixture in press (contact pressure of ten psi) and heat platens to 350°F.

CAUTION
A constant contact pressure of 10 psi must be maintained at this time.

11. When press reaches 350°F apply pressure of 150 psi and hold constant for $\frac{1}{2}$ hr.
12. Cool platens while maintaining pressure of 150 psi.
13. Release pressure, remove fixture, and dismantle.
14. Clean surfaces of multilaminate to remove foil, kraft paper, and mold release.
15. Register laminate to drill template ($\frac{1}{4}$-in. dowels) and drill holes designated for the laminate (No. 65 drill, 0.035 in.).

NOTE
Drill template fabrication by use of $\frac{1}{16}$-in. copper-clad glass base material. Register from registration fixture. Coat material with Shipley AZ top. Print layer that is to be used for drill pattern, develop, and etch.

16. Deburr with a Vibro sander.
17. Roto-brush with Scotch Bright.
18. Plug registration holes with copper bolts. Immerse laminate in concentrated H_2SO_4 for 45 sec. Agitate with an up-and-down motion and water-rinse thoroughly. Dip in hydrofluoric acid (concentrated), agitate for 3 min, water-rinse thoroughly, and air-dry.
19. Make certain that resin and glass fibers have been removed from internal hole surfaces.
20. Proceed with electroless plating.
21. Water-rinse, dip in HCl, water-rinse, and copper-strike.
22. Water-rinse and inspect for complete copper coverage in holes.
23. Scrub laminate with Ajax cleanser and water-rinse.
24. Bake dry for 20 min at 150°F.
25. Use alcohol to wipe surface of laminate.
26. Coat with KPR and air-dry.
27. Bake for 5 min at 200°F.
28. Print pattern on both sides (expose for $2\frac{1}{2}$ min on each side).

CAUTION
Make sure hole centers on positive films have been opaqued closed.

29. Develop in KRP developer for 5 min. Use a vigorous spray to water-rinse the pattern.

NOTE
If pattern washes off or does not develop properly for one reason or another, strip off old resist and repeat steps 23 through 29.

30. Bake dry for 10 min at 200°F.
31. Immerse in Metex solution (swab surfaces to be plated).
32. Water-rinse.
33. Dip in HCl (10% solution).
34. Water-rinse.
35. Copperplate (pyrophosphate bath) at three-quarters of a volt for $3\frac{3}{4}$ hr. (Rotate laminate every hour.)

NOTE
Check hole size before nickel-plating.

36. Water-rinse.

37. Dip in HCl (10% solution).
38. Water-rinse.
39. Nickel-plate (sulphamate bath) for 25 min at 1.4 V.
40. Water-rinse.
41. Dip in HCl (10% solution).
42. Water-rinse.
43. Gold-plate for 25 min. (Use Sel-Rex Autronec CI with adjust-a-volt set at 12.5.)

NOTE

Have laminate cathodic before entering into bath.

44. Rinse in gold drag-out tank, water-rinse, and dry.
45. Strip off resist and water-rinse.
46. Fleck off excess nodules of gold buildup.
47. Etch part (Hunt's S.C.E. solution) for approximately 3 min.
48. Water-rinse, clean with soap and water, and air-dry.
49. Profile the board to proper dimensions and open up guide pinholes.

NOTE

It is advisable to drill test holes outside of the pattern area in the multilaminate when drilling the initial hole pattern. These holes may be cut off the board and sectioned at any time during the plating cycles so that the amount of plating being deposited on the wall of the hole can be determined.

Packaging Computer Circuitry for
Space Applications—A Two-Part Compendium

Donald Shaner and Frank L. Jennings

Light Military Electronics Department
General Electric Company
Utica, New York

As electronic components become smaller and smaller, the mechanical design engineer must devise packaging techniques to assure optimum use of these components. The first paper in this compendium describes precisely how the packaging of digital circuitry for a command storage programmer unit was accomplished. The second paper describes a computer program developed at G.E. to assure an optimum package for this or any similar digital unit. The final product of this combined effort has been a reliable command unit with the impressive circuit density of approximately 1100 transistors, 5000 diodes, 2500 resistors, and 200 capacitors, packaged in 32 separate modules in a total volume of 155 in.[3]

Part I—The Electromechanical Package

By Donald Shaner

THE MECHANICAL PACKAGE

PACKAGING ELECTRONIC CIRCUITS for space applications is an ever-increasing challenge. Circuits are becoming generally more complex, with the evolution of solid state components compounding many of the design problems. The word "miniature" is no longer sufficient in describing the components' relative size; in fact, it is this decrease in size which is really the root of the packaging problem. Cabling and interconnections also become more involved and more vulnerable to failure.

What is needed are designs which reduce or simplify these interconnections, while providing for adequate form factors and circuit needs. These designs, which will result in reliable systems that can perform throughout the life of the mission, must also be capable of ready manufacture without excessive cost or assembly time. The environments of launch, staging, and space itself require still other design considerations, which must be factored into early design decisions.

It is therefore mandatory that early in the design cycle a close, harmonious relationship be established between the circuit design engineer and the circuit packaging engineer, so that each is aware of the other's requirements.

Component selection, for example, must be a mutually arrived at decision. This kind of coordination requires that the packaging often be concurrent with the circuit design development. The "concurrent approach" was used with the digital command storage programmer, described in this paper and in production at GE's Light Military Electronics Department, eight months after award of the contract. As a result, the engineering model was nearly identical to the production unit.

The overall dimensions of this unit are $11\frac{1}{2}$ in. $\times$ $14\frac{1}{2}$ in. $\times$ 7 in. high, and its total weight 31 lb. The circuitry consists of approximately 1100 transistors, 2500 resistors, 5000 diodes, and

200 capacitors. In addition, the unit contains its own power supply, crystal oscillator for all timing functions, and sonic-delay-line digital storage. Its operating characteristics are as follows:

From sea-level pressures to vacuum conditions (nonpressurized)
Vibration: Random—$0.1 \, g^2$/cps from 7 to 2000 cps Sinewave Excitation—20 g rms from 7 to 2000 cps
Shock: 30 g's, 11 msec duration, each of three planes
Acceleration: 12 g's, each of three planes
Humidity: $90\% \pm 5\%$ relative humidity
Temperature limits: $-5°C$ to $+65°C$

Much of this performance was achieved through the use of a lightweight structure in which virtually all of the housing elements are made to serve the dual function of structural members and thermal conductor. The housing and its base (cold plate) are dip-brazed as a single integral assembly of self-jigging, "egg-crated" aluminum plates. No rivets or joining hardware are used in the chassis construction. The circuit modules, which will be discussed in detail later, are mounted in the structure so that the heat is conducted from the modules, along the aluminum plates, to the base plate.

Module retention, Fig. 1, is accomplished by wedging two modules away from each other until they contact the aluminum partitions of their compartment. The wedge uses angles of eight degrees which force against similar molded surfaces of the encapsulated modules. A captive screw which tightens into a threaded insert provides both the mating and "unmating" force required to retain the modules.

Other items are fastened directly to various points in the housing except for the sonic delay lines used to store information. The memory unit, shown in Fig. 2, consists of a drive and read amplifier potted together in a module which is fastened to the sonic delay line to form a memory package with dimensions of $6\frac{1}{4}$ in. $\times$ $4\frac{1}{2}$ in. $\times$ $\frac{3}{4}$ in. The maximum number of stored bits has conservatively been set at 2000. The assembly weighs 1 lb.

A vibration and shock isolation system using silicone rubber pads serves to fixture these items into their compartments. This technique involves preloading the rubber pads, which are placed between the sides of the sonic delay lines and the aluminum panels of the compartments. Tolerances are overcome by selecting the proper thickness of pad to achieve the correct pre-loading of the rubber isolators. Placement within the five-sided compartment is also achieved by other pads positioned between the compartment floor and the line assembly. Metal blocks, which also contain pads, are then forced down on the assembly with a defined force and locked in place.

The isolation system, composed only of silicone rubber strips (or pads) cut from stock sheets of various thicknesses of rubber, effectively reduces the levels of shock and vibration transmitted to the sonic delay line.

Only one thing more had to be done to the package itself. Because high humidity over an extended time may affect stored information, a humidity seal which would release pressure

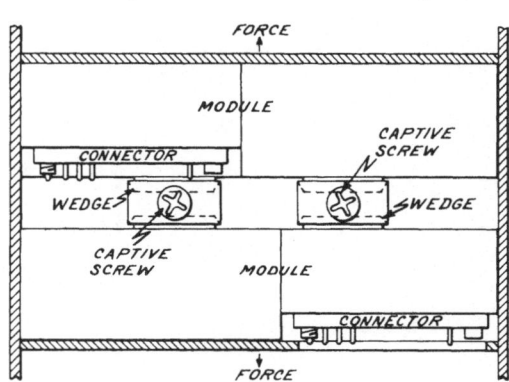

Fig. 1. Module retention by means of wedges and captive screws.

Fig. 2. Memory unit with sonic delay
line and drive and read amplifier module.

when taken from atmospheric conditions to vacuum conditions had to be perfected. This seal
was done as simply as possible. Since dip-brazing effectively sealed the aluminum walls on five
sides of each sonic delay line compartment, the seal on the remaining side had only to be made
by using a new G.E. sealant, called RTV-102. This remarkable adhesive sealant was used to
complete the seal between a laminated glass–epoxy board, placed across the top of each compartment, and the aluminum walls of the compartment.

Cleaning these surfaces was the only preparation necessary as RTV-102 will bond to most
clean surfaces without the aid of a primer. This seal, while guarding against moisture, provided
enough flexibility of movement so that air either remained trapped within the compartment or
escaped slowly when subjected to rapid pressure changes (from atmospheric pressure to the
vacuum of space). No attempt was made to measure this condition as the intent here was for a
humidity seal at atmospheric pressures only.

MODULE DESIGN

The basic circuit element in the logic circuitry is a five-component circuit known commonly as a NOR circuit. One transistor, two resistors, and two diodes make up this basic circuit, shown in Fig. 3.

Additional diodes may be added to form multiple input AND/OR gates. By decreasing
the value of the B^+ resistor, a power NOR is obtained. Two of these basic NOR circuits, when

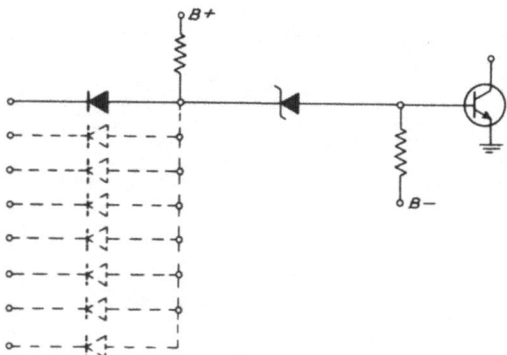

Fig. 3. The basic NOR circuit.

combined with clock-trigger circuitry, form a flip-flop circuit. Buffer NOR's were also used and the flip-flops became known as four-transistor flip-flops.

It was suggested that flip-flops could be built from NOR circuits, although it would require about ten more components per flip-flop. These additional components were not objectionable in view of the fact that the standard NOR circuit was now almost universal throughout the circuit design and a standardization of parts was being achieved.

This standardization was extremely important. It meant that the designer could proceed knowing that a given part would be suitable for several similar functions. Another break-through was in obtaining diodes, resistors, and capacitors with similar physical dimensions, thereby ensuring replacement flexibility.

The basic five-component NOR also became the starting point for actual construction of the physical module. A mounting surface is provided by a 3 in. × 2 in. × 0.05 in. plexiglass board. This board provides for direct placement of 32 NOR circuits with enough space left over for additional supporting components. The board material was chosen for its compatibility with epoxy resins and because it may be readily drilled. There are 412 holes in this mounting board, with 96 used exclusively for transistor mounting. Similar physical dimensioning of the components means that any of the remaining holes are usable for component mounting. Unused holes allow the epoxy–resin encapsulant to bridge through the board, ensuring a stronger module.

Provision for mounting a connector is also provided by an extension of the board on one side. This 2 in. × 3 in. module board proved to be a size that was easy for the assembly personnel to handle, speeding assembly and reducing the total labor cost appreciably.

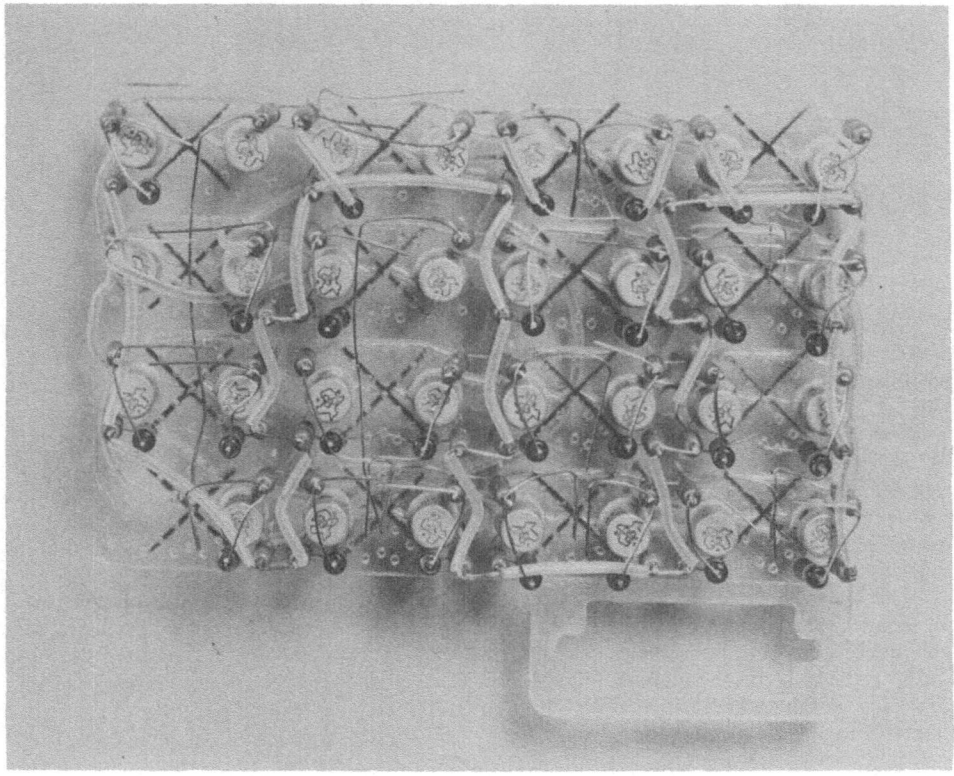

Fig. 4. Production units ready for inspection.

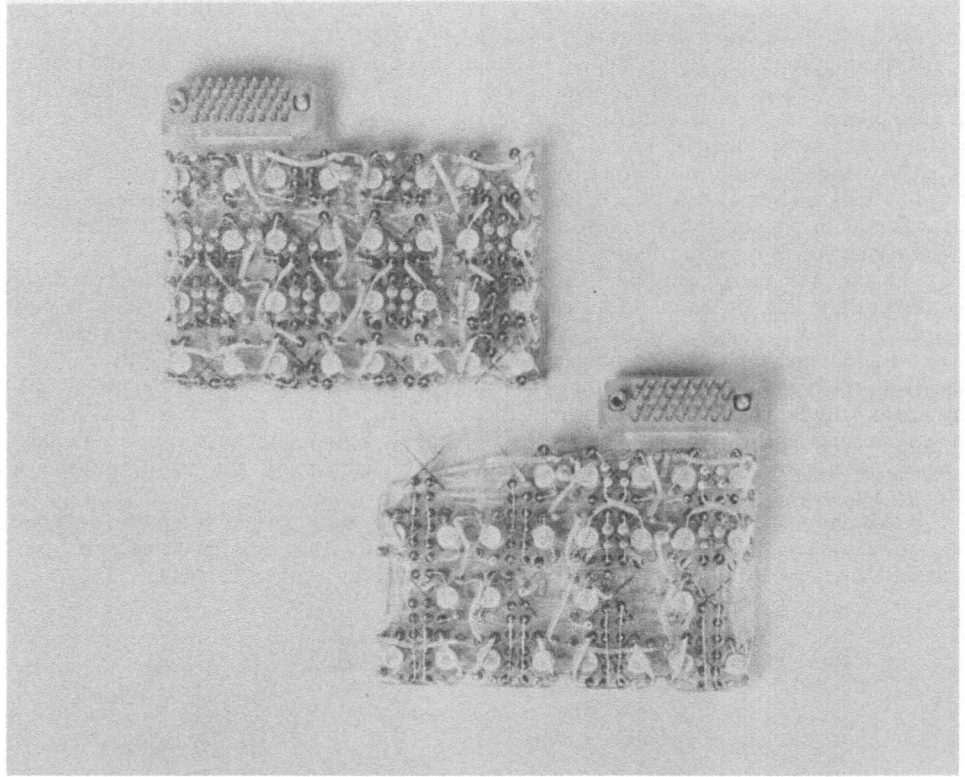

Fig. 5. Typical high-density module assembly used in the equipment.

The actual assembly was relatively simple. Resistance-welding techniques using General Electric square-pulse resistance welders were employed throughout. This highly reliable process permitted rapid assembly of modules with a maximum density of 60 components per cubic inch.

The first assembly step was to insert the transistors into holes already patterned correctly for the leads. The emitter lead of each transistor was welded to a nickel ribbon wire on the opposite side as soon as the transistor was in place. This nickel ribbon wire was continuous and sleeved between welds with tetrafluoroethylene tubing. The next step was adding the B^- resistor and Zener diode. A separate ribbon wire connected their leads and the base lead of the transistor. (See Fig. 4.)

This same principle was used throughout the various levels of assembly. Component leads not requiring further welds were cut back, and the nickel ribbon wiring was used exclusively for interconnection. Tetrafluroethylene sleeving was used as required. The end result is shown in Fig. 5. These are actual production units—fully inspected and ready for encapsulation. The unit on the lower right, which has many NOR circuits omitted, is an example of how this design is adaptable to volume production. Enough units of this particular board justify making a special basic board with fewer than the 32 NOR circuits previously described. On each module it is easy to see the multiple diode inputs and the existence of flip-flop circuits, and yet the basic board pattern of 32 NOR circuits of five components each remains clearly visible. This repeating pattern has permitted assembly personnel, QC inspectors, and troubleshooting personnel alike to become familiar quickly with the design and, as a result, mistakes are prevented or easily corrected.

Assembly personnel find design repetition a distinct help in fabrication. Assembly time decreases with each unit, while reliability and accuracy of workmanship improve steadily. The flexibility achieved by using a basic board of NOR circuits is amazing. So many combinations are available to the circuit packaging engineer that it was possible to build a stockpile of these basic boards during the engineering model design and use them for whatever need the circuit design engineers specified. Circuit changes right up to the point of encapsulation were easily accommodated as splicing of nickel ribbon wire by resistance-welding techniques is accepted practice, and modifications were therefore easily accomplished.

Special circuit needs can also be accommodated by the mounting board as shown in Fig. 6. While digital logic flexibility is not available, need for applications of this type often occur and the module concept was versatile enough to accept this circuit requirement.

An interesting feature of the design was that only one mounting surface was used to support the components. Usually such an approach uses two surfaces in a "cordwood" construction. It was found that the second surface was not necessary. The first board held the components in proper position, and when axial leads were welded to the nickel ribbon wire on the top of each component, the rigidity achieved was more than sufficient to allow handling, testing, and encapsulation. Another important advantage was that repairs, whether by component replacement, or new welds, were now easily accomplished. Even after encapsulation, many replacements were possible.

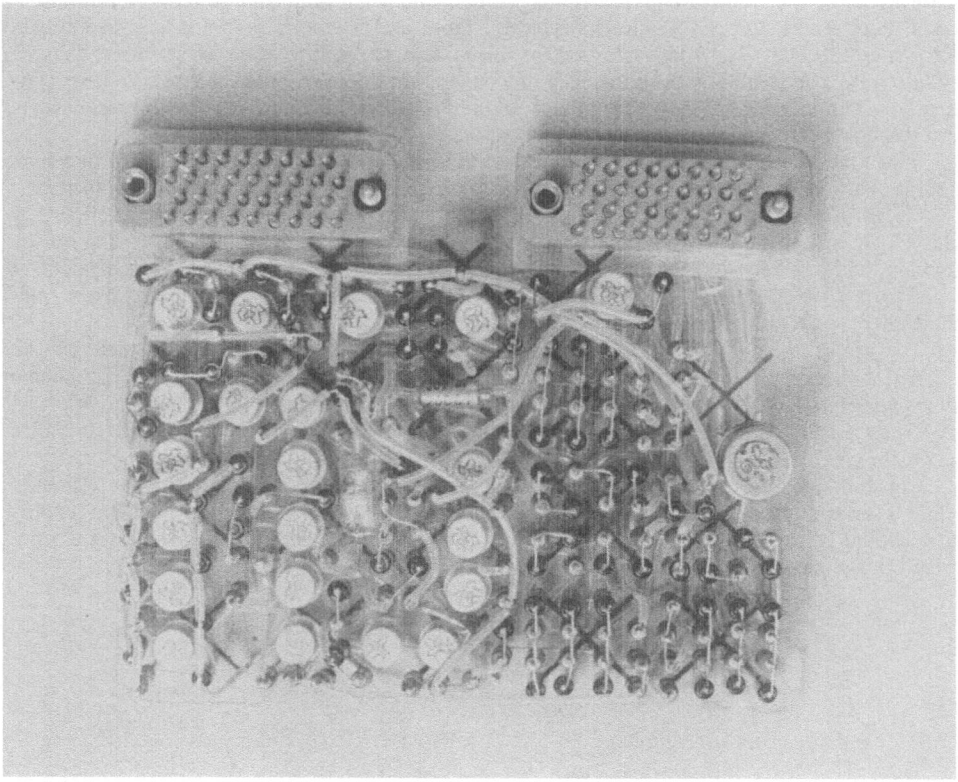

Fig. 6. Mounting board for special application.

ENCAPSULATION

As the environmental conditions were fairly severe, a rigid, but lightweight, encapsulant was required. Heat transfer by conduction had to be used so a foam encapsulant was out of the question. The wedges described earlier also dictated a solid encapsulant. Epoxy resin with phenolic microballon fillers added was used as the encapsulant. The density of 50 pounds/ft^3 made it heavier than foam types but much less than solid resin or filled resin systems. A thermal conductivity of 0.09 Btu/hr-ft^2-°F/ft proved adequate for heat transfer purposes as well. Its rigidity added greatly to the structural integrity of the dip-brazed chassis.

The encapsulated modules were potted in aluminum molds which held dimensional tolerances to ±0.005 in., while assuring full reproducibility of many sets of modules.

Scrapping modules after encapsulation was held to a bare minimum by perfection of repair techniques. Failures could often be traced on logic diagrams to an individual NOR element and its modular location ascertained on the module assembly drawing.

Abrasive blasting techniques at various air pressures were used to penetrate the encapsulant for exposure of the affected components and wiring. Removal of the failed component after exposure was accomplished by using dental-type drills and small high-speed electric tools. Welding techniques were used for easy assembly of a new component in place of the failed part.

CIRCUITRY PLACEMENT

One problem in this design effort was dividing the logic circuitry into module sections with maximum use of each module while keeping interconnections between modules at a minimum. There are many ways one can approach this problem and while some undoubtedly are more successful than others, they all require considerable labor and ingenuity before a reasonable form factor is established. The approach used for this system was first to establish the most suitable module form compatible with the packaging concept, and then to package circuit elements as densely within a module as possible. (This method of accomplishing these two objectives has been discussed earlier.)

This type of module design required a compromise between connector pin limits and high module density factors. Sequential assignment of logic elements to a series of modules isn't feasible, for the intercabling requirements very quickly get out of hand, and the modular density falls off. It is necessary to manipulate these logic elements among modules, depending on the function, and it is here that the difficulty really begins. As the designer assigns these elements it is necessary to keep track of connector pin limitations, dc loading, total component count, and efficiency of both module density and intercabling.

The use of an X–Y chart often proves beneficial. Each NOR or flip-flop element can be assigned a number on the X axis, and every function can be listed on the Y axis. This scheme allows correlation of function in each module against its appearance in another module and establishes a running of necessary connections between functions. This running list simplifies the drafting effort.

Constant juggling to achieve a workable and fairly precise positioning of logic elements within a series of modules requires both ingenuity and hard work. The realization that a computer should be able to do this task more quickly and more efficiently led to the program described in the next paper, "Part II: The Computer Program" by Frank L. Jennings.

Part II—The Computer Program

By Frank L. Jennings

INTRODUCTION

Part I details the design effort in the development of a successful digital command storage programmer. One portion of this effort involved the subdivision of the logic circuitry into modules, a task which was fairly costly in terms of both time and manpower, because most of it was uncreative, trial and error work.

Part II describes the development of computer programs to relieve the engineers of this tedious, but important, layout work.

These programs must be extremely flexible so that they may be applied to any digital equipment design effort. Otherwise, the time required to write and troubleshoot a program for a specific equipment design probably would exceed the time to do the job manually. To make the computer a practical design aid, the programming cost must be amortized over several equipment designs.

While flexibility is mandatory, care must be taken at the same time that a set of programs for general application does not become so watered down that the programs are actually of little use to any one project.

This paper will attempt to relate the development of a set of computer programs that meet these flexibility and utility requirements, to describe what they do, to show how they contribute to our effectiveness as a producer of military electronics equipment, to discuss their limitations and to measure their effectiveness when applied to any design problem similar to the one described in Part I.

When we began to develop a flexible series of computer programs to aid digital equipment design, we turned for advice to General Electric's Computer Department in Phoenix, Arizona. Although the Computer Department produces large-scale computers in contrast to the smaller aerospace equipment that we design and build at the Light Military Electronics Department, we were able to adapt many of their concepts.*

NOMENCLATURE

All of the programs described in this paper depend upon a ten-character logic nomenclature system that uniquely describes each accessible signal and physical logic element in an equipment design. For a logic name to be able to identify its location in the logic complex, it is necessary to subdivide the logic. A computer or other form of digital equipment is subdivided into registers, and a register is subdivided into stages. Finally, a stage is divided into a series of logic elements. An element is defined as the smallest subdivision of digital equipment which still retains its logical identity; i.e., OR gate, NOR gate, flip-flop, etc.

A stage is defined as a collection of interconnected elements. The downstream element of a stage is a clocked element such as a flip-flop, or an element that produces a clock pulse, such as a clock driver, or an element whose output signal is an external output from the system. The remaining elements in a stage form a chain of logic signals which lead to an input of the downstream element. The sole purpose of a stage is to produce one generic logic signal, and all elements within a stage have the same stage name (first six characters of the complete name).

The name of a logic element (physical circuit element) is the same as its logic output signal except for elements with multiple outputs such as combination OR/NOR gates and half-adders. These elements are treated differently.

To act as a meaningful tool for computer design aid programs, a complete logic name must consist of ten characters which are divided first into a group of six and then into a group of four

---

* W. A. Hannig and the staff of the Design Automation section at Phoenix deserve credit for many of the basic concepts used in our programs.

79

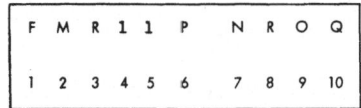

Fig. 1

characters. The first six characters define the stage, and the last four characters define the circuit element (such as a NOR gate) used to implement this part of the logic. The register is identified by the alphabetic characters included in the first four characters of the stage name.

Figure 1 shows how a typical logic name is used to identify a logic signal. The numerals underneath the name refer to character numbers. Character #1 denotes the circuit type being driven by the logic in this stage. In this instance F denotes a flip-flop as the downstream element in the stage.

Together with the first character, the alphabetic portions of characters #2, 3, and 4 define the register. In this case, the register name is FMR. There may be up to ten stages in the FMR register because the fourth character may range from 0 to 9. If more stages are to be grouped in a register, the third and/or second characters may be numeric to permit higher serialization.

The fifth character denotes whether the logic signal is the "true" or "false" output from a flip-flop, or whether the signal is part of a stage that drives the "set" or "reset" side of a flip-flop. In this case, the number 1 tells us the signal is generated in the stage that drives the set side of the flip-flop.

The sixth character tells us that this logic signal is generated in a portion of the machine designated by the letter P. For instance, it may be advantageous to group certain portions of circuitry in major machine units to facilitate maintenance. Thus P may stand for the central processor.

The seventh and eighth characters exactly identify the circuit type that generates the logic signal. NR is used here to denote a NOR gate.

The last two characters are used to differentiate elements of the same circuit type within a stage. Thus, this is number 2 NOR gate within the FMR11P stage.

This nomenclature system is used to communicate with the computer in the following manner: each logic element and signal is assigned a unique name in accordance with the nomenclature rules. The information which is deposited on the logic diagram can now be reduced by hand to an "Element Input List." This list is a tabulation of every circuit element shown on the logic diagram together with all of its input signals.

Figure 2 depicts a small portion of both a logic diagram and the Element Input List (which is formed from the information shown on the logic diagram). Each logic element is listed in the first column. There is space to include up to four inputs beside each element. If an element has more than four inputs, additional lines are used with the first column left blank. This Element Input List, then, serves as the primary input to the entire series of computer programs, and it is also the vehicle for updating design changes.

Note that in the example shown, the elements are listed in an orderly procession. That is, the downstream element, in this case a flip-flop, is listed first, and then elements upstream from it are listed. Also note that all of the "set" stage is listed first and then the "reset" stage. This arrangement is necessary for the successful use of some of the computer programs which process the information. For a large and complex logic design, considerable effort is required to arrange the data in this manner. Since we are dealing with automation, however, an auxiliary program has been built into the main program series to accept data in any order and arrange them so that all elements in a stage are grouped together. In the case of flip-flops, the "set" stage is grouped together and immediately followed by the "reset" stage.

Other inputs, which are very easy to prepare, must be prepared for each program. These data consist of the loading each circuit type imposes upon the element which drives it, the drive capability of each circuit type, the size and power dissipation of each circuit type, etc. In all cases the data are reduced to dimensionless units to provide flexibility so that any design can be handled by the computer.

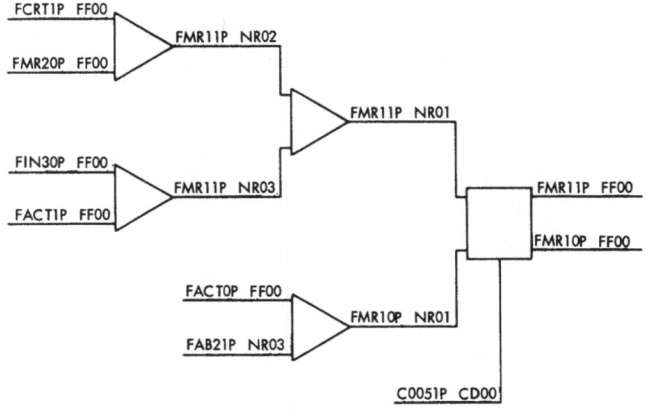

ELEMENT	INPUTS			
	1	2	3	4
FMR11P FF00	FMR11P NR01	C0051P CD00		
FMR11P NR01	FMR11P NR02	FMR11P NR03		
FMR11P NR02	FCRT1P FF00	FMR20P FF00		
FMR11P NR03	FIN30P FF00	FACT1P FF00		
FMR10P FF00	FMR10P NR01			
FMR10P NR01	FACT0P FF00	FAB21P NR03		

Fig. 2

Let's look now at the actual programs to see what they do and to some extent how they work. First, the Element Input List is arranged as described earlier. By searching this list the computer checks the original logic design for closure. That is, it insures that all signals which are listed as inputs exist elsewhere in the machine as output signals, and it checks that all output signals are used somewhere as input signals. The method is simple. The first time an input signal is encountered, the first column is searched to see that the name exists there, and the input columns are searched for each name in the first column.

Each signal name that doesn't meet this check is printed out either in a list of inputs which don't exist as outputs or in a list of outputs that don't exist as inputs. A quick manual check of these lists will show if any signals are included which are not actual inputs to the machine or outputs from the machine.

Given a list which tells the loading that each circuit element imposes and its capability to drive loads, the computer further checks the logic design for dc loading. The "fan-out" of each circuit element is printed along with its loading, and a notice is printed whenever an overload is possible. The fan-out of an element is merely a list of all elements which receive an input signal from that element. Although this may sound fairly simple, this fan-out may be quite large, especially with designs which are fused onto integrated microelectronics circuit elements.

No matter how large the memory capacity of a computer may be, there is a certain limit to the amount of data that may be stored within it. In order to avoid limiting the size of a design that could be automated, the closure and dc loading check programs were written to accept Element Input List data in discrete groups. Each group of data, called a block, consists of a limited number of lines from the Element Input List. In our case, we found we had to limit our block size to 2000 lines.

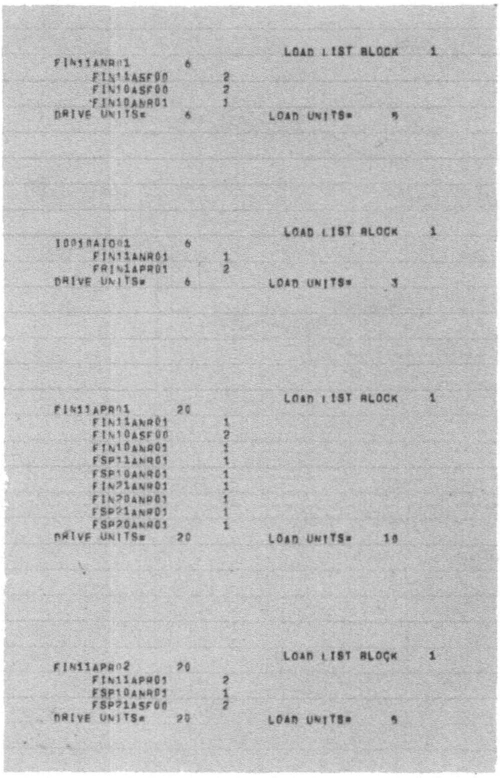

```
                                    LOAD LIST BLOCK    1
FIN11ANR01        6
        FIN11ASF00        2
        FIN10ASF00        2
       'FIN10ANR01        1
DRIVE UNITS=       6        LOAD UNITS=      5


                                    LOAD LIST BLOCK    1
1001RAIO01        6
        FIN11ANR01        1
        FRIN1APR01        2
DRIVE UNITS=       6        LOAD UNITS=      3


                                    LOAD LIST BLOCK    1
FIN11APR01       20
        FIN11ANR01        1
        FIN10ASF00        2
        FIN10ANR01        1
        FSP11ANR01        1
        FSP10ANR01        1
        FIN21ANR01        1
        FIN20ANR01        1
        FSP21ANR01        1
        FSP20ANR01        1
DRIVE UNITS=      20        LOAD UNITS=     16


                                    LOAD LIST BLOCK    1
FIN11APR02       20
        FIN11APR01        2
        FSP10ANR01        1
        FSP21ASF00        2
DRIVE UNITS=      20        LOAD UNITS=      5
```

Fig. 3

A portion of the printed output data of the dc Load Check program is shown in Fig. 3.
Imagine the work involved if the engineer had to cross-check the data from two or more blocks
to be certain that the loading on an element as shown for one block of data was not merely a
partial listing. Consider also that the output as shown here may require 400 sheets of paper.
Here is another instance, as with arranging the Element Input List, where design automation
may only be synonymous with design work shifting.

Therefore, to make this program truly useful for any application, a subprogram which
stored the total loading for each element in each block on magnetic tape was written. When all
blocks are completed, this program processes the data on the tape to produce a compiled list
showing the total loading on each element. Another advantage to the compiled load list is that,
although the designer probably wants to see the complete fan-out for some elements, he can
scan the compiled list quickly for overload elements. Figure 4 is a page from the compiled list.
Notice how much more readily the overloaded elements can be identified.

The program which checks closure and calculates dc loading also produces a magnetic tape
which contains the Element Input List. This tape, then, serves as the major input to subsequent
programs. When a design modification is made, the tape is updated.

Once basic packaging philosophies are worked out, one of the product design engineer's
greatest problems is the distribution of circuitry among subassemblies within the equipment.
Subassemblies may be printed wiring boards, modules, matrices, mother boards, or whatever
basic construction unit is to be used. The problem is to distribute these circuit elements in such
a way that the interconnections between subassemblies can be made with the pins available
within reasonably-sized connectors. Sometimes a cut-and-try technique is the only solution

to the problem. The task is especially difficult for a complex design with large fan-outs whose logic diagram is contained on several sheets of paper.

A "gross placement" program now automates the entire procedure. This program uses as its inputs a tape of the Element Input List, data for the size and power dissipation of each circuit type, and subassembly size and maximum power dissipation capabilities. Distribution based upon power dissipation is optional. Another option is the segregation or integration of major machine units.

The program first sorts all elements by "segments." A segment consists of all elements within a stage, and, in the case of flip-flop stages, includes both the "set" and "reset" stages of a flip-flop. All of the elements within a segment are always located on the same subassembly. This course is followed to meet the memory limitations of the computer and to keep running time reasonable. Segments are further grouped by register, and registers are grouped by major machine unit. Another step taken to meet computer memory and running time limitations is to form a pool of segments from which a subassembly is to be chosen. The pool is filled with segments whose combined total size would fill five subassemblies. Selection of segments for the pool is always made first within a register.

Once a pool has been formed, selection of segments for a subassembly is made as follows: That segment which has the most connections to other segments within the pool is placed on the subassembly. Then a list of candidates is formed from among all segments that are interconnected to the segment on the subassembly.

Each candidate is then scored on the basis of the number of connections to the segment in the subassembly, minus the number of connections which go elsewhere. That segment which has the highest positive score is checked to see that its addition to the subassembly will not exceed size and power dissipation limitations. If it is too big, the candidate list is searched for

LIST OF INPUTS AND LOAD UNITS

INPUT		DU	LU	
FIN11A	NR01	6	5	
I0010A	I001	6	3	
FIN11A	PR01	20	10	
FIN11A	PR02	20	5	
FIN11A	PR05	20	6	
FCT11A	PR01	20	5	
FSP21A	SF00	6	5	
FSP10A	SF00	6	6	
MO0310	PR01	20	2	
FIN11A	PR03	20	7	
FIN11A	PR04	20	4	
FSYN0A	SF00	6	3	
FPU20A	PF00	20	3	
FPU10A	SF00	6	3	
FIN10A	NR01	6	2	
FSP11A	NR01	6	2	
FSP11A	SF00	6	2	
FSP11A	NR02	6	3	
FCT11A	EF01	20	5	
FSP10A	NR01	6	2	
FRIN1A	PR01	20	7	
FIN21A	NR01	6	2	
FIN10A	SF00	6	1	
FIN20A	NR01	6	2	
FIN11A	SF00	6	1	
FSP21A	NR01	6	2	
FSP20A	NR01	6	2	
FSP20A	SF00	6	3	
FSYN1A	NR01	6	2	
FSYN1A	SF00	6	1	
I0020A	I001	6	1	
TCP11A	ST01	6	4	
I0030A	I001	6	6	
FRIN0A	NR01	6	2	
FCT11A	NR01	6	2	
FCT10A	NR02	6	9	****OVERLOAD****
FCT11A	PF00	20	4	
FCT60A	SF00	6	5	
FRIN0A	SF00	6	3	
FCT50A	PF00	20	6	
FCT40A	PF00	20	5	
FCT30A	PF00	20	5	
FCT20A	PF00	20	6	
FCT10A	PF00	20	6	
FCT10A	NR01	6	2	
FCT31A	PF00	20	4	
FCT41A	PF00	20	4	
FCT61A	SF00	6	1	
FCT21A	PF00	20	3	
FCT51A	PF00	20	3	
FST210	PR01	20	4	
FPU11A	NR01	6	2	
FPU11A	NR02	6	5	
I0041A	I001	6	3	
FAC300	PR02	20	9	

Fig. 4

```
                    CONNECTIONS TO SUB-ASSEMBLY SERIAL NO.    1

              INPUT SIGNALS                        DESTINATIONS

              FIN11A  PR03                         FCT11A  EM01
              FRIN0A  SF00                         FCT11A  EM01
              FSP20A  SF00                         FCT11A  EM01
              I0030A  I001                         FCT10A  NR02
              FST210  PR01                         FCT10A  NR02


              OUTPUT SIGNALS                       DESTINATIONS
                                            CIRCUIT ELEMENTS     SUB-ASSY

              FCT11A  PR01                    FIN11A  PR01          0
                                              FPAR1A  NR02          0
              FCT11A  EM01                    FSP11A  NR02          0
                                              FPAR1A  NR02          0
                                              FVER1A  NR01          0
                                              FVER1A  NR01          0

UNUSED CONNECTOR PINS=      23

SIZE RESERVE=     0     POWER RESERVE=     0


                    PARTS LIST FOR SUB-ASSEMBLY SERIAL NO.    1

                              FCT11A  PF00
                              FCT11A  NR01
                              FCT11A  PR01
                              FCT11A  EM01
                              FCT10A  PF00
                              FCT10A  NR01
                              FCT10A  NR02
                              FCT10A  NR02
                              FCT10A  NR02
                              FCT21A  PF00
                              FCT20A  PF00
                              FCT31A  PF00
                              FCT30A  PF00
                              FCT41A  PF00
                              FCT40A  PF00
                              FCT51A  PF00
                              FCT50A  PF00
                              FCT61A  SF00
                              FCT60A  SF00
```

Fig. 5

the smallest segment. If the smallest segment will fit, it is checked against power dissipation limitations. Similarly, if power dissipation limitations are exceeded, the candidate list is searched for the segment which dissipates the least power. If this segment does not exceed power dissipation capability, it is checked against size limitations. In any of these cases, a segment is added to the subassembly if possible, and a new pool is formed to repeat the procedure.

If size or power dissipation has not been a factor, the segment is added to the subassembly, and a new candidate list is formed. This time formation of the candidate list and selection of that segment to consider for addition to the subassembly is based upon connections to both segments in the subassembly. This procedure is continued until the subassembly is filled, whereupon the pool is augmented, and the process starts anew.

The printed output from this program includes a parts list of elements in each subassembly, an accounting of unused space and unused power dissipation capability for each subassembly, and lists of input and output signals with their destinations for each subassembly. Figure 5 shows this output. The bulk of these data, together with data on internal interconnections within each subassembly, is also stored on magnetic tape to be used as input to the next program in the series.

The next, and last, program arranges the circuit elements in each subassembly in relation to each other in such a way that the cabling function approaches a minimum value.* The

* This program is an improved version of a program that was originally developed by Dr. Brewster Gere of Hamilton College and R. H. Glaser for one particular project, the Computer Detector for the W2F-1 Radar system (now E2-A). The improved version includes the ability to handle restraints such as preassigned locations and is applicable to any design, whereas the original program was written specifically for one project with its own design philosophies. The original program is discussed in "A Quasi-Simplex Method for Designing Sub-Optimum Packages of Electronic Building Blocks (Burroughs 220)" by R. H. Glaser which was presented at the Design Automation Survey Meeting of AIEE Design Automation Committee in Phoenix, Arizona, on February 22, 1960.

cabling function is the total length of wires, printed wiring runs, or other interconnection media between elements within the subassembly.

CONCLUSION

This is where the chronicle of the family of computer programs to aid design stops. These programs are the result of over a man-year of effort. There is still a need for additional programs to perform such tasks as providing interconnection data in the form of wiring lists or printed wiring layouts; to check delay times between clocked logic elements; to minimize logic; to implement boolean equations; to simulate logic and produce punched paper control tapes for automatic checkout equipment; and generally to perform many other tasks that now consume large amounts of the engineer's and the draftsman's time. In fact, several companies have developed computer programs to perform various of these functions.

Now, let's see what we have bought with the effort that was required to develop these programs. Since we have a *flexible* series of programs, they can be applied to all digital logic designs. A fair estimate of the manpower requirements for any one job of moderate complexity is close to that required to develop all of the programs. Thus, cost dividends start to accrue as soon as the programs are used for a second design project.

The time necessary to check a design against loading rules and to organize the layout of circuit elements within an equipment can now be reduced to a fraction of that previously required. For those in the military electronics business especially, where equipment requirements seem to change daily, the programs pay for themselves by materially enhancing quick reaction capability.

The use of a computer saves money and time in two ways. Besides reducing direct design labor, it doesn't produce human errors so that rework labor, scrap material, and test and troubleshooting labor are also affected.

Since the placement programs tend to minimize interconnection lengths, neater looking packages with less chance of electrical interference result.

When modifications are necessary, updating is greatly simplified when a computer is available.

Besides freeing the engineer or draftsman from boring and time-consuming jobs so he can devote his time to more creative effort, the use of the computer in design provides a powerful tool for experimentation. Alternate designs now can be checked-out rapidly; time just did not exist in the past to do this sort of thing. Cut-and-try approaches to such things as optimum subassembly size can also be tried in short order.

Other advantages accrue just from the use of the nomenclature system. This system lends organization that aids identification of a signal or circuit element without recourse to the logic diagram. Troubleshooting in the factory or the field is thus simplified to a degree.

Few systems exist without limitations and pitfalls. And this system of computerized design aids has some of both. The price paid for flexibility is a limitation to the amount of customizing that can be done. Thus, there are instances when manual editing and manipulation of the computer's output are necessary. Usage and experience, however, can lead to improvements of technique, and many times special computer programs can be written to handle special cases.

One pitfall that may trap the user of these programs stems from the ease with which changes can be incorporated into a design. We can all find ways to improve things about us, and electrical engineers can be like flighty women unable to refrain from little design improvements here and there. Thus, it is conceivable that a design may be held from final engineering release while the designers are making "one more change."

Now let's see how these programs worked when applied to the design discussed in the preceding paper. Preparation of the input data, including punching cards and checking the punched cards, required three days. Actual total running time on the computer was less than three quarters of an hour. This compares to approximately three weeks effort by one and one-half engineers and two design draftsmen to perform the same tasks without benefit of a computer!

The closure checks and dc load checks did not reveal any errors; however, these checks were performed by the computer after the equipment had been built, tested, and perfected.

The original design located the circuitry for four channels on six modules each, and the timer circuitry on six and one-third modules. The computer program also used six modules for each channel but managed to distribute the timer circuitry among six modules with no overflow on a seventh module.

One hundred thirty-four connector pins were used in each of the four channels in the original design. The computer-produced design used 113 connector pins for each channel. The original design employed 152 connector pins for the timer circuitry, while the computer-produced design required 158 connector pins for this circuitry.

An electronic computer can be used very effectively as an aid to the design of digital logic equipment. It is not a panacea at this stage, but it does help to do the work within a short time span and at greatly reduced cost, provided that it can be programmed to apply to more than one design effort. While providing these advantages, it permits the use of highly skilled manpower for truly creative work. These are significant facts in an industry where reaction time and cost are fast becoming the very keys to survival.

A Cost and Performance Analysis of Encapsulation by Transfer Molding

ROBERT F. ZECHER

Hull Corporation
Hatboro, Pennsylvania

This paper contrasts the transfer molding method of direct encapsulation of electronic components and assemblies with earlier methods, i.e., potting, casting, shell and pellet, etc. The specific areas covered include several types of devices, comparing previous encapsulating processes with transfer molding as to quality of encapsulation and as to production costs. A brief run-down of the transfer molding approach, including development work, cost study, material selection, tooling design, and production technique, is also included. Slides and sample parts will accompany the talk.

THE PAST TWO or three years have seen a significant increase in the use of transfer molding as a process for encapsulating electronic components and devices. This is due to several factors, perhaps the first of which is the need of the components manufacturers for higher production rates coupled with lower costs. Hand in hand with this need has been the development of increasingly superior molding compounds from a handling and a performance standpoint. Last, but not least, has been the sophistication of the marketing programs in electronics, which have placed a greater degree of importance on such items as esthetic appearance, in terms of shape, color, markings, etc., and on increased dimensional stability, for ease of handling, for automatic loading, and for minimum space requirements.

In general, transfer molding offers economies and increased production benefits because of the very essence of multicavity molds. Stated simply, it takes no more time to cycle a mold containing 100 cavities, and hence 100 devices, than it does to cycle one or two. Therefore, the only limiting factor is in the loading of the devices, and the question is whether or not that can be accomplished in the same time as the cure cycle for the particular compound chosen. Likewise, once a compound has been evaluated for performance and a suitable cycle evolved from the development stages on a particular device, this process can be readily transferred over to production of significantly greater quantities with no reengineering of the encapsulation process required.

The performance of units so encapsulated is in most cases equal to or better than the previous methods of encapsulation using thermosetting resins. By far the bulk of this encapsulation is with the epoxy molding compounds, and this is a direct carry-over from the extensive use of liquid epoxy resin systems for casting and potting. Eminently successful additions have been the silicone plastic molding compounds which possess several unique properties, and which have not been available as potting compounds in a comparable formulation. The same is true of the diallyl phthalates. Evidence of this general performance level of the molding compounds is given by the fact that many transfer-molded encapsulated devices are meeting MIL Spec requirements for moisture and for mechanical strength, although the basic compounds themselves have not yet been accepted as a group by the military.

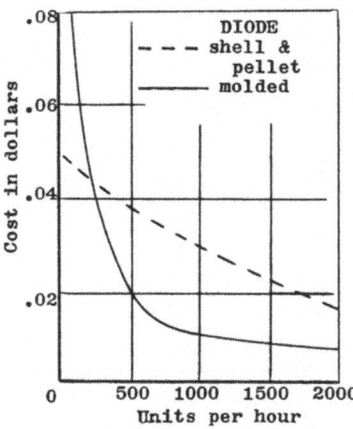

Fig. 1. Encapsulation costs of shell and pellet-molded diodes.

Fig. 2. A 60-cavity diode mold, loading frame, and molded units.

Of equal importance with the chemical properties of the molding compounds is the processing uniformity of encapsulation by transfer molding, which results in a generally increased reliability over older methods using liquid resins. Transfer molding presents an opportunity to eliminate much of the human error in the mixing and subsequent processing of liquid resins.

TRANSFER-MOLDED DIODE

Consider first the case of a simple silicon diode previously encapsulated by means of the shell and pellet technique. This process is a familiar one in which the diode is inserted into a molded shell, usually epoxy or diallyl phthalate, and an epoxy preform is placed in the open end of the shell. This assembly is then passed through a curing oven where the epoxy preform melts, flows down into the shell, and eventually cures, giving a completely encapsulated device. The advantages of this system over previous methods were the absence of a requirement for extensive capital equipment, the uniform size and shape of the final package, the elimination of preweighing and mixing a two- or three-component resin system, and the adaptability to automatic handling. The chief disadvantage is the cost of the combination of shell and preform and the labor involved in individually handling each device up to the point where automatic equipment can be used.

Figure 1 shows the comparative costs, at various levels of production, for the molded diodes versus those encased in epoxy shells. The curves show how the molding quite clearly has an edge in the higher production levels, chiefly due to the reduced labor involved, although a large portion of the cost of packaging a diode in a shell is the shell itself. In order to avoid excessive labor costs if all of the units were hand assembled at the higher production rates, automatic machinery must be considered for handling the diodes and shells, and a conveyorized oven system is probably also required. Figure 2 shows a 60-cavity diode mold complete with a loading fixture containing the finished molded diodes. With a single operator and an overall cycle of under 2 min, it is easily understood why the low cost of molding at these higher production rates is so attractive.

The performance of the two types of diodes is, of course, quite similar, inasmuch as the epoxy shells are molded of similar or the same material as that used in the direct encapsulation. Of course there is always the possibility of air entrapment under the pellet or incomplete cure of the pellet, but likewise, in the molding operations voids can occur due to incomplete fill or improper venting. However, given two successfully encapsulated units in the two different methods, the electrical and mechanical performance is the same. Units encapsulated in these two methods have both passed the MIL Spec 202 method 106A humidity test, which is a ten day humidity test conducted at 158°F, and 98% relative humidity. It is also important to note that the pressures involved in the transfer-molding operation apparently do not have any adverse effect on the performance of the diodes, as units have now been observed in service for four and five years with no apparent degradation due to this factor.

"WRAP-AND-FILL" CAPACITOR

A second comparison of the transfer molding method with an older packaging technique pits the "wrap-and-fill" process for encapsulating mylar capacitors against molding. Figure 3 shows the comparative costs at various production levels of the two processes. The "wrap-and-fill" technique lends itself to automation likewise only in the very high production rates, and so far this automation has only pertained to the wrapping portion. Filling is still largely done by hand, using a liquid epoxy resin system, and the usual cumbersome procedure of weighing the proportions of the resin and hardener, mixing the material, vacuum degassing the material, and dispensing it from syringes or other devices, is to be contended with. Essentially, the performance of the molded capacitor may be somewhat superior to the wrapped unit in that the polyester film tape is not as moisture proof as would be thought. Then there is also the possibility of moisture leakage through the epoxy end seal, either due to poor adhesion of the epoxy to the tape or to the leads, or due to the presence of a void in the epoxy. The molding

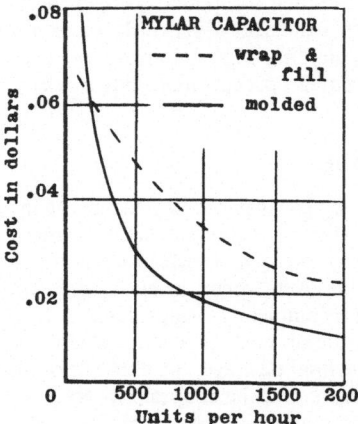

Fig. 3. Comparative encapsulation costs of "wrap-and-fill" and molding processes.

also gives the additional advantage of a permanent polarity indication which can be molded right into the unit, and which does not depend on the durability of an ink marking system.

SOLENOID COIL

An area of even more obvious cost benefits, as well as performance benefits, is in the encapsulation of solenoid coils. Figure 4 shows the cost curves of a vacuum-potted solenoid coil versus a molded coil at various levels of production. Generally, the cost of potting such units runs rather high as individual molds are required for each coil, vacuum potting is usually desirable to eliminate voids which would cause failures in the presence of higher voltages, and the cure cycle of the liquid epoxy resin systems is relatively long even at elevated temperature (1 hr or even more). This means that as the higher production rates are approached, the cost becomes excessive both as to the investment in molds and also as to the amount of labor required for cleanup and for the assembly of the unencapsulated coils in the molds.

At even moderate production rates it becomes advantageous to look into automatic metering, mixing, and dispensing systems for proportioning the resin and for casting under vacuum. A conveyorized oven system may be necessary quite early in the production picture. At a production rate of 250 coils per hour, such a combination of dispensing equipment and oven might well indicate an investment of up to $50,000.

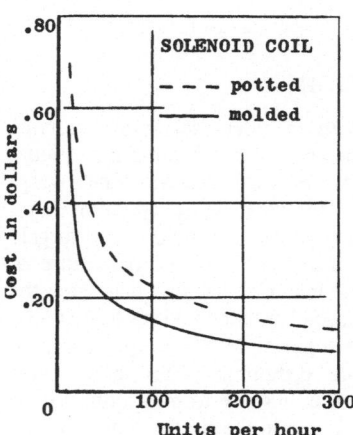

Fig. 4. Comparative encapsulation costs of vacuum-potted solenoid coil and molded coil.

Fig. 5. Shuttle press.

Molding, on the other hand, seems to have an edge all across the board, again primarily due to the short cure of the molding compound and the reuse of each individual cavity in the tool rather quickly. In the case of high voltage coils, preimpregnation of the coil may be required where molding is to be performed, but this can be done fairly inexpensively on a batch basis. The use of a special press, such as the shuttle press shown in Fig. 5, can bring about further economies at high production levels, because the loading of solenoid coils, particularly those with flexible leads, has a very important bearing on the overall cycle time. The shuttle arrangement allows the operator to unload and reload one bottom half of the die while the other section is in the press undergoing the cure cycle of a batch of coils.

MODULE PACKAGE

Of vital interest in today's packaging picture is the module, or other circuit, consisting of a number of individual components or devices. Figure 6 is a comparison of the cost of potting a typical module in a shell with liquid epoxy resin as opposed to molding the unit directly. There

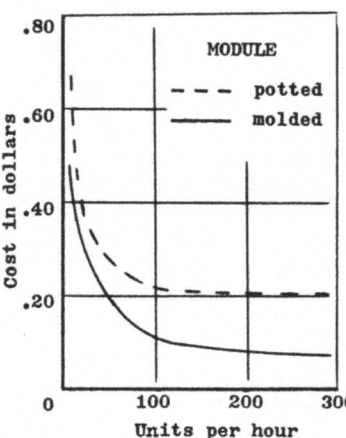

Fig. 6. The cost of potting a module in a shell with liquid epoxy resin compared with the cost of molding the unit directly.

has been a degree of concern over the capability of such a device to withstand the molding temperatures and pressures involved, but experience, in both limited and full production, has shown this problem to be overemphasized. As in the case of the molded shell in the diode picture, the cost involved in potting the module in a shell is largely tied up in the molded case. Because of the tremendous variety of sizes and shapes of modules, most of these molded cases must be custom made, making their cost still more prohibitive except on a high-volume basis. As in the case of potting the solenoid coil, vacuum may be required to insure a void-free encapsulation of the unit, and cure is generally done at elevated temperature and requires at least an hour or more.

Figure 7 shows a two-cavity module mold and some finished units, and represents a production rate of up to 50 units per hour. In high production molding of both the solenoid coil and the module, the cost includes the auxiliary equipment necessary to handle large quantities of material, including preforming and preheating equipment. As with the solenoid coil, the loading operation for modules can be greatly enhanced by the use of a machine such as the shuttle press of Fig. 5.

Performance of the transfer-molded modules in several production installations has met and exceeded the performance of identical units potted in shells with a liquid epoxy resin. In addition to possessing almost identical finished properties to the liquid resin systems, epoxy molding compounds have several advantages during the processing which leads to a lower reject rate upon final testing, and no doubt to a lower failure rate in service. In fact one company reports a reject rate, in relatively high production of transfer molded modules, of less than 0.5%. Probably the most important factors are the one-component material associated with transfer molding and the more easily controlled cycle conditions, whereas the liquid epoxy systems require proportioning, mixing, deaerating, pouring, and finally curing.

These four cases are fairly representative of types of components and circuits which lend themselves to encapsulation by transfer molding. However, they are by no means all inclusive, and it is interesting to note that the use of transfer molding in electronic applications does not stop with total encapsulation. Figure 8 shows the operator removing molded potentiometers

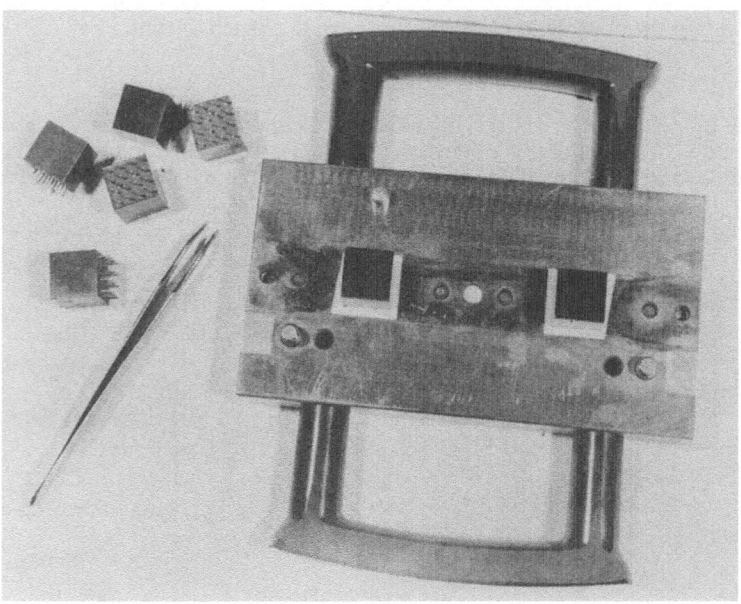

Fig. 7. A two-cavity module mold and some finished units.

Fig. 8. The removing of molded potentiometers.

in which the carbon ring has been left exposed. Similar applications exist in the molding of pilot light assemblies with the bulb left exposed, in the partial molding of bases and headers on circuits, in molding motor and generator stators and rotors with the laminations exposed, and so forth. Since, then, most basic components and simple circuits can be transfer molded, what are the basic steps an engineer can take to investigate this process from a cost and performance standpoint?

EVALUATION OF TRANSFER MOLDING

In approaching the question of whether or not to use encapsulation by transfer molding, there are several considerations. The chief advantages to the use of transfer molding are of course the shorter curing cycles and the easier handling of a material in one-component form. This means simply that a single cavity can be repeatedly used to produce parts instead of being tied up for several hours while a liquid resin is curing. The use of one-component formulations effectively reduces operator error. All of the cost pictures shown above include a large outlay for equipment, which is reflected in the high cost of producing at low rates. In the case of the larger devices such as the module and the solenoid coil, this high cost of equipment is offset by the usually complicated potting procedure required to produce void-free encapsulations. However, it is possible to begin molding on a fairly modest scale, providing attention is paid to the tooling design. Then the break-even point probably comes at a production rate which involves more than one operator in the older method. Since one press with one mold and one operator can effectively produce up to the maximum production rates shown in the four charts, it can be seen how easily the labor becomes an important part of the cost picture.

In all of the above cases the lower-to-moderate production rates are achieved by the use of removable hand-type molds, which are relatively inexpensive. It is important to consider that a production tool has to have a certain minimum quality (and hence cost) or cycle times will be excessively long and reject rates excessively high. Aluminum or soft steel tools are often used in development programs or for prototype work, but when a mold is to be put into continuous operation for the production of hundreds or even thousands of parts, the hardened, chrome-plated, and polished tool generally has the edge. Other important features such as knockout pins, loading frames, placement of heaters, whether a mold is to be removable or permanent, and so forth, all have to be taken into consideration. It is important to note that a large multi-cavity die which can produce just a few cycles more per day than a less expensive tool can easily overcome its initial cost in just a few weeks or months of production. On the other hand, a poor mold can be a costly headache from both a lower production rate and higher reject rate standpoint.

COST DETERMINATION

It is a fairly simple matter to determine the cost of encapsulating a particular device at an established production rate. The volume or weight of material per device can be determined, and a suitable waste factor can be added from the size and layout of the mold. A large production mold will have as little as 20% material waste, a simple one- or two-cavity hand mold may waste 50–100%. The cost of epoxy molding compounds runs 50–100% higher than the average liquid epoxy resin system, but this cost can vary greatly as a function of such factors as filler and color. As a rule, the waste involved in handling a liquid epoxy resin system is considerably higher than that involved in molding, so the cost of materials may pretty well balance out.

A transfer press and a mold are normally the only capital equipment needs for the encapsulation of smaller devices, but preforming and preheating may be necessary for larger ones. Naturally, additional molds are required for each component or each size. A transfer press may be fairly inexpensive, some being little more than a small press frame and a couple of air cylinders, but the requirements for even moderate production or lab work usually demand a hydraulic press with semiautomatic controls and a moderate degree of versatility. Such a press is usually written off as a piece of capital equipment over a period of five years, since it does have resale value at any time.

Tooling, on the other hand, can be written off in a variety of ways, depending on the extent of the program, the type of company and product, and the type of tooling desired. Naturally it is very expensive to write off a fairly good sized tool over a short program of perhaps a few months of production. It is also interesting to note that for a fairly large mold of many cavities the price may equal or top that of the press. However, when the production rates are figured into the cost of a mold over a reasonable period of time, the economy factor is easily seen. For instance, a 60-cavity mold written off over one shift-year of production costs only about 0.05 cents per part per \$1000 of mold cost.

The labor cost in transfer molding is most attractive, inasmuch as one operator can usually handle a press and large tool all by himself, including cleanup and deflashing of the finished parts. With a well designed mold and loading-fixture arrangement, and a semiautomatic molding press, the only human time factors involved are the cleaning of the mold of flash and other debris, and the loading of the alternate frame or loading fixture with unencapsulated devices during the cure cycle. The molding compounds are handled in their powder form or as preforms and offer no mixing, pot-life, or dermatitis problems. The tooling should be designed to provide easy loading, and alignment of the devices should be accomplished as nearly automatically as possible by the closing action of the press. The semiautomatic cycle provides close control of the cure time, and naturally, the other variables, heat and pressure, remain fixed.

PERFORMANCE

The second factor to be considered is, of course, performance, and here there are two areas to be evaluated with regard to the transfer-molding approach. One is naturally the finished properties of the encapsulated units. The second is the advantage of the transfer-molding method during the actual processing as it affects the properties and performance of the devices. There is a great deal of misconception about the latter as regards the temperatures and pressures involved in transfer molding. Although in most cases the molding temperatures recommended (from 175 to 350°F, depending on the compound) may exceed temperature limitations imposed by the military, they are normally applied only for a very short duration. It has been found that such limitations apply more to continuous or long-term exposure, and transfer molding has been used on many such temperature-sensitive devices with no apparent loss of performance. This is especially true of the semiconductor devices.

The same is largely true of pressures. It should be noted that although the compounders advertise transfer-molding pressures as low as 50 psi, and in many cases molding can be accomplished at these low pressures, the average pressures run much higher in practice. This usually comes about through a gradual increase in pressure during the development of a cycle in an effort to gain better molded properties of the material. Of course the pressure cannot be increased to the point where destruction or damage to the devices occurs, but it is not uncommon to use 1500–2000 psi in the encapsulation of a welded circuit module which is generally considered to be susceptible to damage at much lower pressures. The reasoning behind the acceptability of these higher pressures is simple enough: the material enters the cavities as a liquid and must completely fill the cavity and surround the device before the pressure builds up. Therefore, it is like squeezing on an egg with a uniform pressure from all sides. Controlling the rate of flow of the material into the cavity can play a significant part in avoiding damage to the components also.

MOLDING COMPOUNDS FOR ENCAPSULATION

Table I shows a cross section of several epoxy molding compounds, a silicone molding compound, and a diallyl phthalate (all encapsulation grades), compared with a typical liquid epoxy resin system. From the properties given in the table it is easy to see that there is quite a variance in certain specific areas among compounds. There are a dozen different manufacturers offering from two to a dozen or more compounds each—quite a variety of materials available for the packaging engineer to select from.

Epoxy molding compounds, either A or B staged, are available in a variety of colors, with outstanding electrical or mechanical properties, with soft, medium, and stiff flow, in short and longer curing forms, in preforms, and, more recently, in varying prices. Since they possess most of the properties of the liquid epoxy systems, they are being used to replace potting and casting.

TABLE I

Comparison of Resin Properties

Properties	Molding compounds					Epoxy casting resin
	Epoxy A	Epoxy B	Epoxy C	Silicone	Dap	
Specific gravity	1.74	1.6	2.1	1.86	1.79	1.49
Tensile strength (psi)	3350	9000	—	3500	5500	10,000
Mold shrinkage (in./in.)	0.006	0.004–0.005	0.009–0.011	0.007	0.004–0.005	0.50*
Flammability	NB	SE	NB	SE	—	SE
Heat distortion (°F @ 264 psi)	282	280	291	750 (cont.)	350	158
Thermal expansion (in./in./°C $\times 10^{-5}$)	4.75	2.4	5.7	—	4.1	3.7
Water absorption (percent—24 hr)	0.096	0.05	0.061	0.34	0.25 (48 hr)	0.21
Dielectric strength (volts/mil)	400	370	—	340	300	—
Dielectric constant (short time—dry)	3.78 (60 cycle)	3.6 (1 Mc)	4.75 (100 cycles)	3.6 (1 Mc)	4.3 (1 Mc)	4.21 (1 kc)
Dissipation factor (1 megacycle)	0.013 (60 cycle)	0.018	0.0105	0.002	0.009 (wet)	0.0265 (1 kc)
Volume resistivity (ohm-cm)	1.26 $\times 10^{16}$	—	1.43 $\times 10^{14}$	5 $\times 10^{14}$	1 $\times 10^{13}$	3.99 $\times 10^{14}$

* Indicates percent shrinkage during cure.

Silicone molding compounds have no counterpart in a liquid resin system, but they are competing with the epoxies for much of the encapsulation where high temperature performance is a requirement. In fact, silicones have the highest continuous operating temperature of any of these molding compounds, and so have a place all to themselves in that respect. Another application of the silicones is in the molding of circuits where repair or replacement of individual components is desirable. The silicones are attacked by more conventional solvents, and so have an edge in this respect, since it is almost impossible to remove epoxy without some destruction to the circuit.

Two other types of compounds which are coming into some use for encapsulation by transfer molding are the low-pressure phenolics and the low-pressure, soft-flowing diallyl phthalates. The phenolics are well known in the electrical industry and possess a distinct price edge over the epoxies, but have always been too stiff for the encapsulation of devices in a large multicavity mold. The new encapsulation grade compounds flow at pressures roughly comparable to those used in epoxy molding, and so can be used with even the most delicate components. The diallyl phthalates, of course, have always enjoyed military approval, and are well known in the connector and in the potting-shell field. Until recently, attempts at compounding softer flowing diallyls have resulted in generally detracting from their outstanding properties, but the new encapsulation grades seem to perform quite nicely in most respects. Of course the question of military approval on these new diallyls remains to be seen, and perhaps the epoxies will have gained military approval in the near future and render that qualification unimportant.

From the performance specifications given by the molding-compound manufacturers then, the engineer can choose from quite a variety of materials as far as properties are concerned, but, as with most applications involving the use of reactive resin systems, it is wise to perform certain development work to ascertain that these properties are still available when applied to a particular device and a particular configuration. Here again, the advantage of transfer molding over other processes is that a single cavity mold can be used to determine the application of a particular material and to evolve cycle conditions, and the whole process can then be transferred to a multicavity mold with the same results obtainable on a production basis.

Table II is a chart of the outstanding properties available from molding compounds and indicates just what single performance factors are obtainable from the presently available compounds. It should be noted that new compounds are being introduced at a very rapid rate and these properties will soon be surpassed. As with most insulation systems, to gain one

TABLE II

Outstanding Properties Chart

Property	Value*	Material
Specific gravity, low	0.887	Epoxy
Specific gravity, high	2.1	Epoxy
Tensile strength, psi	10,000	Epoxy
Impact strength, ft-lb per in. notch	0.69	Epoxy
Heat distortion, °F @ 264 psi	425	Epoxy
Continuous heat resistance, °F	750	Silicone
Water absorption, percent—24 hr	0.009	Epoxy
Thermal expansion, in./in./°C	21×10^{-5}	Epoxy
Thermal conductivity, cal/sec/cm^2/°C/cm	25×10^{-4}	Epoxy
Low molding temperature, °F	175	Epoxy
Mold shrinkage, in./in., low	0.002	Epoxy
Nonburning	several	Epoxy
Dielectric constant, low	2.52 (1 kc)	Epoxy
Dielectric strength, short time, volts per mil	432	Epoxy
Volume resistivity, ohm-cm	1.26×10^{-16}	Epoxy

* These values are taken from published data of those compounds listed as encapsulating grades (500 psi transfer pressure or less), and indicate the best processing technique suggested by the manufacturer.

TABLE III

Advantages of Transfer Molding

Finished properties	Variety of electrical and mechanical properties Excellent surface appearance Homogeneous, void-free encapsulation Multicolors (interchangeable) Permanent engraving, if desired Dimensional uniformity Lead centering
Cost and processing	One part system—no mixing Short cures (short exposure to heat) High production rate per operator Uniform processing Low reject rate Less material waste Elimination of dermatitis problem Automatic handing
Others	Easily increased production rates No reengineering from development work to production No supply problems on cases, headers, and shells

outstanding property others may have to be sacrificed, and Table II is representative of several different compounds. However, in many cases, two or more of the properties listed are available in one compound.

SUMMARY OF ADVANTAGES

Encapsulation by transfer molding is an approach, then, which may be considered by the engineer for a great variety of devices and circuits where moderate-to-heavy production rates are involved. The selection of materials is now based on a very broad offering by the plastic resin suppliers. The choice of equipment and tooling has likewise expanded, and covers both hand and large production molds, multipurpose and specialized machines, and so forth. The observation of the performance of molded devices over the past few years and the squeeze on manufacturing costs, coupled with high performance requirements, definitely have established transfer molding as an excellent approach to packaging problems. Table III presents a summary of some of the advantages of transfer molding on a general basis, but the engineer must consider his own individual application to evaluate all of the possible benefits.

Packaging of Miniature Diode Assemblies

H. B. Bell and G. A. Doyle

Motorola Semiconductor Products Division
Phoenix, Arizona

Most electronic assemblies are fabricated by connecting discrete packaged components in the desired circuit and then packaging by potting in plastic. Transfer-molding techniques, along with semiconductor surface passivation techniques, have been developed that eliminate the redundancy of this repackaging of hermetically sealed semiconductor packages. This approach has been used to produce a family of miniature diode assemblies. Easier assembly and smaller size, as well as lower costs, have been achieved.

INTRODUCTION

Among the many developments of material for electronic packaging applications, the developments in thermosetting transfer-molding compounds have been of particular interest to electronics manufacturers as a means of economical packaging. With present epoxy and silicone plastics molding technology, it is possible to substitute transfer molding for the potting of electronic assemblies directly. In this manner, manufacturing economy is achieved because a larger number of assemblies can be conveniently handled and processing time can be significantly reduced. However, this change alone will realize only a fraction of the possible cost reduction that can be achieved by using simplified assembly techniques along with transfer-molding techniques.

Most assemblies are fabricated by soldering or welding discrete packaged components into the desired circuit and then packaging by potting in plastic. This system is redundant because it is essentially the repackaging of hermetically sealed packages. Another disadvantage is that each of these individually packaged components occupies an unnecessarily large volume. Furthermore, extensive bending and shaping of component leads is necessary for circuit interconnections. All of these factors tend to make assemblies bulky, complex, and expensive. In some cases these problems can be avoided so that lower cost and smaller size are achieved by eliminating the use of individual prepackaged units, as well as the intricate, expensive, stepwise welding or soldering of individual diodes to form the basic circuit.

DISCUSSION

This approach has been used very successfully in producing a family of Miniature Integral Diode Assemblies (MIDA). In this case, unpackaged device compatibility, interconnecting lead layouts, assembly procedure, and encapsulation technique, as well as their mutual interaction, were considered from the initial package-design concept to final production actuality. The four standard rectifier circuits used in this study are shown in Fig. 1. They are: (1) a single-phase full-wave bridge; (2) a single-phase voltage doubler; (3) a single-phase center tap with common cathode; and (4) a single-phase center tap with common anode. The essential elements of physical location and electrical connection are depicted by the exploded view of the bridge rectifier shown in Fig. 2. Here are indicated the continuous single-element lead forms that

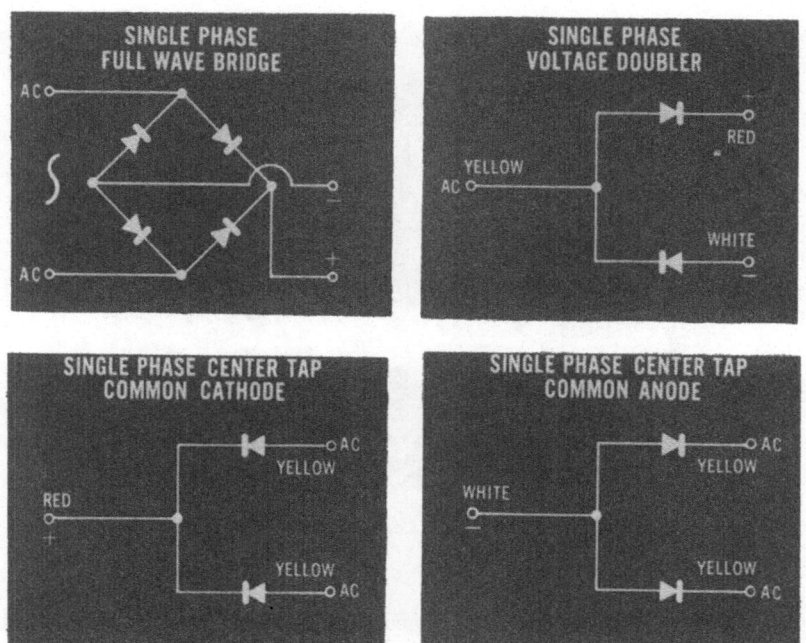

Fig. 1. Rectifier circuits.

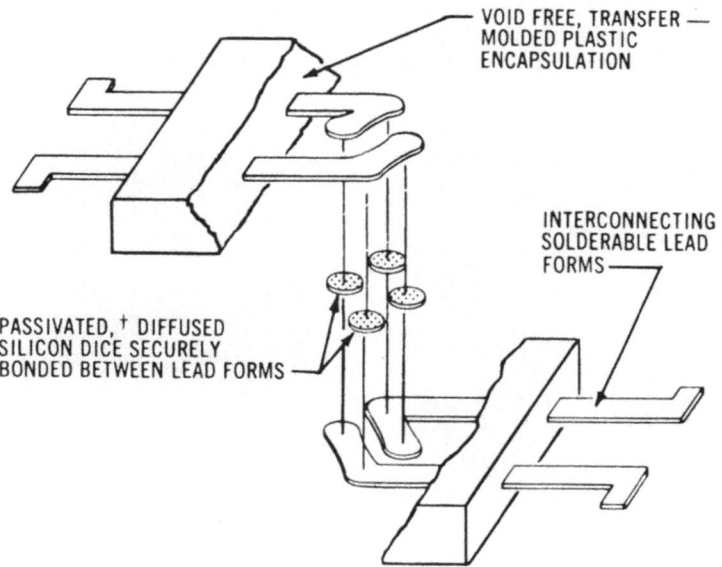

Fig. 2. Exploded view of bridge rectifier.

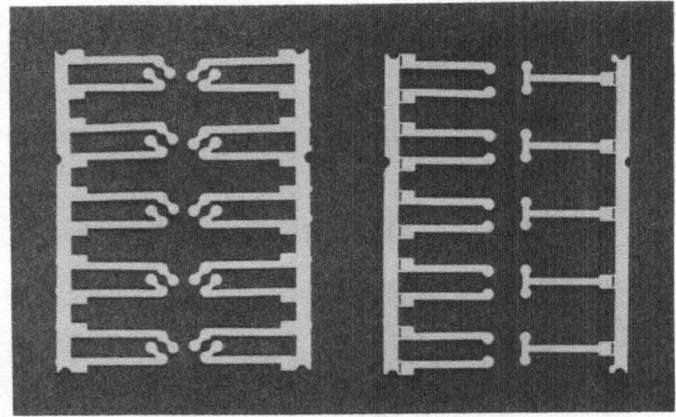

Fig. 3. Lead combs.

provide both interior and exterior electrical connection. Also indicated are the large-area-junction silicon rectifier dice that display good current capability and thermal dissipation. This lead-dice assembly is encapsulated in a thermosetting plastic to yield an efficient, strong package.

PROCESS CONSIDERATIONS

Surface passivation of the silicon diodes is absolutely essential to give the necessary ambient isolation, so that assembly and molding can be done without degrading device electrical parameters. This can be achieved by using a thermally-grown oxide layer [1], an accelerated thermally-grown oxide layer [2], or a vapor-deposited oxide layer [3]. The accelerated thermally-grown oxide layer gives the best passivation and ambient isolation for this particular diode-device geometry. In this manner the glass passivation layer is produced selectively, in the area of the sensitive peripheral junction termination and not on the ohmic contact areas that must

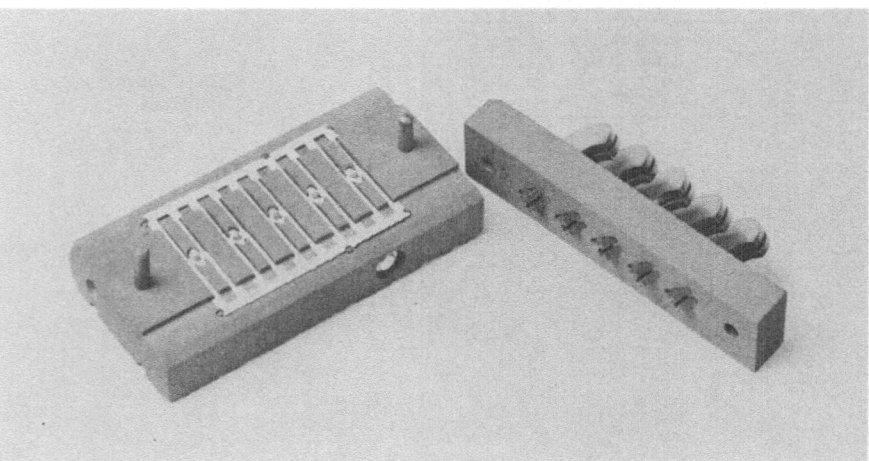

Fig. 4. Assembly-solder jig.

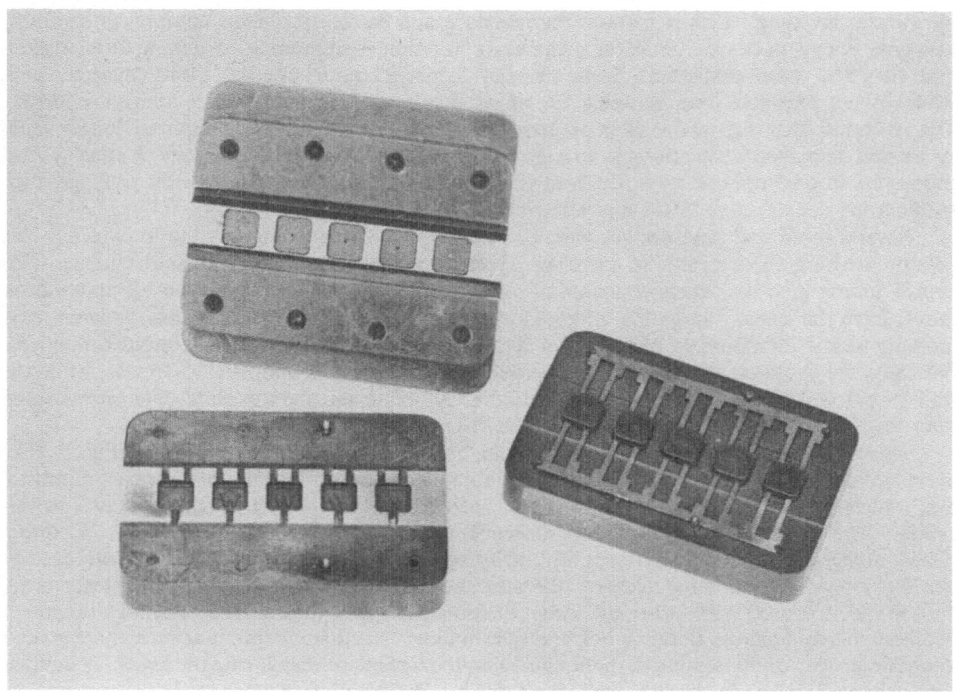

Fig. 5. Mold inserts.

remain solderable. Thus, a hermetic seal is achieved that is as impervious to the penetration of moisture as are conventional hermetically sealed packages.

Assembly of the dice into the desired configuration and provision of the necessary electrical leads is accomplished with a special design lead-comb technique.* The lead comb for the bridge (see Fig. 3) is reversible, so that just one piece part will satisfy the need for both right- and left-hand orientation. In the case of the doubler and center tap, two combs, as shown in Fig. 3, are required. Location and alignment of these combs is achieved simply and accurately by notches in the comb which correspond to locating pins in the assembly–solder jig. Enlarged areas are located on each lead finger. The purpose of these defined land areas is to align opposite dice contacts during assembly. In this manner, the maximum lap-joint bonding area of lead to die is utilized so that good mechanical strength is achieved. Large-area contact also permits excellent transfer of the heat from the die to leads. Designing the leads with a high heat conductivity (copper or silver) and large cross section (10 × 40 mil) then permits excellent heat transfer from device to heat sink. Sizing of the lead clip is based on the best process economy obtainable. This takes into account the assembly considerations of leads, dice, and solder, as well as the molding considerations of handling and locating large numbers of assemblies.

The assembly–solder jig is shown in Fig. 4. The jig loading sequence is simply: (1) lower clip; (2) solder preforms; (3) properly oriented dice; (4) solder preforms; and (5) top clips. The top portion of the jig has movable, metal-weight inserts that apply a constant force on the forty solder joints while in the soldering furnace. Thus, using positive contact and high-temperature solder, a strong and reliable bond is achieved. By using an inert furnace atmosphere and excluding solder flux, most of the surface contamination that can give extraneous paths of conduction is eliminated, thus yielding electrically stable devices.

* Patent applied for.

A compact, void-free, strong plastic package with good electrical properties is achieved by transfer molding. This eliminates the moisture and corona problems common in ordinary cast-type epoxy packages. Several proprietary molding compounds, including both epoxes and silicones, were evaluated. Some epoxies displayed poor viscosity–flow characteristics, while others required long molding cycles or displayed low temperature limits (<150°C). The silicones, although showing good temperature stability (>200°C), required longer mold cycles and displayed poor adhesion to metal leads. For this application, Pacific Resins' S-1311 epoxy molding compound gave the best combination of easy molding, sturdy package, high temperature resistance (175°C), and electrical stability.

Several mold and gate designs were considered. Through-gating has the drawback of an adverse washing effect as well as excessive direct force that can damage lap solder joints. End or side gating involves excessive material loss in runners, as well as a problem of material removal from the leads. Pin gating, although requiring some runner material loss, provides easy molding with a minimum of blemishes or flash. For pin gating a three-plate mold is required. Basically, the mold cavities are located between the bottom and middle ("X") plate, while the transfer pot and runners are located between the "X" plate and the top plate. The gates extend from the runners down through the "X" plate into the top of the cavities.

The mold inserts (removed from the chase for easier viewing) are shown in Fig. 5. The mold cavities are produced by well-known hobbing techniques. Polarity marks, part numbers, and trademarks are included in each cavity to provide for convenient orientation and identification. This eliminates the need for subsequent orientation and marking steps. Chrome-plated, highly-polished mold surfaces not only produce a package with superior appearance but also make plastic ejection much easier. Because the parting line is on the face of the lead comb, the top insert is used with either the three- or four-lead lower insert. The assembly alignment notches are used again to obtain quick and easy loading. Lead-slot depth is accurately regulated to obtain proper cavity sealing without damaging the devices or disfiguring the leads. A molded clip is shown in place in the four-lead lower insert. The molding sequence is as follows:

a. The soldered assembly is located by pins in the lower mold plate. Heated platens maintain the correct mold temperature.

b. The molding press is closed and clamped.

c. A premeasured amount of molding compound is charged to the transfer pot.

d. The transfer ram is brought down on the molding compound. A timer is activated by the ram so that the correct amount of time is provided for melting, flowing, filling, and partial curing of the plastic.

e. When the molding time is completed, the ram is withdrawn and the platens are opened. The gates are broken off when the top and "X" plate start to separate. As separation continues, the gates, runners, and pot cull are ejected. Then, as the bottom and "X" plate separate, the molded comb is exposed and removed.

Usually the mold cure time is regulated to obtain sufficient plastic "hot strength" for safe ejection.

Complete curing of the plastic is then assured by a post-cure step. Post-cure time and temperature depend largely on the extent to which plastic polymerization had progressed while in the mold. After post-cure the individual units are separated from the comb by shearing off the interconnecting strip.

TESTING AND EVALUATION

Testing under various stress conditions was done to verify performance and stability. In Fig. 6 is shown a typical output curve-trace of the full-wave bridge on room-temperature power aging. Drawing a load of 1.2 A with a 92-Ω series resistance resulted in an approximate junction temperature of 160°C. Sample lots have passed 1000 hr of room-temperature power aging without failing. Elevated-temperature power aging was also done. In this test a 300-mA current was drawn by the units while in an oven set at 125°C. Sample lots have satisfactorily passed

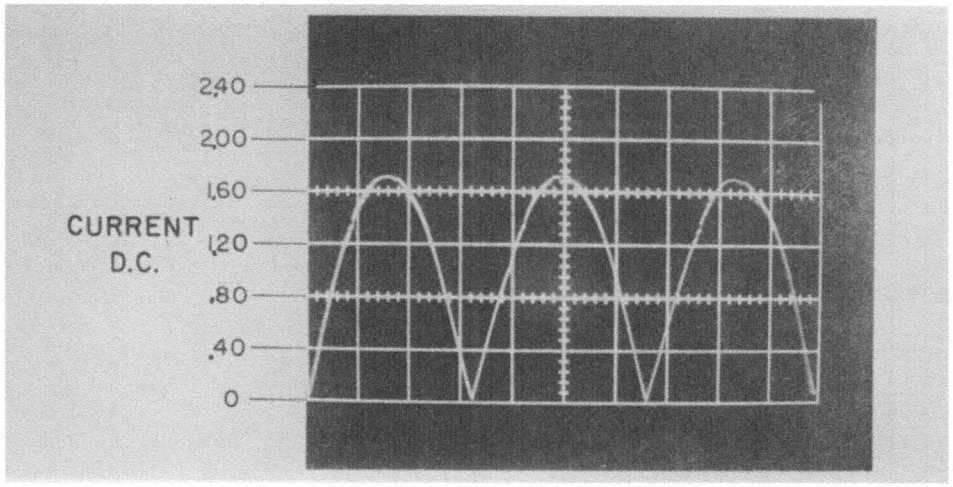

Fig. 6. Output characteristic.

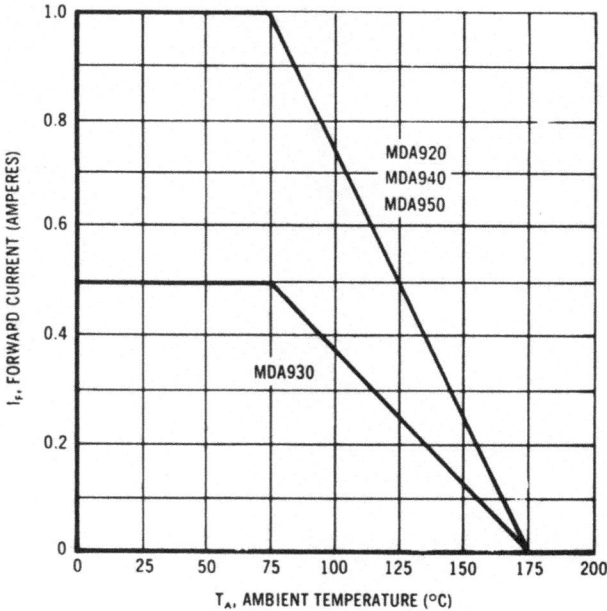

Fig. 7. Derating curve.

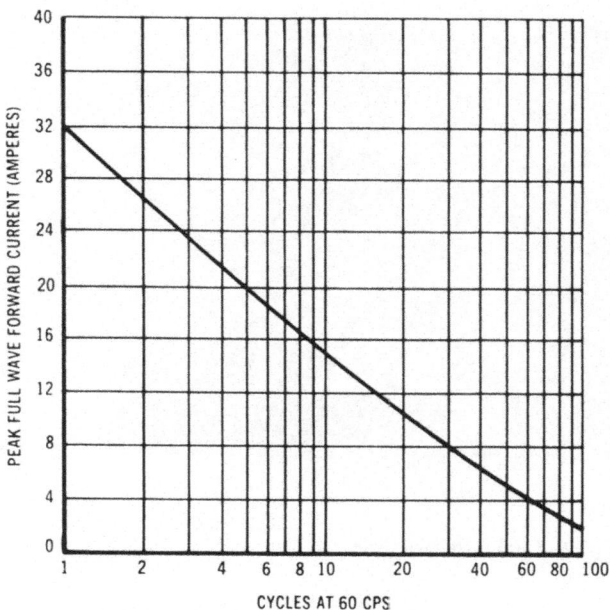

Fig. 8. Maximum forward surge current.

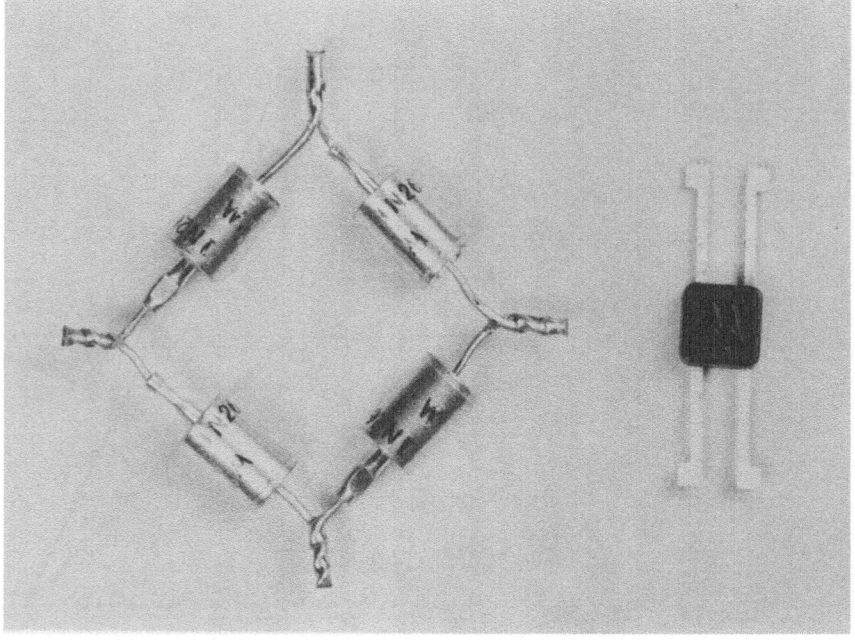

Fig. 9. Discrete and MIDA full-wave bridge rectifiers.

500 hr of this testing. Steady-state humidity testing at conditions of 90–95 % relative humidity and 40°C for 240 hr was done on several lots without failure.

Current ratings and capabilities for these devices were developed. In Fig. 7 is shown the derating curve for a maximum junction temperature of 175°C. With a 75°C ambient, the full-wave bridge and center-tap rectifiers are rated at 1 A while the voltage doubler is rated at 0.5 A. Surge-current capability is shown in Fig. 8. The maximum forward current, at 60 cps, ranges from 32 A for 1 cycle to about 4 A for 60 cycles.

The most important application advantage of these devices is size versus rectified output. A crude comparison of a discrete-component bridge rectifier and the MIDA full-wave bridge rectifier is shown in Fig. 9. In actuality, the MIDA package requires only about one-tenth the volume that is required to contain the discrete devices.

Several simple applications of the MIDA full-wave bridge rectifier were tried for improving operation of small universal motors such as those used in electric shavers or food mixers. In each instance the increased current from the rectified AC improved the starting torque and extended the motor power and speed range. Because of its small size the device can be conveniently located inside the motor case to improve performance without increasing size.

Excellent results were also obtained using the MIDA full-wave bridge rectifier as the small-power DC supply for instrumentation and communication applications.

REFERENCES

1. M. M. Atalla, E. Tannenbaum, and E. J. Scheibner, "Stabilization of Silicon Surfaces by Thermally Grown Oxides," *Bell System Tech. J.* **38**, 749–784 (1959).
2. D. A. Kallander, S. S. Flaschen, R. J. Gnaedinger, Jr., and C. M. Lutfy, "Accelerated Thermal Oxidation of Silicon," Indianapolis Meeting of Electrochemical Society, May 1961.
3. D. Peterson, "Evaluation of Vapor-Plated Oxide Films for Capacitor Dielectrics," IEEE Transactions. of the PTGCP, Vol. CP-10, No. 3, September 1963.

Transfer Molding of Silicone Compounds for Environmental Protection of Electronic Modular Systems

Frederic J. Lockhart

Dow Corning Corporation
Hemlock, Michigan

Encapsulation of modular electronic circuitry by transfer molding has introduced a new packaging concept to the electronic industry. The combination of silicone molding compounds with modular circuitry has fulfilled the basic requirements for environmental protection of modular systems. Characteristics and advantages of these compounds, transfer molding techniques, and material processing are described in depth in this paper. Detailed information includes electrical characteristics, molding temperatures and pressures, and mold-design data.

INTRODUCTION

THE ART OF TRANSFER MOLDING thermosetting plastic materials has been known and practiced by the plastics industry for many years. Until recently, this technique was employed almost exclusively for the manufacture of molded structural parts. Today, with the electronic industry seeking new encapsulation methods capable of achieving and maintaining high production rates, transfer molding has gained added stature. Electronic components encapsulated by the transfer molding technique are more reliable and less expensive than conventionally encapsulated components. Although transfer molding is applicable to the encapsulation of diodes, resistors, transistors, capacitors, and modules, only modular assemblies will be discussed here.

The major requirements for an encapsulating material for modules are: long-term temperature stability from -65 to $300°F$; thermal and mechanical shock resistance; moisture resistance; excellent electrical properties; flame resistance; fungus resistance; radiation resistance; repairability. (Note that these properties must be maintained, under service conditions, for the life of the system.) In addition, the encapsulant must flow readily to ensure a void-free package, it must have high thermal conductivity, and must provide resiliency to relieve stress on delicate encapsulated components. Exotherm during cure must be minimized to prevent damage to temperature-sensitive components.

Undoubtedly, there are other parameters, some of which are still unknown, which could be considered in the choice of an ideal encapsulant, but for the purpose of this paper, only those factors encountered during the actual encapsulant development program will be considered.

THERMOSETTING SILICONE MOLDING COMPOUNDS

Thermosetting silicone molding compounds have been available for eighteen years. Until a few years ago they were employed only for those electrical and mechanical applications where high-temperature properties were required. Today, because of their unique combination of

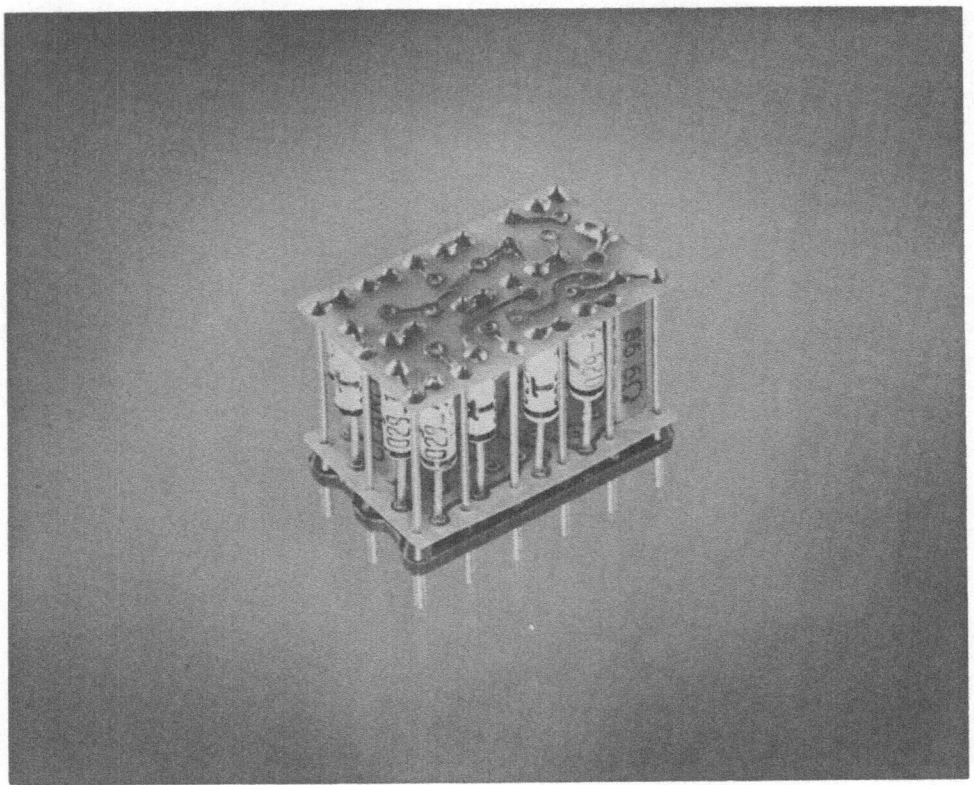

Fig. 1. A standard cordwood module as developed by the Burroughs Corporation. Note the molded header.

physical and electrical properties, as well as greatly improved handling characteristics, silicone molding compounds are used extensively in the encapsulation of delicate electronic components.

The silicone thermosetting molding compound described in this paper was designed specifically for the encapsulation of cordwood modules. A granular, glass and silica filled compound, it can be molded in 1–3 min at 300 to 350°F and at pressures as low as 200 psi. Tests, currently underway, indicate that this compound may also be employed in the encapsulation of "rat's-nest" modules and in micromodular construction (Figs. 1 and 2). While not designed for components other than modules, this compound has also proved excellent for the molding of resistors, capacitors, and semiconductor devices. The multipin headers shown in Figs. 1 and 2 were molded from the same compound as the module body. This was done to ensure package continuity and compatibility and to gain maximum sealing effect at the header–module body junction.

In this particular module design it was desirable to provide a header containing the connector pins. This separate part is not a requirement, and the leads or connector pins can be molded integral with the components.

The first molding compound developed specifically for encapsulating cordwood modules is Dow Corning* 306 molding compound (formerly Dow Corning M–9–1070 molding compound). Tables I, II, and III indicate the typical physical, electrical, and performance properties of this material in as-molded and post-cured conditions. Note that the dissipation factor of

* Registered trademark.

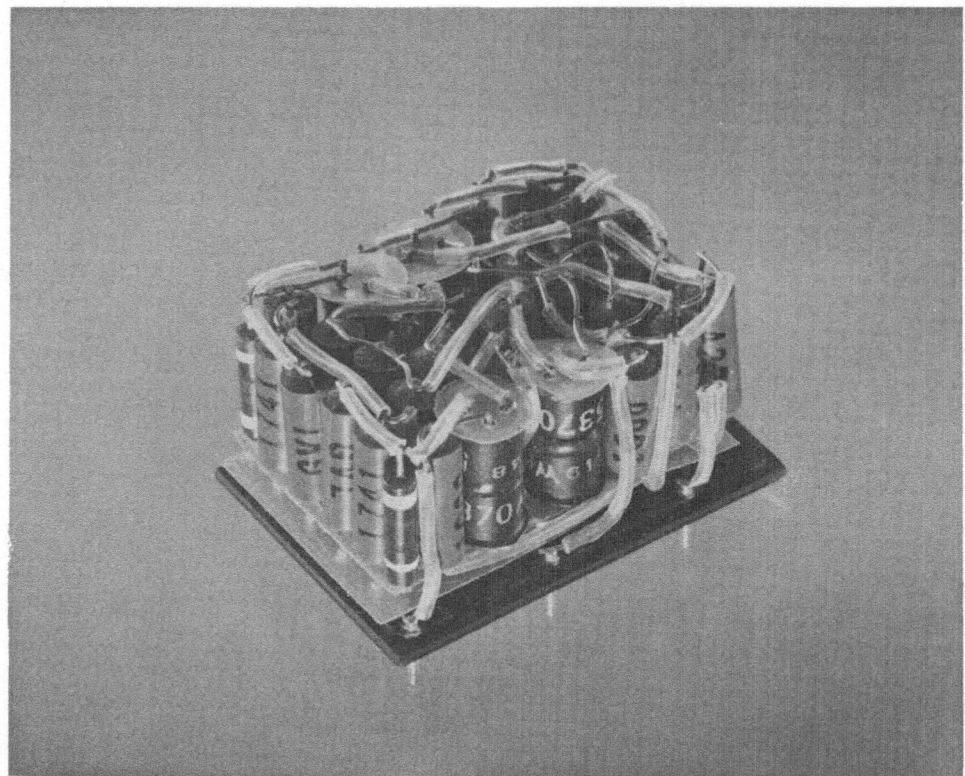

Fig. 2. A "rat's-nest" type module. Note the molded header.

Dow Corning 306 molding compound does not exceed 0.004 after water immersion or when measured at 150°C (300°F). This low dielectric loss along with excellent arc resistance assures that circuit performance will not be compromised in extreme operating conditions. This assures maximum reliability. Note also that Dow Corning 306 molding compound is inherently self-extinguishing. No flame retardants that degrade electrical properties are required to obtain this property.

Unlike organic molding compounds, the physical and electrical properties of Dow Corning 306 molding compound show no appreciable changes after exposure to more than 1000

TABLE I

Typical Physical Properties

ASTM test method	Property	As molded	Post-cured 2 hours at 400°F
D790	Flexural strength, psi	7000	6000
D695	Compressive strength, psi	10,900	10,600
D790	Flexural modulus, psi	1.5×10^6	1.4×10^6
D638	Tensile strength, psi	3600	2800
D256	Izod Impact Strength, ft–lb per in. notch	0.31	0.34
D785	Rockwell hardness (M scale)	71	77
D792	Specific gravity	1.95	1.95

TABLE II

Typical Electrical Properties

ASTM test method	Property	As molded	Post-cured 2 hours at 400°F
D495	Arc resistance, sec	190	260
D149	Electric strength,* volts per mil		
	Tested in air 23°C, greater than	160	165
	Tested in oil 23°C, greater than	370	400
D150	Dielectric constant at one megacycle		
	Condition A†	4.25	4.12
	Condition D†	4.28	4.17
	Tested at 150°C	4.81	4.27
D150	Dissipation factor at one megacycle		
	Condition A†	0.0033	0.0026
	Condition D†	0.0035	0.0030
	Tested at 150°C	0.014	0.0039
D257	Volume resistivity, ohm-cm		
	Condition A†	6×10^{13}	6×10^{13}
	Condition D†	6×10^{13}	6×10^{13}

* Tested under $\frac{1}{4}$ inch ASTM electrodes, 500 volts per second rate of use; specimen $\frac{1}{8}$ inch thick.
† Condition: Condition A—As received.
  Condition D—After 24-hr immersion in water at 23°C.

megarads of gamma radiation. For space and other applications where radiation will be encountered, there are probably no better electronic packaging materials available than silicone.

The unique thermal stability of silicone molding compounds is shown in Tables IV and V and Figs. 3, 4, and 5. As indicated in Table IV, the electrical properties of a typical granular transfer molding compound show no significant changes after 1000 hr aging at 250°C (482°F). Even after aging at 372°C (700°F) for 1000 hr, all of the electrical properties are still within acceptable ranges for most applications.

TABLE III

Typical Performance Properties

ASTM test method	Property	As molded	Post-cured 2 hours at 400°F
	Thermal conductivity, cal/sec/cm^2/°C/cm	1.0×10^{-3}	1.0×10^{-3}
	Thermal expansion, in. per in. linear coefficient per deg C at		
	−60	3.1×10^{-5}	
	0	3.3×10^{-5}	
	60	4.6×10^{-5}	
	120	5.5×10^{-5}	
	150	5.5×10^{-5}	
	200	5.5×10^{-5}	
D648	Heat distortion temperature, °C		450
	Thermal vacuum performance 24 hours at 100°C and 10^{-6} torr vacuum, wt. loss, %	0.224	0.095
	Radiation resistance usable after, megarads at deg C		2000
D635	Flammability	Self-ext.	Self-ext.
D570	Water absorption, %	0.056	0.103

TABLE IV

Electrical Properties of a Granular Silicone Molding Compound After Heat Aging

ASTM test method	Property	Hours at temperature											
		482°F				572°F				700°F			
		250	500	750	1000	250	500	750	1000	250	500	750	1000
D495	Arc resistance, sec	244	255	265	258	303	238	301	304	301	284	304	285
D150	Dielectric constant												
	10^6 cps, dry	3.53	3.49	3.52	3.66	3.47	3.47	3.49	3.50	3.54	3.5	3.5	3.7
	10^6 cps, wet*	3.58	3.54	3.55	3.71	3.58	3.62	3.64	3.70	3.97	4.5	4.5	4.5
D150	Dissipation factor												
	10^6 cps, dry	0.0025	0.0027	0.0027	0.0027	0.0033	0.003	0.0036	0.0041	0.0077	0.0092	0.007	0.013
	10^6 cps, wet*	0.0048	0.0033	0.0033	0.0063	0.012	0.018	0.016	0.021	0.040	0.13	0.050	0.048
D257	Volume resistivity, ohm-cm,												
	dry	1×10^{15}	7×10^{13}	1×10^{14}	4×10^{14}	6×10^{14}	3×10^{13}	1×10^{14}	2×10^{14}	3×10^{12}	3×10^{11}	3×10^{11}	2×10^{10}
	wet*	1×10^{14}	9×10^{11}	5×10^{13}	2×10^{14}	6×10^{13}	1×10^{13}	6×10^{13}	1×10^{14}	2×10^{11}	2×10^{11}	3×10^{11}	5×10^{9}
D149	Electric strength,† volts/mil	354	347	392	434	377	322	403	421	375	355	317	253
D257	Insulation resistance, ohm-cm												
	dry	2×10^{14}	2×10^{14}	3×10^{14}	5×10^{14}	6×10^{13}	3×10^{14}	6×10^{12}	1×10^{12}	5×10^{12}	5×10^{10}	6×10^{9}	5×10^{9}
	wet*	1×10^{13}	7×10^{13}	1×10^{13}	6×10^{12}	1×10^{13}	6×10^{12}	6×10^{12}	9×10^{12}	4×10^{9}	3×10^{9}	4×10^{6}	8×10^{6}

* "Dry" values were obtained on specimens conditioned in standard laboratory atmospheres: 77°F, 50% relative humidity; "Wet" values obtained on specimens after water immersion: 24 hr at 77°F.
† Tested under oil, using standard ¼-in. ASTM electrodes, 500 volts per second rate of rise; specimen approximately 125 mils thick.

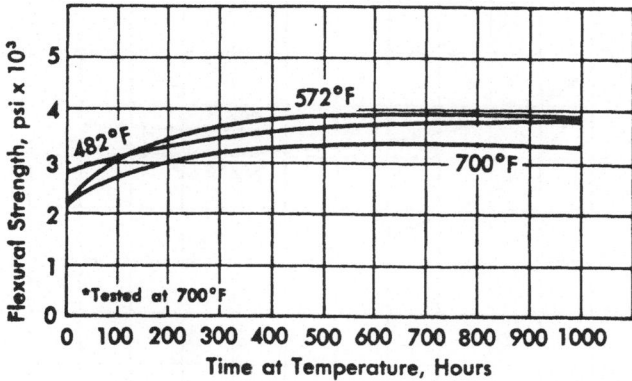

Fig. 3. Flexural strength *vs.* temperature of a granular silicone molding
compound. Samples aged and tested at temperatures indicated.

Similarly, the mechanical strength and physical integrity of the silicone molding compounds
do not change significantly after aging at high temperatures even when tested at these operating
conditions. To indicate the relative thermal stability of silicone and organic compounds, the
changes in flexural strength of several materials after aging for 250 hr at various temperatures
is shown in Fig. 5. These data show that the silicone compounds have operating capabilities
far exceeding those of organic compounds.

Maintaining both electrical and mechanical properties and moisture resistance after long
exposure at high temperatures assures maximum reliability. Even when the electronic equipment
will not operate at very high temperatures, silicone packaging provides the extra safety factors
needed in aerospace and military applications as well as in demanding industrial uses.

THE MOLDED MODULE

In conjunction with the defense and space group of MECD, Burroughs Corporation,
Detroit, Dow Corning developed and perfected silicone transfer molding compounds to meet
the exacting requirements for cordwood module encapsulants. Quantitative requirements for
encapsulating compound qualifications were evolved as indicated in Table VI. These tabular
data are based on Burroughs specification No. MPX 20190.

TABLE V

Change in Size of Molded Part After Removal From Molded Cavity

	Mold shrinkage	2 hr, 250°C	1000 hr, 175°C	1000 hr, 250°C	1000 hr, 310°C
Fibrous silicone	−0.08%	−0.06%	−0.06%	−0.03%	−0.01%
Granular silicone	−0.72%	−0.75%	−0.68%	−0.75%	−0.85%
Fibrous diallyl phthalate	−0.50%	−0.46%	−0.46%	Destroyed	
Fibrous phenolic	−0.01%	−0.01%	−0.10%	Destroyed	
Fibrous epoxy	−0.11%	−0.09%	−0.17%	Destroyed	

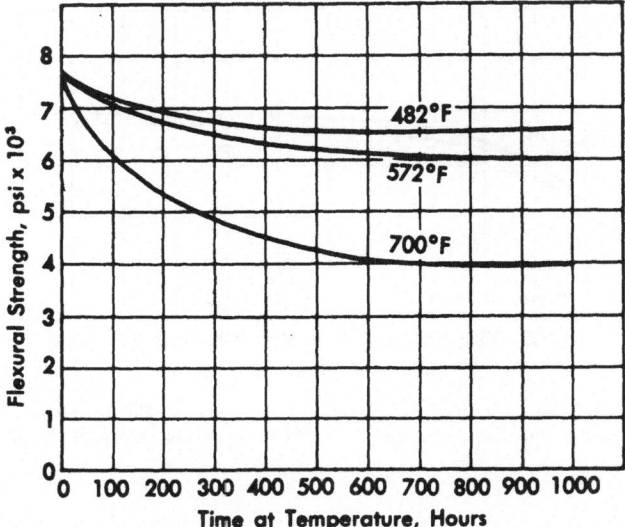

Fig. 4. Flexural strength *vs.* temperature of a granular silicone molding
compound. Samples aged at temperatures indicated and tested at
room temperature.

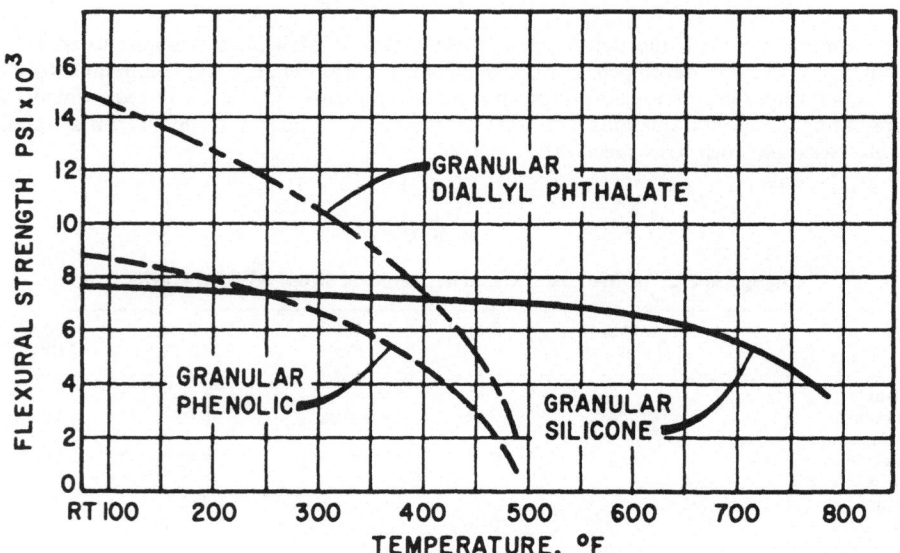

Fig. 5. Effects of heat aging on flexural strength of silicone *vs.* organic molding compound. All
samples aged 250 hr at temperature (tested at room temperature).

TABLE VI

Quantitative Requirements for Qualification Tests

Property	Units	Required value		Test method	
		Maximum	Minimum	ASTM standard	Other
Arc resistance	Seconds	—	175	D-495	—
Moisture absorption	Percent	0.09	—	D-570	—
Impact strength	Foot-pounds per inch	—	0.17	D-256 (Method A)	—
Hardness	Rockwell "M"	75	—	D-785	—
Thermal conductivity	Calories per second per square centimeter per degree C per centimeter	—	7.0×10^{-4}	C-177	—
Dielectric constant at 1 megacycle					
Condition A		4.5	—	D-150	—
Condition D		4.5	—	D-150	—
Dielectric strength	Volts per mil in air	—	255	D-149	—
Dissipation factor at 1 megacycle		4×10^{-3}	—	D-150	—
Volume resistivity	Megohms per mil	—	3×10^{13}	D-257	—
Linear coefficient of thermal expansion	Length per unit length per degree C	7.5×10^{-5}	—	D-696	—
Thermal shock resistance at −55°C to +85°C (−67°F to +185°F)	Cycles	—	5	—	MIL-STD-202, Method 107A, Condition A; 5 min max. between cycles
Flammability		—	Passes	D-635	—
Specific gravity		1.97	—	D-792	—
Fungus resistance		—	Passes	—	MIL-E-5272, Fungus resistance test procedure I
Storage life	Months	—	3	—	—
Test spiral	Inches	—	15	—	—
Color		—	—	—	—

Fig. 6. Cross-sectional view of encapsulated module.

Fig. 7. Three sizes of encapsulated cordwood modules.

In addition to aiding Burroughs Corporation in determining specifications for the molding compound, Dow Corning worked closely with Burroughs' engineering department, press manufacturers, and tool makers to establish the best processing techniques in relation to tool design, mold cycles, plastic flow characteristics, etc. Such problems as insufficient flow of the molding compound at low pressures, insufficient hardness (Rockwell M scale), and excessive molding time were encountered. Deficiencies in properties of the compound were resolved by formulation, compounding, and testing in the Dow Corning application engineering and development laboratories. Application problems were resolved by redesigning the module molds.

Properly molded modules possess extremely good density characteristics. The pressures used in the final stages of molding assure that no voids exist in the encapsulated modules. Component distortion and dislocation is negligible as indicated in Fig. 6.

A unique property of silicone molding compounds is repairability. To effect a repair, the molded plastic can be removed entirely by dissolving it with a chlorinated solvent. After the faulty component is replaced the module can be remolded. This feature of Dow Corning 306 molding compound is a major benefit in packaging expensive modules.

Dow Corning 306 molding compound exhibits low shrinkage during molding and after post cure. The shrinkage figures given in Table V indicate the small changes in dimensions of a molded part after exposure to high temperatures. This property results in minimum pressures on the electronic components. Glass enclosed diodes, film-type resistors, and transistors are not damaged. The added processing expense of precoating with silicone rubber usually required when potting in epoxy compounds is not required.

During the development of the module and its encapsulant a need for color coding the modular system became apparent. In addition to the standard black compound, red, blue, orange, dark green, brown, and chartreuse colors are now available.

MOLD DESIGN

The successful application of silicone molding compounds depends on the correct press and proper mold design. Presses employed in module molding development were 10- and 25-ton top-transfer types with variable ram speeds and pressures. Typical commercial presses are shown in Figs. 8 and 9.

Both the header and finished package molds were constructed of hardened tool steel of approximately 54 to 56 Rockwell hardness. The molds used in this project were not plated as they were designed only for prototype production. Production molds should have chrome-plated surfaces to ensure surface smoothness in the finished module.

Proper gating of molds is extremely important. Gate design in the header is especially critical because of the extremely close tolerance required for accurate pin location. To obtain uniform shrinkage of the molded part the gate for the header mold was located at the side rather than at one end of the header. A Teflon pin jig was used for pin location and release from the steel mold.

The gates in the header and finished module molds were of the ribbon type, 10 mils deep and $\frac{1}{2}$ in. wide (Fig. 10). Gate sizes and locations are variable and dependent upon these criteria: inherent flow characteristics of the encapsulating compound; degree and type of preheating; speed and pressure of the transfer ram; number, size, and location of pins in the header; number, size, and location of components in the module.

MOLDING PROCEDURES

Preforming of the compound prior to preheating and molding ensures that a uniform mold charge will be used and eliminates the loose-powder problems normally encountered by the press operator (Fig. 11).

Although not absolutely essential to the molding operation, electronic preheating improves the flow characteristics of the silicone molding compound and reduces the total mold cycle time.

Fig. 8. A 10-ton Hull standard top-transfer-type press with variable ram speed.

Fig. 9. A 25-ton Hull standard top-transfer-type press with variable ram speed.

Fig. 10. Typical prototype transfer mold design for encapsulating modules.

The correct degree of preheat required to give the best flow characteristics under a given molding time, temperature, and pressure can be easily determined by the operator after a brief training session.

Typical molding and processing data for silicone molding compounds designed for module encapsulation are as follows:

Molding temperature, °F	300 to 350
Molding time, min	1 to 5
Transfer molding pressure, psi	200 to 1500
Post-cure, hours at °F (optional)	2 at 400
Spiral flow, in.	16*

CONCLUSIONS

Transfer molding of cordwood modules with silicone molding compounds offers a number of advantages:

1. Rigid, void-free packages provide maximum environmental protection.
2. Stable electrical and mechanical properties under severe operating conditions assure highest reliability.

* Hull spiral mold, conditions 800 psi, 350°F.

Fig. 11. Preheating preforms of molding compound prior to molding.

3. It permits repair or recovery of expensive modules.
4. It offers uniformity of package for assembly in system.
5. It permits versatile design or unique package construction at minimum cost.
6. Design parameters are broadened, permitting reduction in size and weight.
7. It eliminates the need for precoating or cushioning normally needed to protect against stress on components.
8. It allows maximum utilization of manpower and manufacturing equipment by increased production rates and reduction in the number of rejects.

There are many press and tool manufacturers who will cooperate with electronic engineers and material suppliers in designing molds, materials, and presses to satisfy specific design problems for electronic packaging. By using transfer molding techniques to apply silicone molding compounds in high-volume electronic encapsulating applications, unequalled performance and environmental protection can be obtained at minimum cost and with fewer rejects.

A Feasibility Study in Miniaturized Packaging*

T. J. WILLIAMS

Sandia Corporation
Sandia Base, Albuquerque
New Mexico

AND R. R. ROGERS

Sprague Electronic Company
Visalia, California

This paper deals with three major problems in packaging a complex logic module: mechanical layout; interconnection techniques; and encapsulation. The first two sections independently consider mechanical layout and interconnection techniques. The third section combines information from the first two, and derives a method for design evaluation. The fourth and final section develops and applies encapsulation methods.

INTRODUCTION

DURING THE EARLY PART of 1963, several magnetic logic applications were evolving at the Sandia Corporation. The expected mechanical similarities between logic elements prompted a study of general methods for their assembly. To this end, a contract was entered into with the Sprague Electric Company to evaluate both proven and novel packaging schemes. In scope the program was concerned only with such things that would be production feasible by early 1964. The object, of course, was to develop basic concepts and technologies which could then be applied as specific requirements emerged.

APPROACH

A straightforward approach was pursued in order to accomplish the stated goals, i.e., take a logic element known to be representative of the class of component in question and examine it in great detail.

Figure 1 shows the schematic of a 23-bit gated-voltage shift register selected for this purpose. Each bit consists of a square loop magnetic core, two diodes, and one capacitor. Components were chosen to reflect maximum developments in miniaturization consistent with availability.

Figure 2 compares components used in this program with those in a similar state of development as of late 1962.

The program was planned so that physical properties of piece parts, interconnection modes, and encapsulation were studied independently from the mechanical layout of components. Results from the former studies were then applied to the layouts, allowing a quantitative

* This work was supported by the United States Atomic Energy Commission. Reproduction in whole or in part is permitted for any purpose of the U.S. Government.

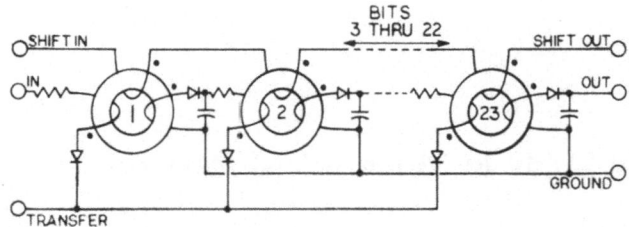

Fig. 1. Twenty-three-bit shift register schematic. Diode: T.I. silicon "Micro Diode"; capacitor: Sprague 165D "Tantalex"; core: Dynacore Mo Permalloy tape, 0515 bobbin; resistor: Cuprothal 294 resistance wire, input windings on coils.

evaluation of each package configuration. Designs were rated as a function of volume efficiency, manufacturing feasibility, and interconnect characteristics. Finally, conclusions drawn from design analysis were supported by actual pilot assemblies and environmental testing.

MECHANICAL LAYOUT

As may be obvious, there is a limited number of methods for the grouping of discrete components within a complex subassembly. The generalized methods chosen for extensive investigation were referred to as planar, cordwood, build-on, modular, and hybrid. The first task was to illustrate the application of each method and derive the variation most suitable for the 23-bit shift register under investigation. The following observations were made as a result of this initial work:

Planar

The structure is typified by a printed circuit board layout. It becomes immediately obvious that a literal interpretation of the planar structure is not suitable for the application. This occurs because the surface-area–to–volume ratio is necessarily large and the package becomes unwieldy. A slight modification yields a more compatible form factor while retaining most of the planar characteristics (see Fig. 3).

Advantages of the planar style are accessibility, ease of inspection, and repairability. The two major disadvantages are poor volume efficiency and difficulty in mechanizing the coil winding relationship between individual bits.

Cordwood

A cordwood package is defined as one in which components are placed in a three-dimensional arrangement with longitudinal axes parallel. Because of geometric anomalies introduced

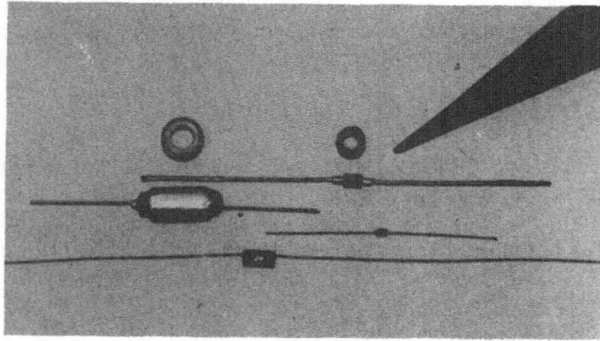

Fig. 2

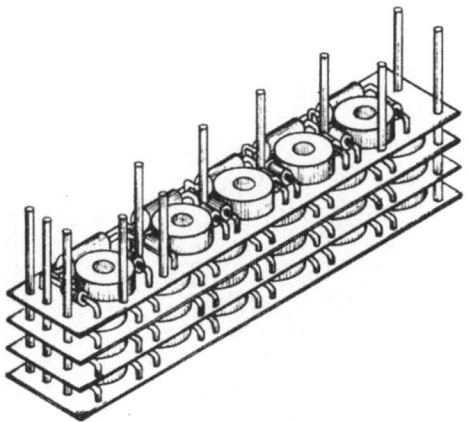

Fig. 3

by the toroidal core, the magnetic shift register cannot adhere exactly to the above definition. Figure 4 illustrates the cordwood style as applied here.

Although this package has high volume efficiency, it also has some serious drawbacks. The scheme is inherently difficult in the alignment of components between polyester-film layers. This problem is amplified by the number of and the small size of the components in this particular package. In addition, repairability and inspection access are severely limited.

Build-On

Probably the highest density of all is realized by this method. The approach is simply to start with a single component and pack other parts around it, making electrical connections at the same time. In this manner no supporting structure is required and all nooks and crannies are filled. Several considerations allowed a rapid dismissal of the build-on techniques:

a. A single size-dominant part is required to form the nucleus of the structure. None exists in the 23-bit register.
b. Success of the assembly is highly dependent on operator skill and judgment.
c. The number of individual components is excessive.
d. The manufacturing process cannot be controlled to an extent commensurate with desired reliability.

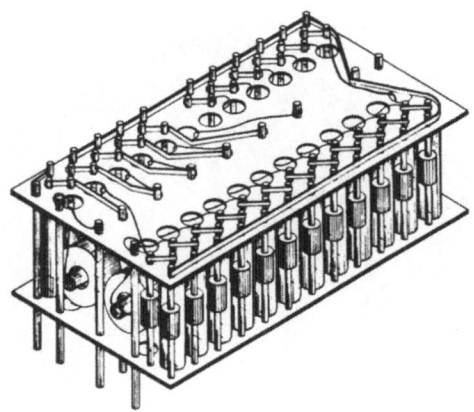

Fig. 4

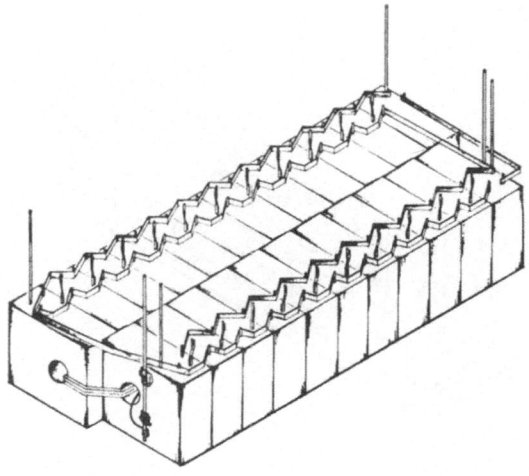

Fig. 5

Modular

By packaging each bit individually many advantages are apparent. The design does not depend heavily on component similarity or periodic symmetry. Subassembly testing is greatly simplified and component changes are easily accommodated. Also, repairability is enhanced if not reduced to the point of interchangeable bits in the final assembly. On the other hand, the scheme results in increased cubic volume and more interconnections. Figure 5 shows a modular assembly.

Hybrid

As the name implies, hybrid describes those packages which employ combinations of the preceding styles. The best example of a hybrid design is the honeycomb structure. One layout of this type is shown in Fig. 6. As can be seen, the Honeycomb module utilizes component arrangement principles of the cordwood package but allows better accessibility. Also, the odd-shaped coils are handled more conveniently, and components are supported throughout the manufacturing process. The design does suffer from one consideration. It is rather inflexible to component change. The honeycomb is most efficient when components, or groups of components, are repetitive in form factor (i.e., 12 capacitors in a row). The insertion of an odd-sized component not only disturbs the package symmetry but also requires retooling of the support structure.

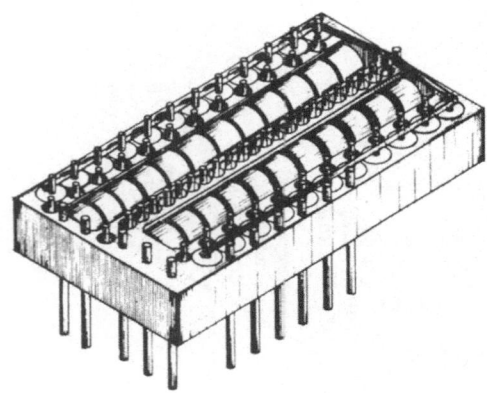

Fig. 6

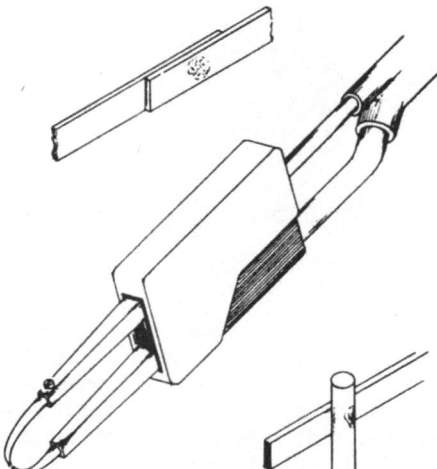

Fig. 7

INTERCONNECTION TECHNIQUES

Concurrent with the mechanical layout studies, an investigation was made of interconnection methods. Since interconnections typically occupy an appreciable volume (on the order of 25%), the objectives were to scale-down common techniques and evaluate some less obvious methods. Although many techniques are possible, those considered practical for this program may be grouped together under the general descriptions of welding, soldering, and two-dimensional.

Welding

The welding evaluation included series resistance, parallel-gap, ultrasonic, and percussion. More advanced techniques such as electron-beam and laser welding were not studied.

In an attempt to secure the maximum flexibility in a semistandard technique, tweezer welding was tried. Figure 7 illustrates the hand-held welder. Ironically, the very attraction of

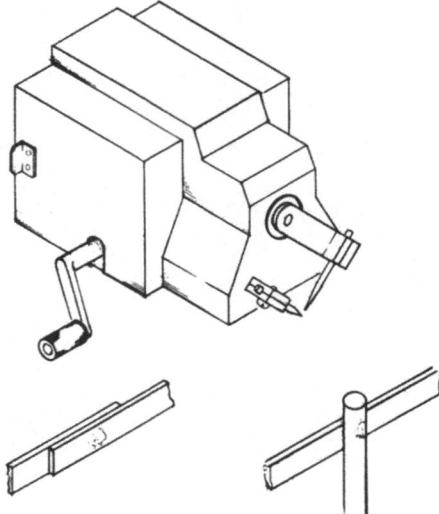

Fig. 8

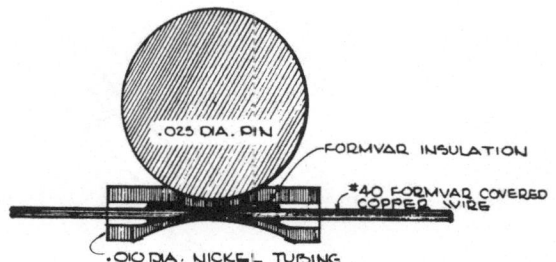

Fig. 9

tweezer welding works against it in this application. The hand operation requires too much skill; it is difficult to maintain electrode alignment, visibility in work area is poor, and in general its unique maneuverability produced nonuniform registration with the work. Finally, tweezer welding did *not* produce a denser termination matrix than did other more standard methods.

Another method of series welding utilizes the pincer-type head (see Fig. 8). The process is a common one and will not be described in detail. Suffice it to say the technique is amenable to this miniaturization program. Theoretical weld density is defined solely by electrode and weldment geometry. In the cross-wire joining of 0.010-in. weldments, a feasibility of 33 welds/inch was demonstrated as compared to the theoretical limit of 42 welds/inch.

Because of the numerous coil terminations in the shift register, special attention was given to magnet wire. Two techniques were used to successfully weld the copper wire to nickel terminals. The first is simple cross-wire welding in the pincer head. Insulation is removed and the weld schedule is adjusted in a normal manner to produce the weld. To obviate removal of insulation, the second technique employs a nickel tube as shown in Fig. 9. Both methods produced welds with no indication of embrittlement or other deleterious effects. The conclusion is that copper magnet wire can be terminated by welding, but much more experience is required to assure process control.

Parallel-gap welding is shown in Fig. 10. This technique is especially compatible with a planar-type interconnection. Experiments show best results occur when a flat ribbon

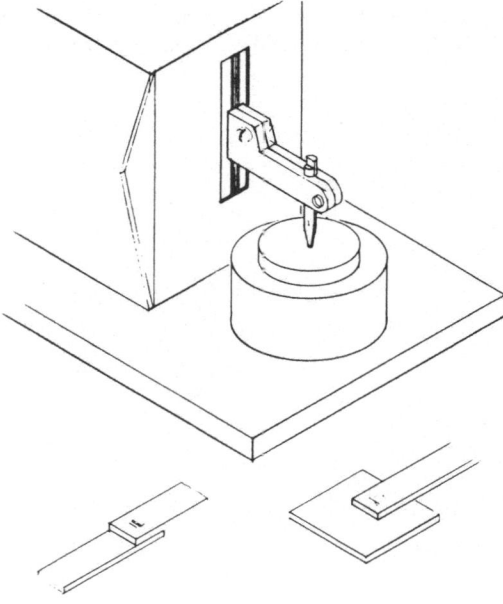

Fig. 10

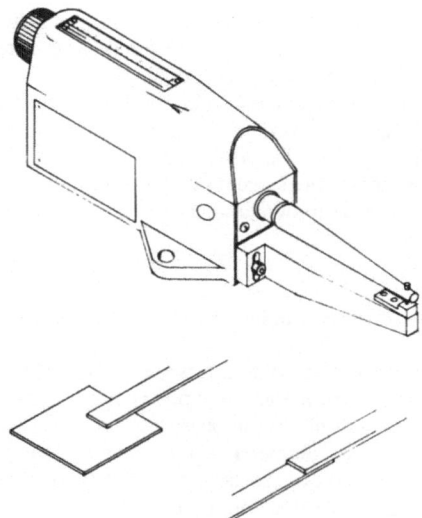

Fig. 11

termination is made to a low-conductance (e.g., Kovar or nickel) circuit pad. Heat balance and pulse energy control are more critical than with series welding. Number 44AWG magnet wire, nickel ribbon, and nickel wire were successfully welded to gold-plated nickel-clad circuit board. However, attempts to weld the same materials to copper-clad board failed.

Ultrasonic welding was considered from the standpoint of being useful in physical situations similar to the parallel-gap method. In addition, this method holds promise for a greater flexibility in accepting combinations of materials. Since the process is entirely mechanical, possibilities exist for bonding nonconducting materials. Attempts to weld insulated magnet or resistance wire were unsuccessful. Although the other more common terminations are feasible, the ultrasonic process did not appear to offer any advantages over parallel-gap. In fact, physical requirements, notably positioning work on the anvil, actually impose some serious limitations. A typical ultrasonic welding installation is shown in Fig. 11.

One variety of percussion welding was briefly considered. The mechanical arrangement is shown in Fig. 12. In this process, a radio-frequency voltage ionizes the air between weldments,

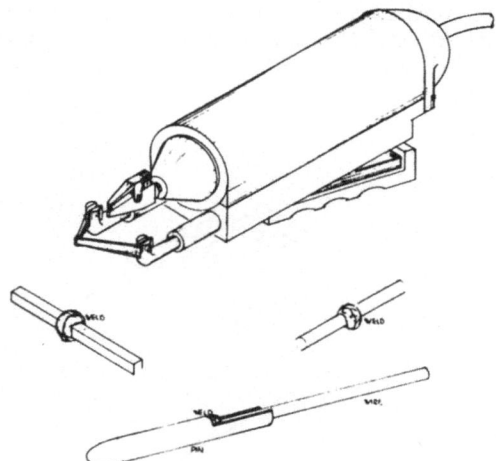

Fig. 12

starting an arc which is sustained by a high-voltage capacitor discharge. When weldments have reached the proper energy state, they are forced together, completing the weld. This method is well-suited to butt welding, but was judged generally inapplicable to module-type work.

Worthy of brief mention is a miniature arc welding machine capable of welding No. 44 magnet wire. A machine was demonstrated, and it successfully bonded magnet wire to copper-clad circuit board, to No. 24 gold-plated brass, and to No. 24 nickel wire. For module work, the machine was impractical because of the required opaque face mask and a rather bulky setup. We are given to understand, however, that a new machine for transistor assembly is under development.

Soldering

Besides miniaturized standard methods, studies included controlled-energy and ultrasonic soldering.

A clever innovation in soldering is the pulse-dot system. Heat to the solder joint is controlled by timed pulses from a variable power supply. Electrical energy is supplied to the heating element through a small nichrome resistance loop. Solder is carried to the joint in the form of small (0.005- to 0.062-in.-diameter) solder balls adhering to the heating element by means of a paste flux. Adequate solder joints were made with No. 36 and No. 44 bare magnet wire to copper, gold-plated, and solder-plated circuit boards, and also to 0.015-in.-diameter unplated nickel wire. On the larger connections, such as to a 0.062-in. printed circuit pad, a pulse of up to 10 sec duration was required to produce adequate solder flow.

Limits of conventional soldering density were tested using the smallest tool available. This was a 12-watt iron with a 0.030-in.-diameter tip; No. 44 AWG bare copper wire was soldered to 0.010-in.-diameter nickel, spaced as closely as 0.040 in. This was accomplished by using a 15X binocular microscope, 0.020-in.-diameter 63-37 solder, and carefully prepared tweezers for wrapping the magnet wire. One of the greatest problems is the resilience of the copper wire. A toothpick was required to hold the wire in place while soldering. Also, the iron tended to overheat and oxidize after a few minutes use.

A third method of soldering is by means of ultrasonic agitation. The object is to eliminate fluxing. Heat is still applied to the joint, but mechanical rather than chemical means are used to break up surface contamination. One way to apply ultrasonic energy is to attach a transducer to the conventional soldering iron tip. Another scheme, for dip soldering, is to immerse the transducer in a solder pot. Solder adhesion to magnet wire, nickel wire, and copper-clad circuit board was tested. Results were marginally successful, but the method was not pursued because it appeared to offer no advantages to the program.

Planar Interconnection Methods

Of greatest potential with regard to volume economy are the two-dimensional, or planar, interconnection methods. Here it is possible to form a component termination which requires essentially no volume.

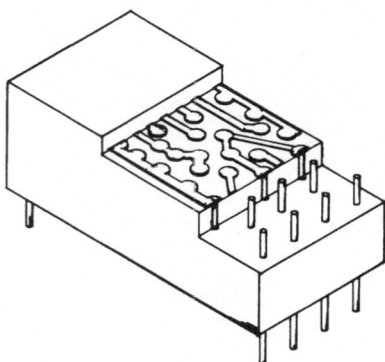

Fig. 13

The first method investigated was a conductive epoxy resin, filled with metallic particles, typically, aluminum or silver. It may be used to either "glue" or "paint" a termination or circuit run. No size advantage is gained in bonding conventional joints such as wrapped wire or cross-wire. One advantage, however, is that in some cases the epoxy may form a bond when conventional soldering or welding is incompatible with materials being joined. Of more particular interest is the possibility of producing the interconnection circuit pattern and component termination simultaneously. This was attempted by encapsulating magnet wire in an epoxy block and then milling a plane perpendicular to the wire axes. This left wire cross sections exposed. These were then interconnected by means of a conductive epoxy coating. Conclusions are that adhesion is not sufficient to form a reliable bond of this type; it is very difficult to obtain the required conductive path detail, and resistivity of the conductive media is orders of magnitude higher than that of metallic conductors.

A more sophisticated method for accomplishing the two-dimensional effect is by actually plating the circuit in place. This allows the use of precision photolithography in conjunction with metallurgical component interconnections. The feasibility of this approach was tested by encapsulating copper, nickel, and Kovar wires in an epoxy block. The block was then "faced off" by milling and a circuit pattern was deposited over exposed conductor butt ends (see Fig. 13). Continuity of the conductors to plated circuit was then monitored over a temperature range of −55 to +105°C. The success of this experiment prompted additional metallurgical analysis to insure the type and quality of interconnect bond. Results indicate the feasibility of this technique. However, a great deal of skill and care is required for the initial copper reduction deposition and the subsequent mechanics of photo-resist application and etching.

APPLICATION OF DESIGN TECHNIQUES

Having established a base of design tools, the task now is to apply the most suitable combinations to a particular 23-bit shift register. Starting with the premise that volume, flexibility, and reliability are important to the same degree, one may assume that no single design will optimize all three. Therefore, the approach taken was to select the most applicable techniques, combine them to produce workable designs, and then evaluate the designs such that trade-offs are illustrated in a quantitative manner.

A review and careful analysis of the preceding layout methods show that advantages of the build-on are overcome by its disadvantages. Likewise, ultrasonic welding, parallel-gap welding, arc welding, ultrasonic soldering, pulse-dot soldering, and conductive epoxy were shown to have some less than optimum characteristics when applied to the selected layouts. Manipulation of the remaining design methods yielded four prototype packages shown in Fig. 14. At the

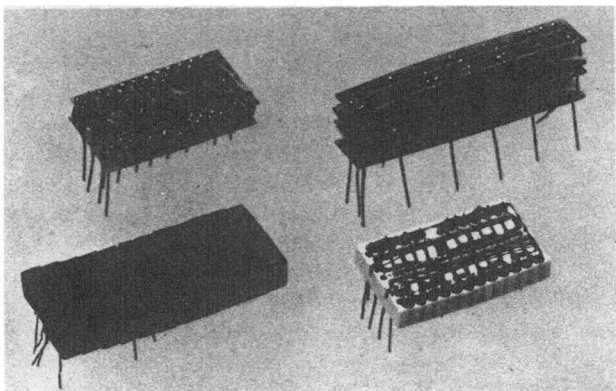

Fig. 14

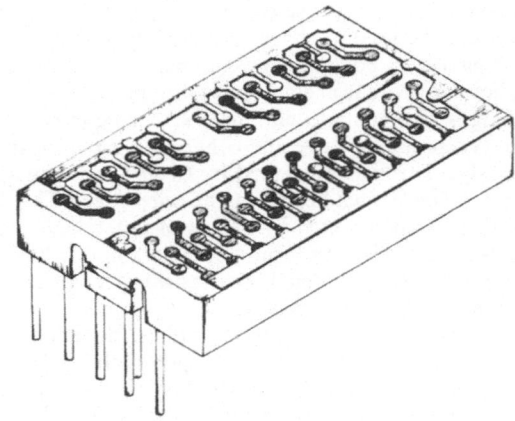

Fig. 15

time of this writing, a fifth design was approaching the prototype stage. This package utilizes plated circuitry and is shown in Fig. 15.

Prior to assembly, each package was analyzed and rated as follows:

a. Volume efficiency—rating based on geometric calculations.
b. Design flexibility—rated on ability to accommodate a variety of component form factors and accept design changes.
c. Interconnection merit factor—rating calculated from number of connections and facility with which each is made.
d. Manufacturability—rating considers component handling, support and preconditioning, and assembly encapsulation and repairability.

Since most of these characteristics require human judgment, a scheme was devised to objectively evaluate each one. For each characteristic, a number was calculated depicting a gradation from "optimum" ($= 1$), through "acceptable" ($= 2$), to "marginal" ($= 3$). Note that feasibility of all operations has previously been demonstrated, therefore no "unacceptable" rating exists here.

As an example of how the numerical rating was accomplished, consider the interconnection merit factor for a honeycomb package (see Fig. 6). Here the data in Table I apply. The factor, Q, was assigned after evaluating the environment associated with each connection type. This consisted of work area and inspection accessibility, stress applied as a result of the interconnect operation, exposure of uninsulated conductors, and conductor cross-overs.

TABLE I

Connection	Number of connections (N)	Technique	Quality (Q)	$N \times Q$
Coil to cap	23	Solder	1	23
Coil to cap	8	Solder	1	8
Coil to diode	45	Solder	1	45
Coil to diode	27	Solder	3	81
Cap. to bus	46	Weld	1	46
Diode to bus	46	Weld	1	46
Term. to bus	8	Weld	1	8

$$\Sigma N = 203 \quad \Sigma N \times Q = 257$$
$$\text{Merit factor} = \frac{\Sigma N \times Q}{\Sigma N} = 1.3$$

Data in Table I provide an immediate indication of any design weakness. For instance, the marginal coil-to-diode rating was caused by a series of conductor cross-overs. This situation was subsequently remedied by relocating the transfer diode.

Following techniques similar to those illustrated above, a master comparison chart was derived. Table II summarizes individual characteristics of each package. Note that for all these figures of merit, smaller numbers indicate superior characteristics.

From Table II, one is now in a position to estimate the relative weights of volume efficiency, flexibility, and reliability. (It must be understood here that "reliability" does not imply a rigorous mathematical definition. Rather, it is an estimated quality factor based on interconnection merit and manufacturability.) A quick glance at Table II shows that of the five packages listed, only three possess outstanding characteristics. The plated circuit module is preferred by virtue of its minimum volume, modular because of flexibility, and honeycomb because of manufacturability.

One must now weigh the advantages of each design against its disadvantages. When this is done, the honeycomb package is seen to offer a very reasonable compromise between all characteristics of interest. The program as executed actually called for pilot assembly of the three outstanding modules. Methods studied during this phase verified the estimated ratings for each package.

ENCAPSULATION

Now that the internal structures have been defined, a final consideration must be given to encapsulation. Experience shows that occasional catastrophic failures associated with encapsulation do occur. The object of this section is to explore the physical relationship between components and the impregnating system.

Design requirements indicate the use of a rigid thermosetting resin, at least externally. Furthermore, simplicity suggests a single-stage potting procedure. However, experience with rigid resin systems warns of certain risks when internal components are not isolated from mechanical effects of the resin. The problem is to show just what effects are at work in the resin and then estimate their magnitude. From these data, one may derive an approach to utilize the beneficial resin attributes without suffering the deleterious ones.

Two resins were chosen for evaluation. They represent a class of general purpose epoxy resins and were selected for low temperature coefficient, operating temperature, viscosity, and general handling characteristics. Both were silica filled; one was self-extinguishing and the other not.

If components are to be ruptured by a resin, then a relative displacement of the resin must be accompanied by adhesion and/or a lower yield strength on the component. A qualitative but very informative test was performed to check adhesion. Groups of diodes were partially submerged in a thin film of each resin. After cure the diodes were pulled from the resin by means of a force gage. Adhesion was such that the glass envelope consistently yielded before being pulled free. Another experiment demonstrated mechanisms causing resin displacement as well as yield strength and adhesion.

Here a glass microscope slide was coated on one side with epoxy, cured, and then cycled from -55 to $+125°C$. Bowing of the slide as a result of cure shows that significant stress and adhesion are developed in this process. Actual fractures in the glass during thermal cycling point out differences in yield strength and expansion coefficients. A more quantitative

TABLE II

	Volume, in^3	Flexibility	Interconnect merit factor	Manufacturability
Planar	0.40	2	1.2 ($N = 267$)	1.6
Cordwood	0.31	3	1.5 ($N = 213$)	2.2
Modular	0.37	1	1.0 ($N = 259$)	1.6
Honeycomb	0.28	2	1.3 ($N = 203$)	1.4
Plated circuit	0.21	3	1.9 ($N = 196$)	2.5

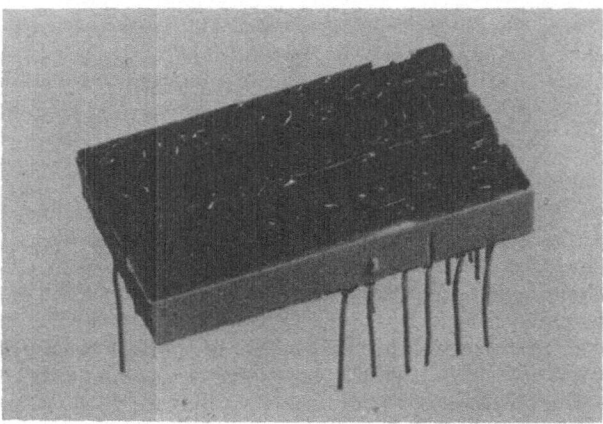

Fig. 16

experiment performed by encapsulating strain gages confirmed these large-magnitude stresses during cure and thermal cycling.

From the above experiments one may conclude the necessity for component protection. Not only must some antiadhesion agent be employed, but a "cushioning" effect must also be introduced. Tests on silicone mold release and RTV silicone rubber show both capable of preventing adhesion. The latter was chosen for its handling properties and potential as a buffer against the final encapsulant. Unfortunately, the RTV compounds have temperature coefficients an order of magnitude greater than other materials in the package. If the quantity of RTV is appreciable with respect to package volume, the outer envelope can actually bulge and rupture. In order to provide mechanical reinforcement of the assembly, prevent epoxy adhesion to components, provide a mechanical buffer, and allow for stress relief under temperature excursions, the following procedure was adopted: (a) Apply a very thin coating of agent to all components (Fig. 16); (b) deliberately entrap a certain amount of void space by wrapping with PTFE tape (Fig. 17). The assembly is then cast in the final rigid epoxy system as shown in Fig. 18.

Although environmental testing substantiates the above encapsulating hypothesis, the process itself leaves something to be desired. A more adroit way to accomplish a similar result is by prepotting with glass microballoons. First the assembly is molded with a mixture of hollow

Fig. 17

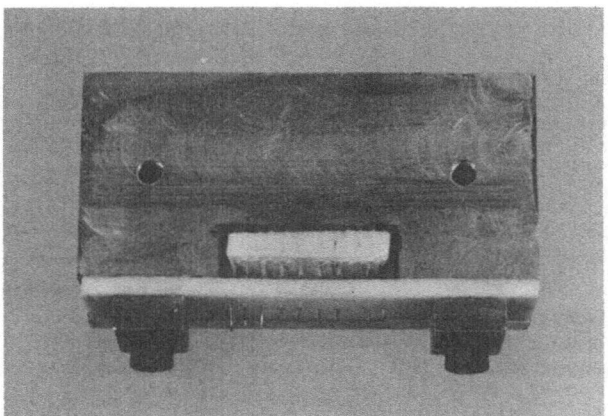

Fig. 18

glass spheres carried in an organic binder. The extremely small size of the spheres (approximately 100μ) in conjunction with an organic solvent produces a very low-viscosity fluid. Hence penetration is thorough. As a result of low shrinkage and low temperature coefficients, the assembly may be completely filled, eliminating a large judgment factor. Repair is also greatly facilitated because the microballoon encapsulant system is easily and safely dissolved. Before final epoxy casting the microballoon-binder surface is treated with a sealer to prevent epoxy penetration. This system was tried and found to be quite satisfactory.

CONCLUSIONS

1. A 23-bit magnetic shift register can be designed with a volume reduction of approximately 10 to 1 over what was considered practical a year ago. This is accomplished by the simultaneous application of ultraminiature components, refinements in interconnection methods, and careful mechanical design. Further volume reduction may be possible, but only at a great sacrifice in production feasibility.

2. By methods not discussed in this paper, volume efficiency was investigated. It was discovered that the real problem in reducing volume by an order of magnitude is not in increasing the ratio of component volume to gross volume. Rather, the object is to apply smaller components and still maintain a constant volume ratio. Future reductions in size can be expected to depend upon smaller components and a proportional decrease in the volume required for interconnections and supporting structure.

3. The optimum mechanical design cannot be synthesized. Analysis of a family of solutions yields the most suitable design approach. Selection of the most satisfactory package design can be made on a semiquantitative basis by weighing factors of volume, manufacturability, and reliability.

4. Imposing more than one major functional requirement generally precludes use of the optimum technique for each individual requirement. The best package is then a compromise between interrelated functions and techniques. For the 23-bit magnetic shift register, the honeycomb offers the best balance between volume, flexibility, and reliability.

ACKNOWLEDGMENTS

Welding Equipment

Sippican 1A power supply
Sippican 14 weld head
Lincoln 110 tweezer weld head
Weldmatic 1016C power supply
Weldmatic 1032 weld head
Weldmatic MA2166 electrode holder
Sonoweld W-1040-TSL welder
Cannon Model 100 power supply with percussive/arc hand tool
Birdsell Model 520 inert gas arc welder

Soldering Equipment

Circon PDS-LK Pulse-Dot System
American Beauty B-2000 $22\frac{1}{2}$ W iron
Hexon Hornet 1-110 12 W iron
Sonosolder 25 W power supply
Sonosolder S-0-HN-56-4 solder head
Alpha Solder and Flux, variety
Kester Solder and Flux, variety

Encapsulation

National Engr. Castiplast 894 and hardener No. 29
Hysol 4169 and hardener No. 3471
Furane Epocast 12-55 and hardener HN-94-10
Emerson–Cumings Stycast 3020, catalyst No. 9 and No. 11
Dow Corning RTV 601
Dow Corning 20 mold release
Bacon Industries syntactic foam SF-1
Bacon Industries FA-8 adhesive with BA-5 activator
"Insulation" Feb. 1964, Lake Publishing Company

Photo-Etch Process

Kodak Ortho Resist
Kodak Photo Resist
Shipley Azoplate AZ-17
Shipley Deposition Process PI 501

Components

Texas Instruments–TI 257 Diode
Sprague Electric–Type 165D Tantalex Capacitor
Sprague Dynacor 051S6.5A1 Core
Phelps–Dodge Soldereze Magnet Wire
Cuprothal 294 Resistance Wire

Interconnection and Organization of Functional Electronic Blocks

Norden Division, United Aircraft Corporation
Norwalk, Connecticut

Norden has developed a packaging concept specifically designed to organize and interconnect logical groupings of microelectronic integrated circuits, both thin-film and semiconductor circuits, and hybrid combinations of both. The basic submodule building block approach consists of stacking functional electronic blocks (FEBs) and casting them rigidly within an insulative three-dimensional supporting structure. The FEB leads are flush and exposed on the surface planes of the submodule thus formed. The exposed FEB terminations are interconnected on a batch process basis simultaneously rather than sequentially by vapor-phase nickel deposition. Subtractive photo-etching techniques define the conductive interconnect pattern. The submodule permits handling and assembly into the next level of system organization by conventional well-known methods. Wire bound split-pin connections were evaluated as an optional method of interconnecting submodules. The technique provided a practical demonstration for making this type of permanent connection, a connection that could be severed and remade a finite number of times, on 0.050-in. centers. The program has demonstrated the feasibility of utilizing this submodule building block approach for efficient packaging densities, preservation of device reliability and economy in manufacture, and produces an optimum solution to the problem of interconnection and organization of functional electronic blocks.

INTRODUCTION

THE ANTICIPATED LARGE-SCALE utilization of microelectronic integrated circuits challenges the equipment design engineer to make full use of the inherent advantages of these circuits: outstanding performance, high reliability, low cost, and minimum weight and volume.

A look at microelectronic research and development to date indicates that most of the effort expended has been in the direction of fabricating integrated circuits. By comparison, little effort has been directed toward developing more effective ways of packaging these circuits into complex systems. Methods of providing structural support, interconnections, and thermal control of microminiaturized circuits have been, for the most part, extensions or adaptations of methods presently in use in conventional component packaging.

Norden has developed a packaging concept specifically designed to organize and interconnect logical groupings of microelectronic circuits. It utilizes a submodular building block approach, and is based on the performance, size, and configuration requirements of the miniaturized circuits. This development effort was the result of a Norden study program sponsored by the Air Force, and Norden's continuing effort to extend its aerospace electronic packaging capabilities.

The program's objectives were to organize microelectronic circuits, which are substrate mounted and essentially two-dimensional in form, into three-dimensional submodules performing a system function. The two-dimensional circuits, known as functional electronic blocks (FEBs), have excellent volumetric efficiency. The program called for preserving this efficiency while at the same time providing a method for interconnecting all of the FEBs within a

133

submodule simultaneously on a batch process basis. The submodule was to exhibit good heat-transfer characteristics, be readily maintainable, and be easily integrated into the next level of system organization.

DEVELOPMENT OF ORGANIZATIONAL DESIGN CONCEPT

Choice of Basic Circuit Elements

The elements chosen for the packaging studies were FEBs. These units are the most flexible and volumetrically efficient electronic packages available today. An FEB is any microelectronic integrated circuit that can be fabricated in two-dimensional form. It can be a thin-film network, a semiconductor network, a group of uncased components mounted on an insulative substrate, or a hybrid combination of any or all of these. It can be mounted in a flat package or deposited on a thin flat supporting substrate. It has terminations that extend out of the sides of, and are coplanar with, the package or substrate. The terminals are usually flat ribbon leads spaced on fixed centers.

FEBs can be fabricated in a variety of sizes and shapes. They can be as small as 125 mils by 125 mils, and sizes as large as 375 mils to 500 mils square are in common use today. There is no theoretical upper limit on the size of an FEB. As the technology progresses, the trend will be to produce even larger FEBs with greater functional complexity.

In summary, the common features of FEBs are their two-dimensional form factors and disciplined terminal lead-out configurations. Any type of circuit can be contained within an FEB. An FEB therefore possesses great circuit flexibility from the standpoint of available fabrication methods and performance characteristics obtainable.

Packaging Philosophy

A submodular building block approach was chosen because it satisfied the operational requirements of both present and future equipment designs, and permitted utilization of simple, reliable, and economical fabrication techniques.

A submodule is the lowest organizational level contained within a system. Each submodule performs a discrete circuit function, and can be readily tested independently of the rest of its system. Defective submodules are discarded rather than repaired.

The submodules can accommodate FEBs of various shapes, sizes, and lead-out configurations. They can contain as few as four FEBs or as many as twenty, and can dissipate heat effectively. Thus while submodules provide the advantages of standardized, repetitive building blocks, they have great packaging flexibility and design freedom.

Basic Submodule Building Block

A submodule consists of a stack of FEBs cast rigidly within an insulative supporting structure. The FEB leads are flush with the surface planes of the submodule thus formed. By depositing a conductive plate on the surface planes, all the exposed terminations are electrically interconnected. Then by photo-etching techniques, interconnecting conductors are defined using photo-masks of the desired wiring pattern.

Figure 1 is a sketch of a typical submodule showing some of the features that can be incorporated. It shows four FEBs arranged so that their leads are flush with two opposite surface planes of the submodule. Deposited conductors interconnect the FEB leads, through-conductors, and cross-over conductors, and L-shaped submodule termination header leads. The header leads originate in the interconnection planes and extend out of the base. They are the input–output terminals for interconnecting submodules into the next level of system organization.

In high-density interconnection systems, cross-overs generally present the most difficult interconnection problem. In the submodule the cross-overs are easily handled by using pre-fabricated matrices within the submodule structure. They permit connections to be made between opposite planes of the submodule, and make it possible to achieve cross-over connections in the same plane as well. The matrices eliminate the necessity of using multilayer wiring for cross-overs on the surface planes, making complex interconnections practicable.

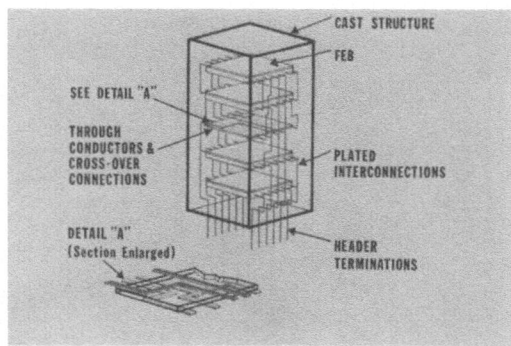

Fig. 1. Submodule design features.

The use of thermal shims can improve thermal transfer between the FEBs and the external heat sink. They can also serve as electrostatic shields between stages when required. The shims can be extended beyond the surface of the submodule to serve as fins for convective heat transfer systems, or can be terminated at the surface of the submodule for conductive heat transfer to an ultimate heat sink. In general, the relatively low power levels of the majority of integrated circuits encountered do not require these auxiliary devices, and a fairly efficient thermally conductive casting resin can maintain the thermal gradients within the submodule at the desired levels.

FABRICATION PROCEDURE

Fixturing Frame

It is necessary to position the leads of the FEBs in the submodule surface planes accurately so that they will register properly with an interconnection photo-mask. Package or substrate lead position accuracies are generally inadequate for the registration tolerances required. Small and fragile leads also present a difficult handling problem.

To overcome these problems a lead registration fixturing system was devised. As shown in Fig. 2, a split cavity mold was designed with grooved locations for positioning leads. A low melting point alloy was used to cast and imbed the leads in a frame with predictable accuracy. The thickness of the frame was closely controlled so that, when stacked, each row of leads would be equally spaced.

Figure 3 shows how the positioning of through conductors, cross-over connections, and header terminals is accommodated in the same common U-frame jig. Having framed the FEBs and associated elements, the next assembly operation consists of stacking the frames on indexing pins. The assembly is completed by adding a silastic header terminal support and end plates. This creates the mold used to cast the submodule.

This fixturing method was used for the original development program. New fixturing for production quantities has been planned to achieve the same results at lower costs.

Casting

An epoxy-based potting material was chosen as the submodule structure because of its compatibility with the FEBs, deposited conductors, and associated processing parameters. The material is stable, impervious to common etchants, and has good electrical insulation properties. Conventional vacuum casting techniques and a long-term curing cycle assure void-free casting of the submodule assembly.

Machining

Removal of the frame fixturing is accomplished by milling. The indexing pins in the frame assembly are used as a reference in machining the submodule to final size.

Fig. 2. Split cavity mold.

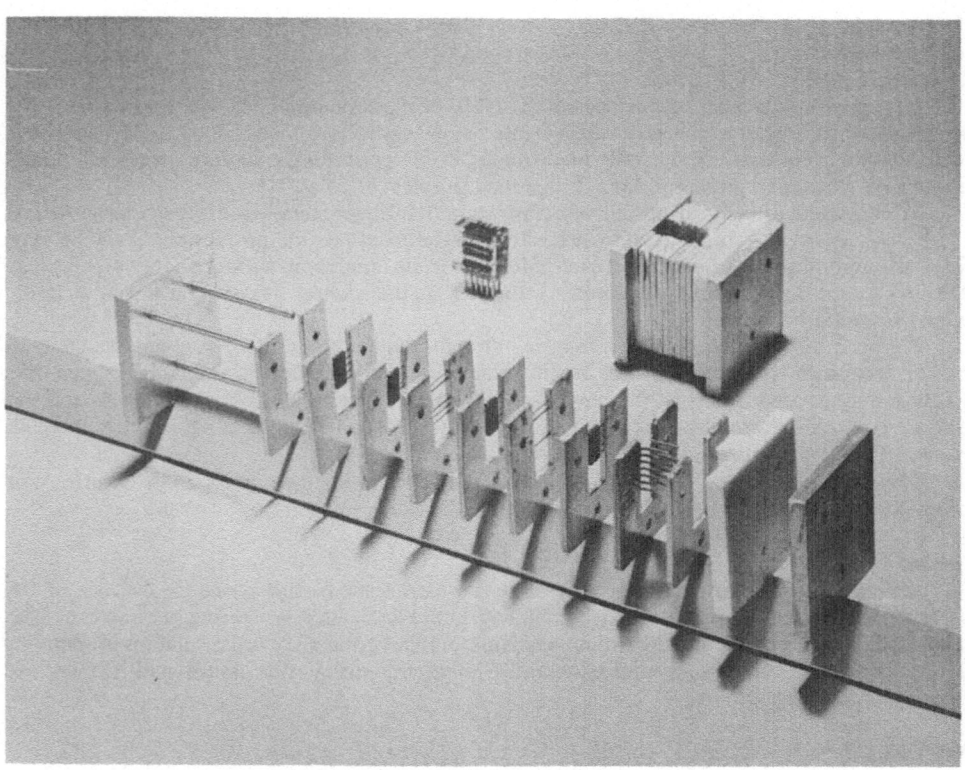

Fig. 3. Frame fixturing and casting mold.

At this stage, the machined submodule is a rugged, reliable assembly, its elements no longer requiring "handle with care" treatment. The subsequent interconnection operations are thus much easier to perform.

Plating

Vapor-phase deposition was selected as the method of depositing interconnections on the submodules because of its tenacious bond and penetrating deposit. An efficient deposition rate and a relatively simple processing procedure assured a reproducible and readily controlled deposition process.

In vapor-phase deposition, the conductive material is present as part of a metal organic compound. The compound is vaporized and the metal deposited on the heated surface planes of the submodule by a surface-catalyzed thermal-decomposition process.

Nickel carbonyl, selected for its low decomposition temperature, produces a nickel deposit on the submodules. Nickel is suitable as a conductive material and is compatible with both the FEBs and the epxoy resin casting compound. For the intended application it can be plated in sufficient thickness to keep its resistance low and its current-carrying capacity large enough to satisfy any submodule interconnection requirements.

Masking and Etching

The desired interconnection pattern is produced on the plated submodule by well-known subtractive photo-etching techniques. The submodule is first coated with a photo-resist material. The resist is then exposed through a photo-mask having the desired interconnection pattern. Etching of the pattern is accomplished with a bubble etcher and a ferric chloride etchant.

Conformal Coating and Functional Acceptance Test

To protect the submodule interconnection patterns from surface contamination and mechanical abrasion, a thin insulative epoxy coating is applied. After curing the conformal coating, the submodule is ready for a final functional acceptance test, which completes the manufacturing cycle. Completed modules are delivered as tested functional building blocks that can be integrated into the next level of system organization by established, well-known assembly and interconnection methods.

Rework

In the event that improper connections are made, through use of an incorrect mask, improper alignment, poor control of deposition or etching, or mechanical abrasion of the interconnect pattern, the deposited interconnection can be readily reworked without damaging the cast submodule or its elements. All that is required to prepare the submodule for replating is the removal of the previously deposited plate by either chemical or mechanical means. The repair or rework is accomplished without imposing any of the stresses normally experienced in reworking more conventional integrated circuit assemblies.

Examples of Submodule Fabrication

Various types and sizes of submodules have been fabricated. Figure 4 demonstrates the sequential stages of batch processing interconnections, and shows the capability of the system to include a variety of FEBs. Submodules on the left are machined with the cross-sectional terminal contact areas flush and exposed in the interconnection plane. The submodules in the center are plated, electrically interconnecting all the previously exposed terminal contact areas. The submodules on the right have been selectively etched to define the interconnecting conductor pattern.

The smaller submodule represents the successful manufacture of a four-stage ripple-through counter. It consists of the logical organization and interconnection of four semiconductor integrated circuits into a submodule that is $0.2 \times 0.4 \times 0.4$ in. Operational units were delivered to the Air Force in August, 1963, and are operational today.

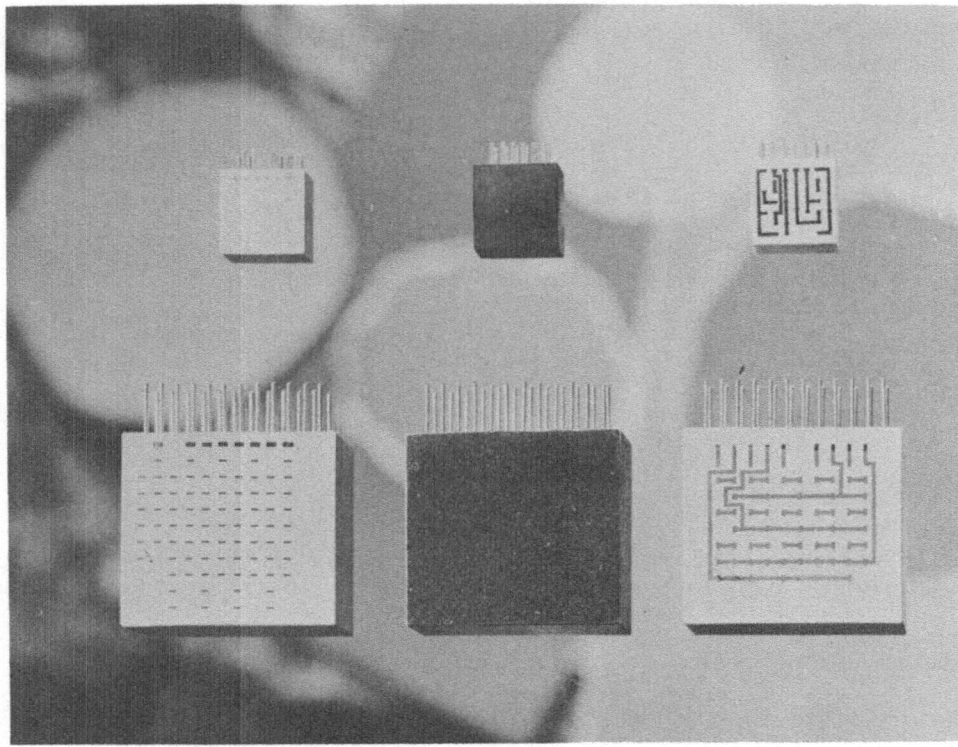

Fig. 4. Sequential stages of batch processing interconnections.

The larger submodule is the logical grouping of thin-film networks deposited on substrates with active components mounted on separate substrates, organized to form a functional decade counter consisting of five flips-flops and associated circuitry. The submodule size is $0.4 \times 0.8 \times 1.0$ in.

SYSTEM INTEGRATION

In developing the system packaging philosophy, a level of system modularity had to be established that was compatible with present system designs and would accommodate a complementary intermix of the more conventional variety of modular fabrication techniques available today. The packaging philosophy also had to consider the organization and interconnection of future systems whenever greater use was to be made of integrated circuits. Overriding considerations throughout were the required reliability and maintenance levels, and the cost of replacing functional submodule assemblies.

The submodule permits handling and assembly into the next level of system organization by conventional methods. Soldering, welding, special connectors, or split-pin wire-bound connections are the various methods available for interconnecting submodules. The most obvious organizational scheme is to mount the submodules on a printed circuit mother board in a manner similar to that used today for packaging high-density cordwood modules. Figure 5 depicts a wire wrap circuit board utilizing the organizational techniques.

An improvement on this method (which reduces the required interconnection density per unit area of the mother board) is shown in Fig. 6. Here vertical welded-matrix interconnect boards are assembled to the basic mother board to accommodate the interconnections between

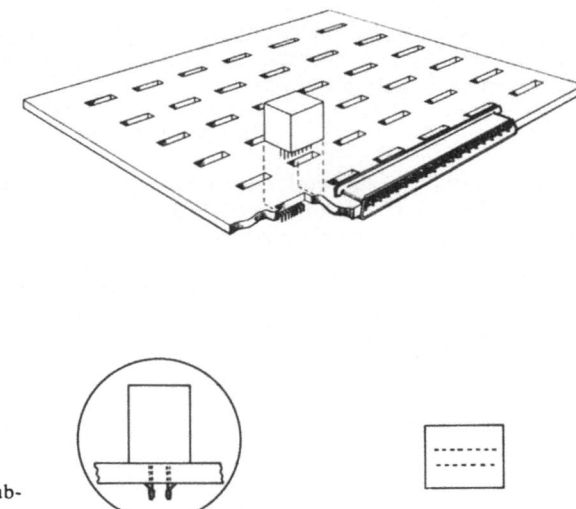

Fig. 5. Conventional integrated submodule assembly.

the submodules in a given row. The required submodule interconnections are thus shared by the vertical boards and the base mother-board. Higher packing densities can thus be achieved without the penalties of higher assembly costs or impaired reliability. The split-terminal wire-bound connections afford a readily maintainable system without introducing the problems normally associated with the resoldering or rewelding of terminal connections.

CONCLUSIONS

The development effort has established the feasibility of the concept, the practical application in design execution, and the manufacturing technique required to produce a basic building block submodule. Submodules are flexible in size and shape and can accommodate a variety of circuits, such as thin-film networks, semiconductor networks, or complementary hybrid combinations of both. They are volumetrically efficient, can dissipate heat effectively by conduction or convection, and are easily maintained and readily integrated into systems. Reliable multiple interconnections are produced simultaneously on a batch process basis. The ability to produce literally hundreds of interconnections on a batch process basis makes the submodule

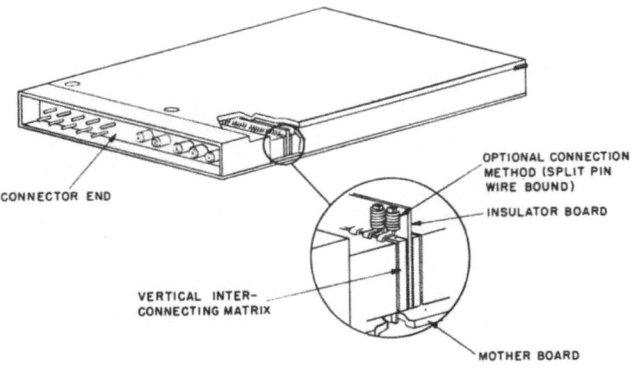

Fig. 6. High-density integrated submodule assembly.

manufacturing technique more economical than comparable techniques in existence today. Thus the submodule building block approach combines efficiency of packaging and economy of manufacture to produce an optimum solution to the problem of interconnection and organization of functional electronic blocks.

ACKNOWLEDGMENTS

A major portion of the work described is the result of a program sponsored by the Air Force under U.S.A.F. contract number AF33(657)-8720 and was administered under the direction of the Molecular Electronics Branch, Electronic Technology Division, Air Force Avionics Laboratory.

The success in establishing an encapsulating and plating system for fabricating the submodules was due to the efforts of Messrs. T. J. Silver and F. R. Corwin of the Norden Materials and Process Section of the Solid State Engineering Section.

Designing for Multilayer Circuits

LAWRENCE D. HUNTER AND ROBERT W. KORB

*TRW Space Technology Laboratories, Thompson Ramo Wooldridge, Inc.
Redondo Beach, California*

The multilayer circuit board offers many advantages in electronic design, particularly for packaging of semiconductor integrated circuits and other microelectronic elements for spacecraft systems. Essentially a multilayer circuit board is a stack of printed circuit (etched wiring) boards forming a matrix with all interconnecting wiring prefabricated and built in. To determine multilayer circuit design requirements, the problems presented by an existing space electronic assembly were analyzed in detail. Redesign, using multilayer circuits and microelectronic parts, resulted in a 15-to-1 reduction in the volume of the package. Consideration was given to layout and spacing, use of miniature mounting posts, fabrication methods, electrical test and checkout, repair and servicing, part-connection techniques, and environmental resistance. A feasibility demonstration model of an electroplated five-layer circuit board was constructed experimentally. Based upon the design problems presented by this unit, practical standards for multilayer circuits have been evolved.

DESIGN STUDY

A MULTILAYER CIRCUIT BOARD STUDY was conducted by TRW Space Technology Laboratories as part of a continuing program to advance the state of the art of electronic packaging design. Because of the complex relationship between performance, design, reliability, processes, and fabrication, the practical problems presented by an actual electronic subassembly for use in space environment were used for properly integrated study. This subassembly, a digital telemetry unit, now performing successfully in an Air Force satellite orbiting the earth, was originally designed (Fig. 1) using 44 welded cordwood modules interconnected on five mother boards.

Fig. 1. Welded-cordwood module version of spacecraft digital telemetry unit (156 in.³).

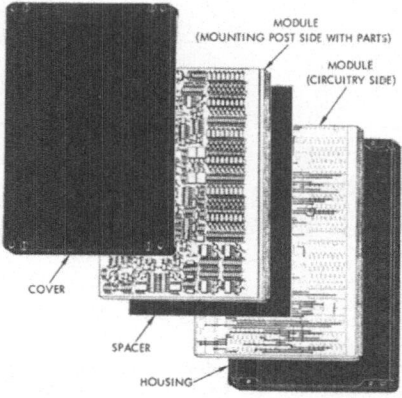

Fig. 2. Microelectronic multilayer circuit version of digital telemetry unit (10.4 in.³).

The orbiting cordwood version is a competent design occupying only 156 in.³ and weighing 4.2 lb.

The redesigned multilayer circuit board version of the same unit (Fig. 2) uses two five-layer circuit boards, each measuring 3.3 by 4.7 in. The volume is only 10.4 in.³ and weight is less than one-third of a pound.

Presently available microelectronic parts and semiconductor integrated circuits were used in the redesign. The multilayer circuits incorporate the design requirements which evolved during the course of the design investigation. When no supplier was found to manufacture boards meeting these design requirements, a multilayer circuit board was fabricated in the laboratory as a feasibility demonstration (Figs. 3 and 4).

Knowledge and information obtained in this investigation are useful for design of micro-electronic assemblies. Feasibility has been demonstrated for such techniques as electroplated circuit board construction, miniature mounting post terminals, close spacing of conductors, and other related design features.

ADVANTAGES

At the present time, multilayer circuits seem the most practical method for meeting the packaging design requirements associated with microelectronics.

Multilayer circuit boards are, in general, of two types. One type is fabricated by laminating several printed circuit boards together. The other type is fabricated by a progressive build-up of electroplated layers of etched circuitry separated by layers of insulation. In the laminated type, interconnections between layers of circuitry are made by plated-through holes. Electronic

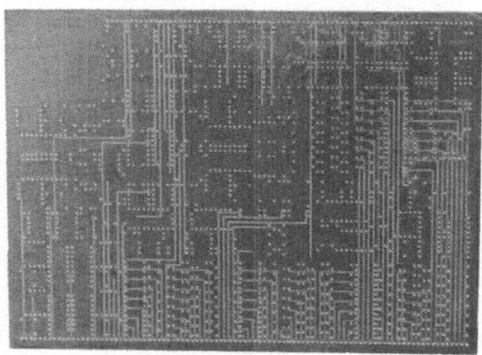

Fig. 3. Multilayer circuit board containing five layers of circuitry (circuitry side).

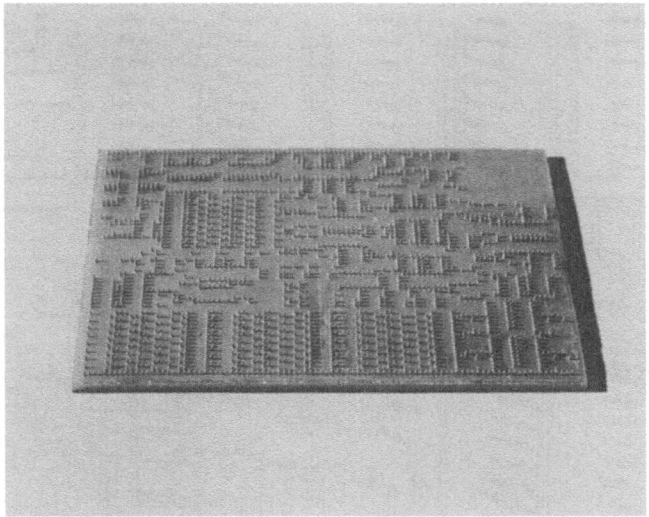

Fig. 4. Multilayer circuit board showing posts for the mounting of 717 miniature electronic parts and semiconductor integrated circuits on 3.3 by 4.7 in. surface.

parts are attached by soldering part leads into these plated-through holes. In the electroplated multilayer circuit board, interconnection of circuitry layers is accomplished as part of the electroplating process. Many variations of both types are possible, each offering some advantages for specific applications.

Basic advantages of the multilayer circuit board may be summarized as follows:

1. More wiring can be packaged in less volume. If the volume of wiring is not thus reduced, the space saving made possible by integrated circuits and other microelectronic devices is meaningless.
2. Hand operations in assembling and interconnection are reduced to a minimum. This materially decreases the chance for error in the handling of microelectronic parts of small size.
3. Production processing of the interconnection wiring results in uniformity and higher inherent reliability.
4. The multilayer circuit board provides a mounting surface for parts, serving as integral structure as well as electrical interconnection.
5. Environmental resistance is easily assured with selection of suitable materials and processes. Stringent vibration requirements are met without difficulty and heat removal is easily accomplished.
6. All electronic parts are easily accessible on a planar surface for testing, checkout, servicing, inspection, repair, and replacement.
7. With elimination of time-consuming assembly and servicing operations, the multilayer circuit board can eventually result in lower costs for microelectronic equipment.

LAYOUT AND SPACING

To determine multilayer circuit design layout and spacing, the problems presented by the digital telemetry unit were analyzed in detail. These resulted in establishment of practical standards generally applicable to other similar microelectronic assemblies.

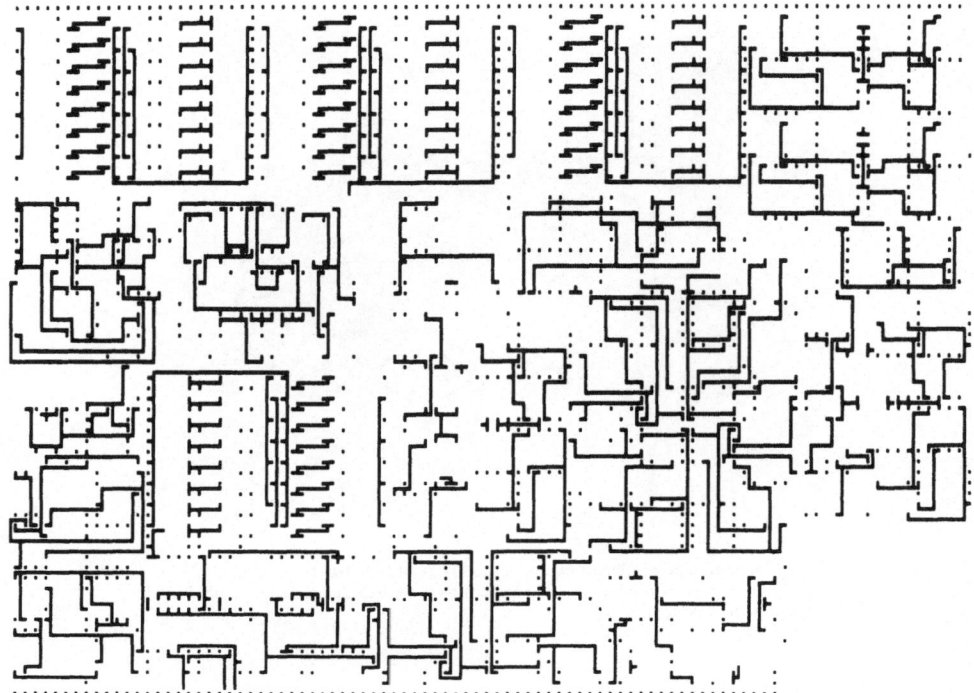

Fig. 5. Multilayer circuit wiring layout (first layer).

Preparation of engineering layout drawings proved to be less difficult than originally anticipated. Because of the small size of the electronic parts, all engineering drawings were made on a scale of ten times actual size. Preliminary layouts were made of each of the functional circuit blocks such as pulse drivers, flip-flops, analog gates, and gate drivers. These small layouts were then fitted together to provide the layout for the overall multilayer circuit board design. In this manner, it was possible to design the demonstration multilayer circuit board containing 717 electronic parts without an excessive expenditure of drafting time.

It was originally thought that ten circuitry layers might be needed, but this many layers proved unnecessary. For the digital telemetry unit, and other similarly complex circuits, five layers of circuitry appear to be adequate. Although there were some modifications in the final design, it was determined that conductors can be generally parallel in each layer, with alternate layers perpendicular to each other (Figs. 5 and 6). The first two circuitry layers provide all connections for the functional circuit blocks, including cross-over connections. The next two layers provide connections between circuit blocks. The fifth, or final, layer is used for external input–output connections. With this separate layer for external connections, an external electrical connector can be included as part of the multilayer circuit board design, eliminating the need for cabling.

A desirable size for a typical multilayer circuit board appears to be approximately 3 by 5 in. A board of this size accommodates between 600 and 800 small electronic parts or as many as 200 integrated circuits. Larger-size boards present problems of complexity in fabrication, testing, and servicing. Smaller-size boards present handling and packaging problems.

A grid spacing of 0.050 in. was selected for the multilayer circuit board design to permit close mounting of most microelectronic parts. This spacing is compatible with dimensions of flat-package semiconductor integrated circuits, which commonly have leads spaced on 0.050-in. centers. For these units, the ability to pass an internal conductor between the leads is a decided

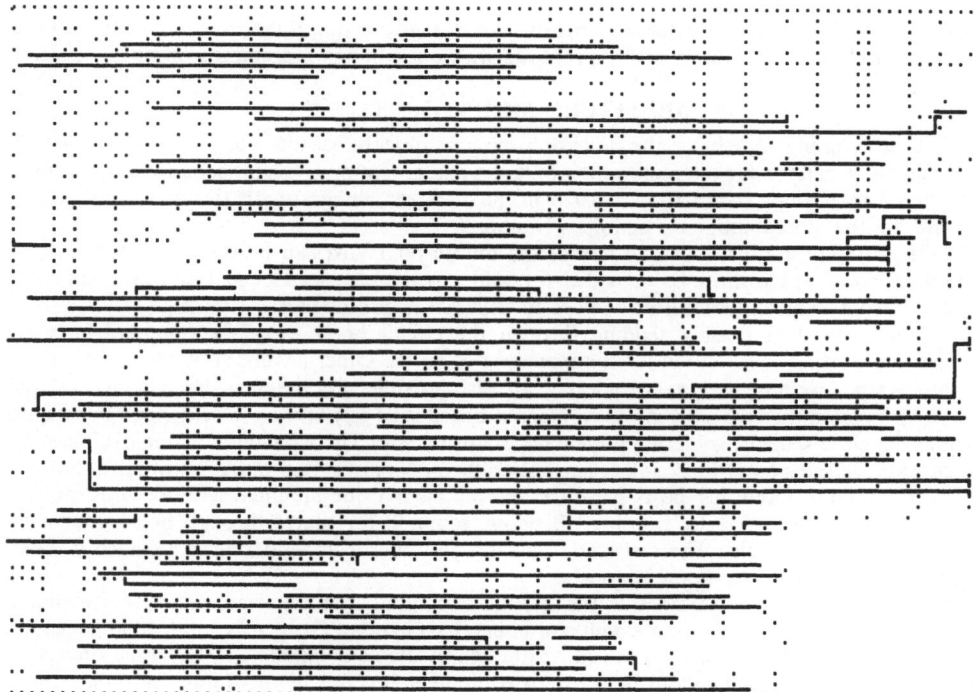

Fig. 6. Multilayer circuit wiring layout (fourth layer).

advantage, avoiding the space-consuming routing of conductors around the outside of integrated circuit packages. A 0.015-in. width was therefore established for conductors. This allows a conductor to pass between two others spaced on 0.050-in. centers with 0.010-in. insulation laterally (Fig. 7). The 0.010-in. insulation is more than adequate and provides a tolerance for such process variations as electroplating build-up.

Arrangement of circuitry on a standard grid pattern offers significant advantages. With a unit of small dimensions, containing a large number of parts, a grid pattern is a decided convenience, providing uniformity, simplicity of drawing, indexing, and time-savings in production and assembly. The chief objection to a grid pattern, in favor of random design with point-to-

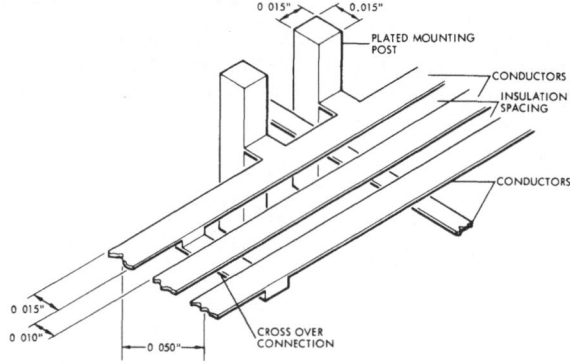

Fig. 7. Internal spacing on 0.050-in. grid within multilayer board, permitting conductor to pass between posts.

point wiring, is loss of density. Loss of density, however, is relatively insignificant and far outweighed by the simplicity achieved. Furthermore, a grid arrangement makes possible future automatic fabrication and assembly processes with operations programmed on an *x-y* grid.

MINIATURE MOUNTING POSTS

In attempting to achieve the desired spacing, it quickly became obvious that the plated-through hole method for interconnection of layers and mounting of parts has serious limitations. Terminal mounting posts 0.015 to 0.020 in. in diameter, integral with the multilayer circuit board, on the other hand, have definite advantages.

If posts are used, the overall size of the multilayer board can be smaller. With plated-through holes, space must be provided for solder pads, part lead bend radii, and sufficient spacing to prevent soldering heat damage to sensitive electronic parts.

With mounting posts extending through the circuit board (Fig. 8), opposed-electrode resistance welding can be used for the mounting of parts. Other advanced methods, including split-electrode welding and microsoldering, are also possible. On the other hand, connection of parts to a plated-through-hole multilayer circuit board is largely limited to soldering. For a unit with such close spacing, soldering presents many problems. These include flux removal, lack of access for soldering tools, loose solder particles, difficulty of inspection, and excessive handling.

In addition, mounting posts permit electronic parts to be mounted without stressing the very fine part lead wires by bending. Bending, required for plated-through holes, may also damage glass part seals. Thermal stress damage, resulting from soldering, is avoided if parts are resistance welded to mounting posts.

Of considerable importance is the specific capability that the mounting-post technique offers for use of flat-package semiconductor integrated circuits. These are contained in packages as small as 0.125 by 0.250 by 0.050 in. Although available commercially for some time, flat packages have been difficult to design into equipment because of mounting and connection problems. Multilayer circuits with miniature mounting posts offer an acceptable design solution.

The use of posts permits parts to be moved and replaced without damage to the circuit board. On the other hand, repeated application of soldering heat to plated-through holes may seriously damage both the circuit board and the associated electronic parts.

More positive electrical connections are provided between circuitry layers with posts and electroplated wiring than with drilled plated-through holes on a laminated-type board. The internal area of interconnection between a conductor and a 0.015-in. post is equivalent to that between a one-ounce copper conductor and a 0.050-in.-diameter plated-through hole.

These advantages are of major importance when the cost of a completed multilayer assembly is considered. The complexity, labor, and parts costs make it exceedingly expensive to scrap a defective assembly. It is therefore advisable, with use of mounting posts, to eliminate sources of trouble known to occur in assembly and servicing by simplifying design and fabrication.

FABRICATION TECHNIQUES

To determine the feasibility of the multilayer circuit design requirements, including layout, spacing, and mounting-post technique, a demonstration board was fabricated in the laboratory. Although it is normally advisable that such boards be constructed by commercial-specialist

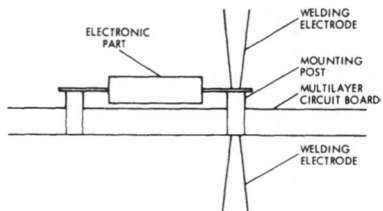

Fig. 8. Opposed-electrode resistance welding of parts to mounting posts on multilayer circuit board.

manufacturers, laboratory fabrication can be accomplished with relatively inexpensive and simple equipment for photo-resist processing, photoprinting, silk-screening, etching, and electroplating.

The following items are of interest to the packaging engineer concerned with designing, directing, or specification of multilayer circuit board fabrication.

Because of the small size of all elements, engineering drawings were made on a scale of ten times actual size. First, each of the circuit blocks in the system, such as flip-flops, pulse drivers, logic gates, analog gates, and gate drivers, were laid out separately. These blocks were then combined in a complete layout drawing for the multilayer board unit. The demonstration board, as designed, contained 717 electronic parts, including semiconductor integrated circuits, on a single module measuring 3.3 by 4.7 in.

The artwork required for photoprinting processing was prepared by cutting the pattern on a transparent ruby-coated stable plastic film (Rubylith). The art was made ten times actual size, using essentially the standard etched printed circuit artwork techniques. Posts internal to the plated structure were drawn with a 0.015 by 0.015 in. square cross section, rather than circular, for more efficient layout.

Actual size negatives required for processing were made on glass plates, and photographically reduced in a single step from the artwork, using a copy camera with a high precision lens. Glass plates were used rather than film negatives both for accuracy and for compatibility with the fixturing which was constructed for precision registration.

An epoxy glass laminate 0.090-in. thick was used as the substrate for the multilayer board, providing sufficient rigidity and strength for the requirements.

An initial step in fabrication required the precision drilling of 1700 holes of 0.020-in. diameter accurately spaced on 0.050-in. centers for the insertion of the mounting posts into the substrate. This was accomplished by making a metal drill pattern, and then drilling the substrate with the pattern pinned to the substrate. The stylus of a high-speed drilling machine is pressed into the pattern holes and a drill comes up through the substrate at 45,000 rpm. Holes are drilled at the rate of 60 per minute. Satisfactory precision is achieved with proper selection of drill and drill bushing.

Considerable effort was necessary to determine a suitable method for bonding the 0.020-in.-diameter copper wire mounting posts into substrate holes. Posts must be bonded securely to prevent any movement during plating or after fabrication which might result in breaking of electrical connections. After a number of attempts, an epoxy adhesive was found to provide satisfactory bonding. Vibration of the substrate board is required as a prior step to remove dust particles remaining from the drilling operations. Precautions are taken to keep the adhesive from the contact surfaces of the posts and to seal the holes against entry of the electroplating solution.

Surface flatness of the substrate board is essential for intimate contact between the negative and the board for good photoprinting. Initially, it is necessary to lap the surface on a granite micro-flat block. During subsequent processing, care is taken to control the uniformity of copper, insulation, and photo-resist coatings to minimize the number of lapping operations.

For proper handling during photoprinting, plating, and other processing operations, it was necessary to construct both a holding fixture for the multilayer board and a registration fixture to accommodate the holding fixture in the photoprinter. The holding fixture is held in the registration fixture under spring tension (Figs. 9 and 10). An initial adjustment is made to align the board with the first negative glass plate; subsequent negatives are in registration with each other. With each change in board thickness during processing, shims are used to assure close positioning against the negative.

It is necessary to protect the post side of the multilayer board on which parts will be mounted to prevent damage during processing. Also, it is necessary to make a reliable electrical connection to each of the posts to carry the plating current during processing. A single-sided copper-clad epoxy–glass board with holes drilled to fit tightly over the posts is recommended for this purpose. Connections can then be made to the posts by dip soldering, thus providing electrical connections and protection at the same time.

The photographic masking coating normally applied was found to be too thin and porous

Fig. 9. Multilayer circuit board holding fixture.

to withstand the necessary two-hour plating cycle, unless carefully cured at elevated temperature. Adding a photosensitive lacquer in a 1 to 5 ratio also was found to provide the required characteristics without reducing the image definition appreciably.

For insulation between layers of electroplating, an epoxy resin with glass-cloth reinforcement was used. Care was taken to match the coefficient of expansion of the insulation to that of the substrate. For circuit layers, insulation is 0.0015 in. including the 0.0014-in.-thick glass-

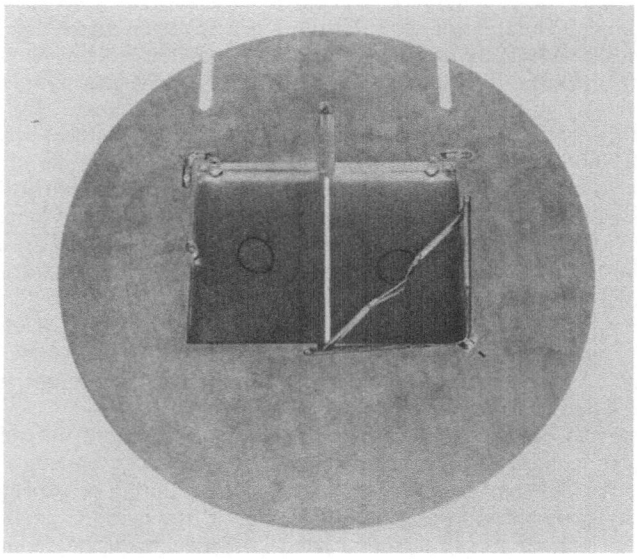

Fig. 10. Registration fixture for use with photoprinter.

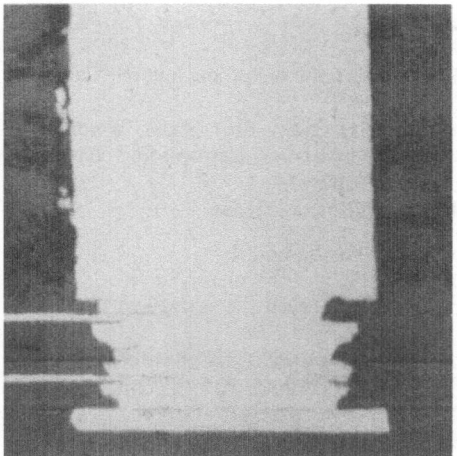 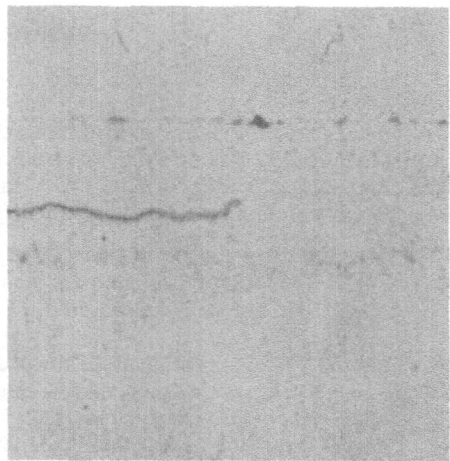

Fig. 11. Microstructure of electroplated post showing solid metal structure with electroless copper remaining only at small portion of interface. Left, cross section of post (100 ×); right, electroless copper interface (1000 ×).

cloth (type 104) reinforcement. For post interconnection layers, insulation is 0.0025 in. thick including 0.0020-in.-thick glass cloth (type 108). Insulation was applied by a vacuum bagging process.

The basic concept of plated multilayer circuit construction is the fabrication of a solid block of plated copper containing all electrical paths separated by insulation (Fig. 7). If this is accomplished, high reliability is achieved. To avoid possible defects in the structure, pending further study, it was decided to remove all electroless copper material deposited as a result of process requirements. Considerable effort was devoted to removing this material, which complicated the fabrication of the demonstration multilayer board. Subsequent sectioning and metallurgical examination of a cross section of an electroplated post (Fig. 11) showed that good bonds were achieved between the copper layers during processing. It should be noted (Fig. 12) that a lateral build-up of the material occurs during the electroplating of each layer. This build-up is controlled by limiting the thickness of each electroplated layer.

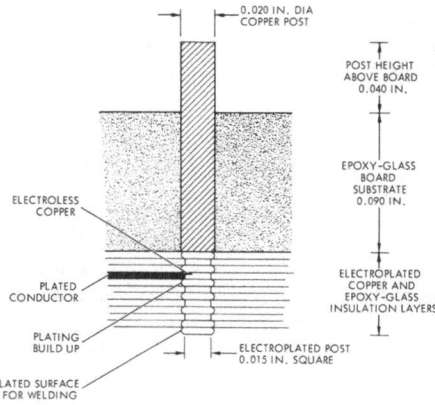

Fig. 12. Diagram of mounting-post construction.

FABRICATION STEPS

Following preliminary development investigation to determine the processing requirements, the demonstration board was constructed.

Initial preparations involved drilling the substrate (G11 epoxy fiber glass), bonding the pins into the holes, making electrical connections to all posts to assure electroplating continuity, and checking registration of the artwork pattern on the substrate.

Fabrication of a single layer of circuitry involves the following steps:

STEP 1. Posts are electroplated so that they project from the board.

STEP 2. Insulation (glass-cloth and epoxy) is applied.

STEP 3. The surface is lapped until the electroplated posts reach the surface of the insulation.

STEP 4. A photo-resist image of the primary post pattern is applied by photoprinting.

STEP 5. Posts are electroplated to a thickness sufficient to connect with the circuit layer.

STEP 6. The photo-resist is removed.

STEP 7. Insulation (glass-cloth and epoxy) is applied.

STEP 8. The surface is lapped until the electroplated posts reach the surface of the insulation.

STEP 9. Electroless copper is applied to the entire surface to conduct plating current.

STEP 10. Copper is electroplated on the entire surface to a thickness of 0.0002 in.

STEP 11. A photo-resist image of an offset post pattern is applied by photoprinting.

STEP 12. The surface is etched to remove both the electroplated and electroless copper interface.

STEP 13. Posts are electroplated to the level of the previous copper layer (step 10).

STEP 14. The photo-resist is removed.

STEP 15. The entire surface is electroplated until the copper reaches a thickness of 0.001 in.

STEP 16. A photo-resist image of the circuitry is applied by photoprinting.

STEP 17. Unwanted portions of the copper layer are etched away, leaving only the copper circuitry.

STEP 18. The photo-resist is removed.

STEP 19. Insulation (glass-cloth and epoxy) is applied.

STEP 20. The surface is lapped until the copper circuitry reaches the surface of the insulation.

For the fabrication of a five-layer board, steps 4 through 20 are repeated five times for a total of 88 processing steps. Although at first glance this procedure seems somewhat complex, it is accomplished by experienced and knowledgeable personnel without difficulty.

Following processing, the multilayer circuit board is removed from its holder, which includes the protective covering on the mounting-post side of the board and the electrical connections between the posts, which carry the plating current. Posts are lapped to the required 0.040-in. height. The board is then trimmed to size and cleaned. Following electrical checkout, the board is ready for the mounting of electronic parts.

ELECTRICAL TEST AND CHECKOUT

Considerable attention was given to design requirements for electrical testing of the multilayer circuit board. Problems in this area indicate that new techniques and procedures are needed for electrical testing, inspection, and trouble shooting. There are problems both for acceptance testing of multilayer circuit boards before attachment of parts, and for the identification of defective electronic parts following final assembly.

For acceptance testing of the multilayer circuit board before attachment of parts, it is necessary to check each miniature post for electrical isolation or continuity with each of the other posts on the board. Maximum electrical resistance of the conductors was established as

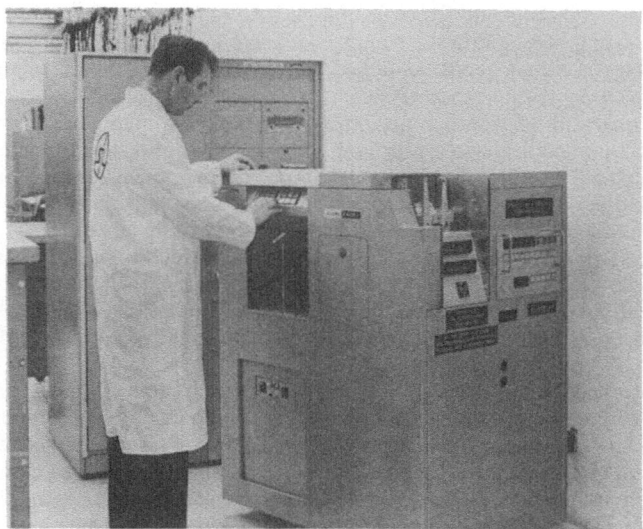

Fig. 13. Automatic circuit tester capable of scanning 600 terminations per minute.

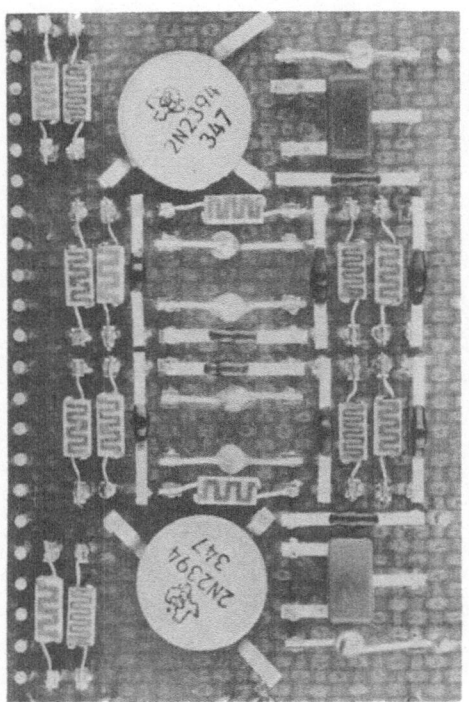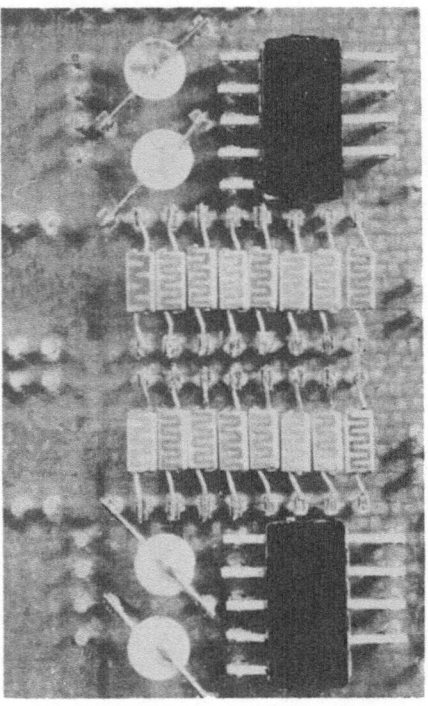

Fig. 14. Microelectronic parts welded to miniature mounting posts on multilayer board (dual blacking oscillator, left, and dual flip-flop, right).

0.1 ohm per inch on the basis of circuit requirements. To test dielectric strength, a requirement was established that adjacent conductors be able to withstand application of 50 volts.

Acceptance testing of a multilayer circuit board is a task requiring automatic techniques. The feasibility demonstration board contained approximately 1700 posts, requiring a total of some 1,444,150 electrical wiring checks if each mounting post is checked with each of the others. This is an impractical and excessively time-consuming task if automatic scanning test equipment and carefully programmed testing are not used. Several types of testers capable of performing this task are available (Fig. 13) utilizing punched tape or IBM-punch-card inputs. Tests can be performed at relatively high scan rates and programmed so that all data for the feasibility demonstration multilayer circuit can be included on 850 IBM punch cards.

The time required for acceptance testing of a multilayer circuit makes it necessary to initiate test planning as early in the circuit layout design phase as possible. In addition to wiring lists, design of test fixtures, connectors, and cabling is needed. The drill pattern used to fabricate the circuit board can be utilized in the design of a test fixture electrical connector, with proper planning.

There are also special problems in functional system checkout of an assembly consisting of microelectronic parts mounted on a multilayer circuit board. Since individual subcircuits cannot be electrically isolated and disconnected, it is necessary to test the entire assembly as a unit. Normally, however, it should be possible to check functioning of the complete assembly with standard test equipment. Each terminal of each electronic part on a multilayer circuit is accessibile for monitoring during electrical tests, thus making available as many electrical test points as are needed.

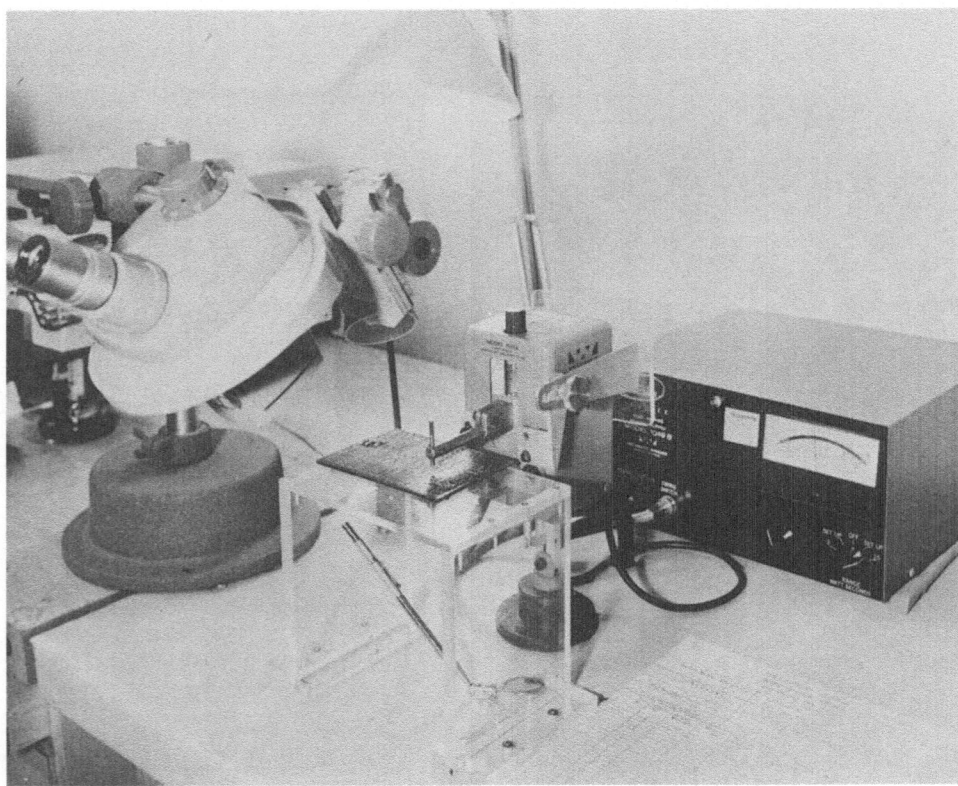

Fig. 15. Setup for opposed-electrode welding of electronic parts to multilayer board.

ELECTRONIC PART CONNECTION TECHNIQUES

As part of the design study, an investigation was made of resistance-welding techniques and methods for attaching electronic parts to miniature mounting posts. The connection techniques were verified following fabrication of the feasibility demonstration board with attachment of typical discrete parts (Fig. 14).

Although opposed-electrode resistance welding is the preferred method of attachment of parts at the present time because of reliability considerations, other methods of bonding may be used with miniature mounting posts. Tests indicate that split-electrode (parallel-gap) resistance welding can be practical also. Other methods are conventional soldering, use of solder preforms, and conductive adhesives.

With opposed-electrode resistance welding, connections are made by placing electrodes on opposite sides of the board and welding each part lead to its post. Welding fixtures are relatively uncomplicated (Figs. 15 and 16). A table of plastic sheet material provides adequate support for the multilayer circuit board. The lower welding electrode extends slightly above the surface of the table to provide positive electrode contact with the bottom of the post. The work is viewed through a binocular microscope under a magnification of ten times.

In developing weld schedules for the miniature mounting posts, weld settings were found to be extremely sensitive. Since poor resistance-welding bonds were achieved initially to the copper posts used in the feasibility demonstration board, a variation in the welding technique was devised. Cut lengths of 0.005 by 0.020 in. Alloy 180 (nickel–copper) ribbon were first welded to the copper posts to serve as an interface material. These tabs provide a cap on the

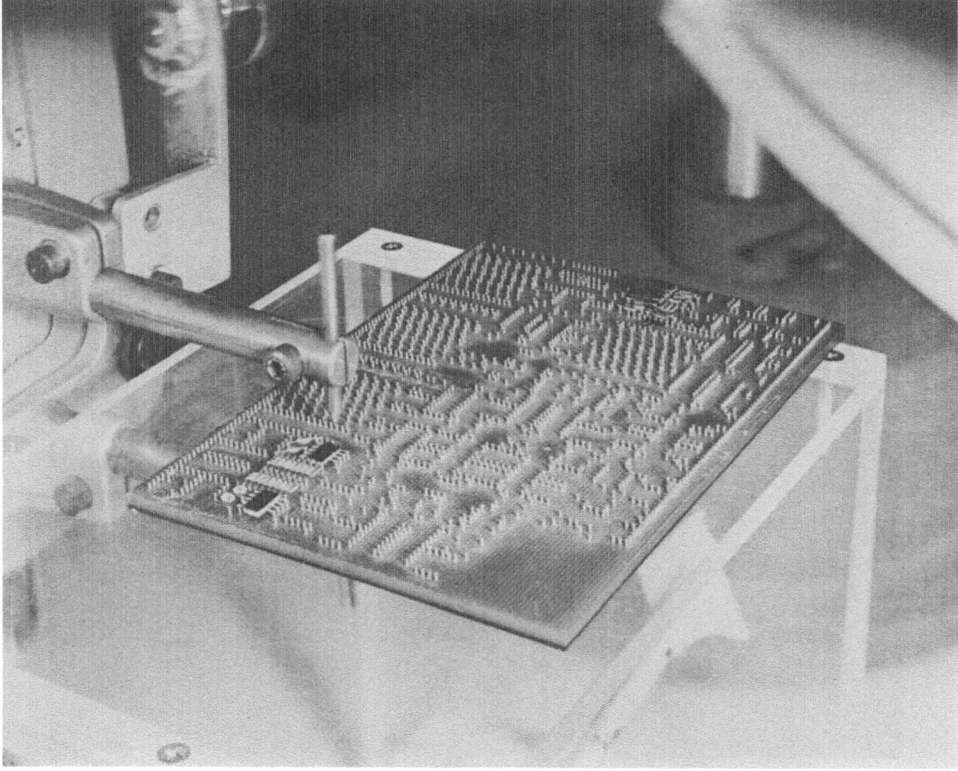

Fig. 16. Plastic positioning table for opposed-electrode welding to multilayer board.

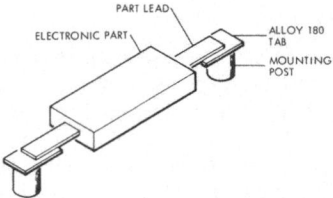

Fig. 17. Sandwich-type resistance weld.

top of the post. It was then relatively simple to weld part leads to the tabs, providing a sandwich-type weld (Fig. 17). Weld schedules were developed for tin-plated copper wire (0.016-in. diameter), gold-plated Kovar ribbon (0.002 by 0.010 in., and 0.003 by 0.019 in.), solder-coated nickel wire (0.010-in. diameter), and gold-plated Alloy 180 wire (0.010-in. diameter). Alloy 180 ribbon (0.005 by 0.020 in.) and gold-plated Alloy 180 ribbon (0.0035 by 0.021 in.) were easily welded to the posts without use of an interface material. In future design studies, use of more weldable material for posts will be considered. The sandwich-type weld, however, provides an acceptable and simple solution to the electrical connection problem (Fig. 18).

The miniature mounting posts are sufficiently rugged to permit replacement of defective or mislocated parts without damage to the circuit. Since posts are 0.040-in. high, they may be clipped off and burnished smooth in preparation for rewelding. Repairs may be made easily three times without excessively shortening the posts.

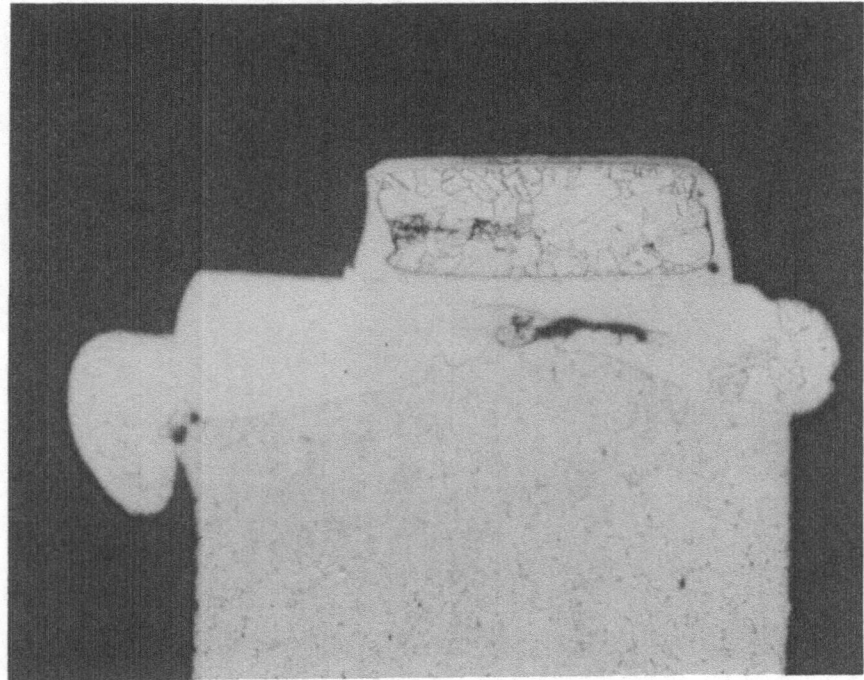

Fig. 18. Microstructure of sandwich weld (gold-plated Kovar ribbon to Alloy 180 tab to copper mounting post). 150 × magnification.

ENVIRONMENTAL RESISTANCE

After electrical test and checkout of a completed multilayer circuit board assembly, a transparent coating of silicone resin can be applied. This coating serves to bond all parts to the board, provide heat dissipation, and otherwise furnish environmental protection.

Multilayer circuit assemblies, thus protected, appear well able to meet all present environmental specification requirements. The thermal properties of the silicone material should provide a suitable heat-conduction path, as well as protecting the individual electronic parts against the effects of shock and vibration. Materials used in fabrication of the multilayer circuit board are selected with thermal expansion coefficients taken into consideration, so that no separation of layers is expected to occur. All materials have satisfactory low-outgassing and sublimation rates, after initial conditioning, for the required resistance to the effects of space vacuum. Adequate environmental resistance is thus designed into the circuit board.

REPAIR AND SERVICING

Concepts for rework, repair, and servicing of the multilayer circuit board are required as part of the initial design, since the costs of parts and labor in such a complex assembly prohibit rejection and scrap of defective units.

Since the multilayer board is essentially a planar surface, all electronic parts are easily accessible for inspection, testing, removal, and replacement.

In the feasibility demonstration circuit board design only one internal connection is made to any one mounting post. It is a relatively simple matter, in the event of a defect, to drill out any post without affecting other circuitry. The post can then be replaced with an insulated terminal and the proper electrical connection to the post can be made by means of a jumper wire on the surface of the board.

If an internal conductor proves to be defective, it is similarly possible to replace it with a jumper wire on the surface of the board, between the posts affected. Minor changes and modifications in circuit design can also be accommodated in this manner by drilling and jumpering if necessary, even following final assembly of a complete unit.

EVALUATION AND FUTURE DEVELOPMENT

As the density of microelectronic packaging increases, it is apparent that considerable design effort is required in all phases of circuit layout, artwork preparation, and interconnection of microelectronic elements. New and unavoidably more complex requirements for testing, and servicing, as well as for fabrication and assembly, must be anticipated and designed into the electronic package.

The multilayer circuit board apparently provides the most acceptable solution to the packaging of microelectronics, but this solution is not without its problems. Nevertheless, the multilayer circuit board technique offers the packaging engineer a compact and versatile interconnection method for integration circuits, thin-film networks, and discrete microparts.

The following design characteristics are considered desirable in multilayer circuits for microelectronic packaging:

1. Parts should be attached to the multilayer circuit by welding. To this end, weldable terminals in the form of miniature mounting posts must be provided.
2. The diameter of the mounting posts should be between 0.016 and 0.020 in.
3. The mounting posts should extend completely through the circuit board to permit welding by opposed electrodes.
4. Center-to-center distance between adjacent mounting posts should be 0.050 in. to accommodate flat-package integrated circuits and permit close spacing of miniature electronic parts.
5. It should be possible to pass an internal conductor between any two terminals on 0.050 in. centers.

6. Terminals should be sufficiently rugged to permit replacement of parts without damage to the circuit.
7. Electrical resistance of conductors should be less than 0.1 ohm per in.
8. Dielectric strength between adjacent terminals and conductors must be sufficient to withstand 50 volts.
9. Multilayer circuit processes should permit a minimum of five layers and a probable maximum of ten layers of circuitry to be fabricated.
10. No more than one internal electrical connection should be made to each mounting post and the mounting post grid should be offset from the connector pattern to permit repair of the board.
11. The multilayer circuit board should be capable of withstanding the rigors of the aerospace environment.

These specifications broadly delineate design requirements for multilayer circuitry for interconnecting flat-package integrated circuits, thin-film networks, and microparts. These requirements have been shown to be technically feasible and well within the capabilities of multilayer circuit manufacturers and suppliers, following a relatively short development period.

The advantages of multilayer circuits in microelectronic design, together with the widespread attention devoted to use of integrated circuits, should combine to produce demand for multilayer boards great enough to encourage rapid development of this technique.

A Flexible Module Concept for Electronic Box Design

FREDERICK L. KOVED

GPL Division, General Precision Aerospace
Pleasantville, New York

Design for the module concept to be presented is under development in the packaging group associated with the GPL Division Microelectronic Facility. The design for the electronics portion of a proposed airborne system starts with component studies and continues through the circuit-board level, striving for an optimum package at minimal overall cost. Versatility of electronic box design is stressed to provide maximum utilization of space, power, and cooling techniques. Three objectives selected to guide the program were weighed in the following order: (1) reliability; (2) maintainability; and (3) weight and volume. Reliability is recognized as a basic factor throughout the circuit design, element selection, and packaging phases of the program. Special attention has been given to mechanical structures and to thermal balance as required in avionic equipment.

DESIGN OBJECTIVES

THE MAINTENANCE PHILOSOPHY to provide the flight line with the ability of locating and replacing defective functional modules is a key factor in the module and electronic box design. After review of several approaches, a plug-in-type connector to be incorporated as an integral part of the module was selected as the best design technique. Analysis of the vendor data presented on a connector selected for the module indicated low failure rate (approximately 0.09 failures per 10^6 hours). The proposed equipment design now employs plug-in features at both the module and board level. This feature enhances the serviceability and maintainability of the equipment. Accessibility is provided to all sections of the electronic box, as sketched in Fig. 1.

MODULE MOUNTING

The module concept for electronic circuit packaging has gained widespread acceptance based on important maintenance and logistic advantages. Accepted cost reductions are possible through the conservation of maintenance manpower, the elimination of stocking redundant functions, higher reliability achieved, and lower engineering design costs.

The GPL plug-in technique presents an approach to module construction which uses standard connector sections fastened to the internal assembly board of the module and molded to form a portion of the mounting surface. Two keying pins are also located in varying positions on the mounting surface to provide mechanical rigidity and avoid error in location of a module type when plugged into a board. Modules and boards are stamped with identifying part numbers to further simplify placement of modules, as shown in Fig. 2.

The strip connectors are standard Amphenol wire form contacts and connectors series 220,221. The delicate male sections are shrouded in the longer plastic segments, whereas the tubular female sections are mounted in an internal board assembly and project above the board

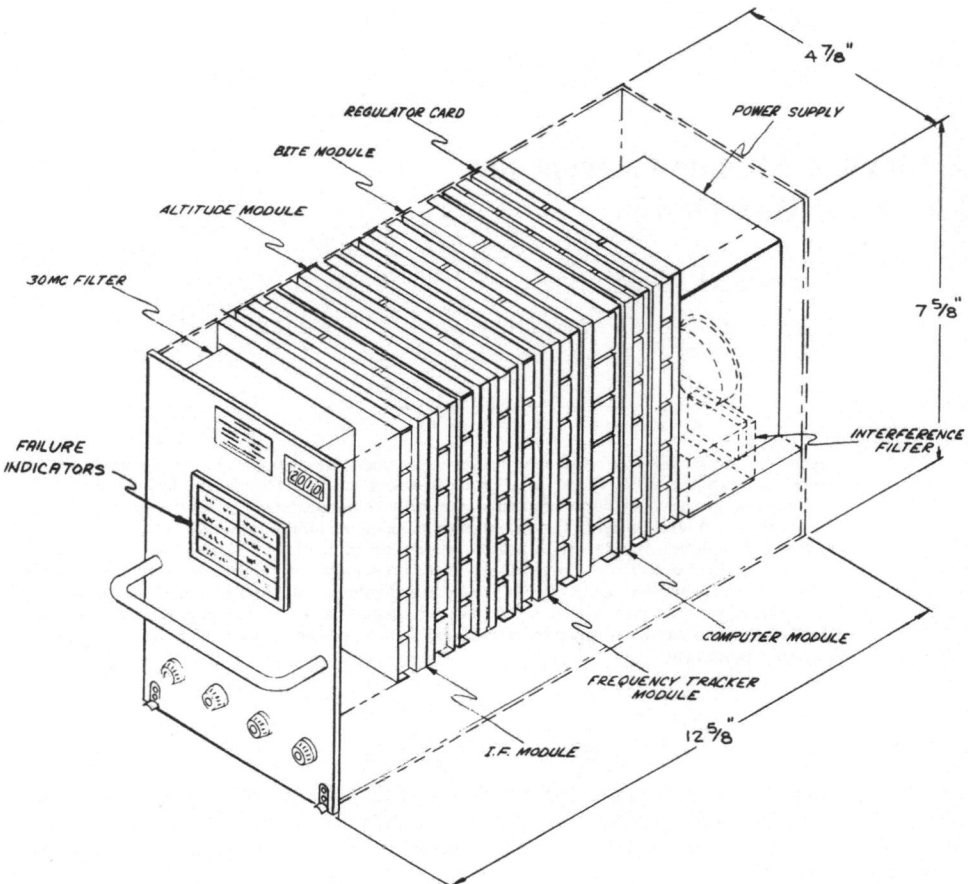

Fig. 1. Electronics unit.

surface as shown in Fig. 3. Studies have shown this arrangement to be satisfactory, rugged, and relatively easy to assemble. The guide pins assure positive alignment of the male and female sections when placing the module on the board. Typical modules are shown in Fig. 2. Connector sections are available in 6-in. lengths, and may be cut to length as required in each module based on the number of pins required.

Details of the unassembled parts of the connector are shown in Fig. 4. The wide plastic strip is the housing for the shrouded male pin, and the narrow section is used for the long female tubular section in the board connector. Pins are placed into the plastic sections as required, using an inexpensive tool available from the vendor. Excellent adhesion has been found with the polycarbonate material of the connector and the di-glycidyl ether of bis-phenol, an epoxy system used to pot the module.

Each module is uniformly dimensioned with a 1.125-in. width and 0.75-in. height. The length will vary with circuit function and allowable component density based on a "worst case" thermal analysis. Preformed diallyl phthalate shells were used for casting the modules. The materials and subassembly parts of a modular unit are shown in Fig. 5.

Emphasis has not been placed in this paper on the internal module construction. Consideration and techniques are being developed for the incorporation of discrete components,

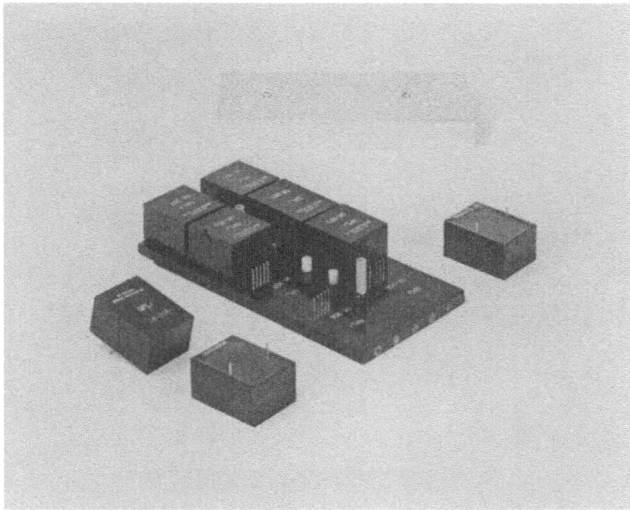

Fig. 2. Detailed view of circuit board and plug-in module.

thin-film devices, and integrated circuit flat packs into units. These have all taken the form of modified cordwood welded assemblies. Programs are being conducted to assure that epoxy resins selected will permit determination of the mode, should failure occur, and allow for factory repair of units. Analysis of thermal gradients internal to the module structure and related to the potting resins and shell material used in the assembly is in progress.

BOARD CONSTRUCTION

A circuit tray or board replaces the conventional single layer printed circuit card and is intended as a mother board for interconnection of system functions. Those designers advocating a welded structure may use an internal assembly of four layers of standard nickel interconnects with mylar backing produced by Amphenol Corp. Vertical tubular female connectors provide interlayer connections in addition to normal contact to the module. An epoxy glass board predrilled provides accurate positioning of the tubular female sections, and is a supporting

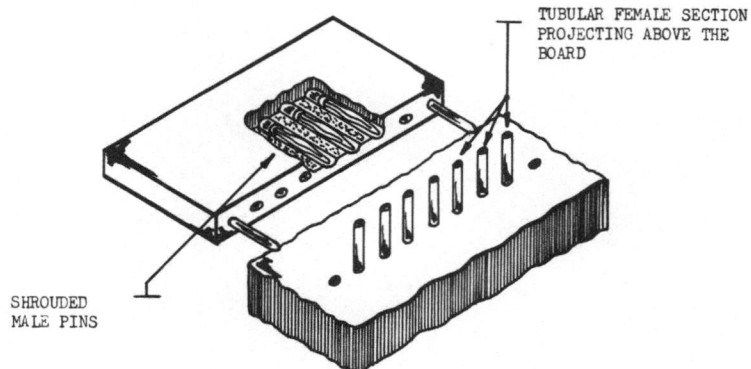

Fig. 3. Cross section of connectors.

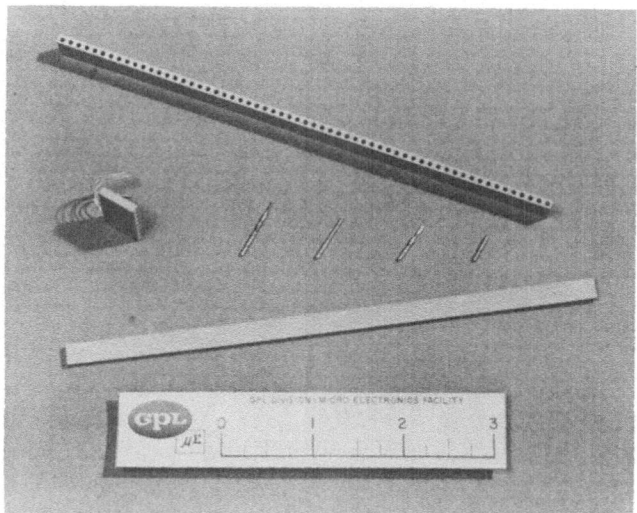

Fig. 4. Connector subsection.

vehicle for the nickel interconnect patterns, connectors, and test jacks, to be described in a later section. The tubular sections may also be used in conjunction with multilayer printed circuit boards drilled to accept the female units. A sketch shown in Fig. 6 illustrates a cross-sectional view of a proposed assembly. Note that this entire assembly is ultimately cast into the tray shape using a 60% filled epoxy resin system with a negative coefficient of thermal expansion.

A connector strip of the type used in the modules with shrouded male sections is used at the end of the board assembly as shown in Fig. 7. Refinement of the design resulted in the connector strip being moved from the top board surface to a recessed position under the board

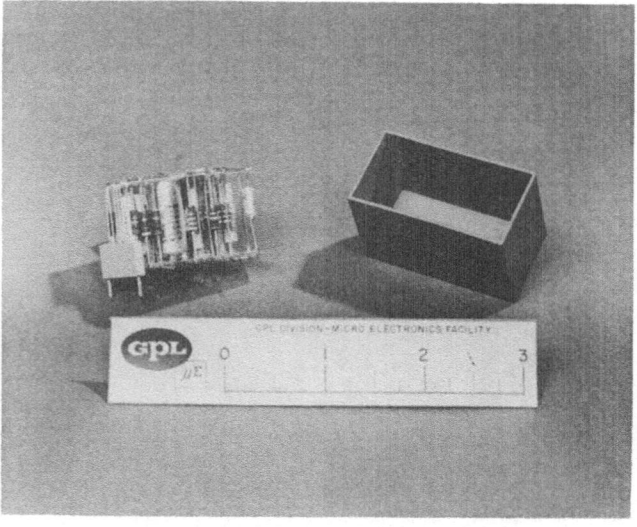

Fig. 5. Typical module internal assembly.

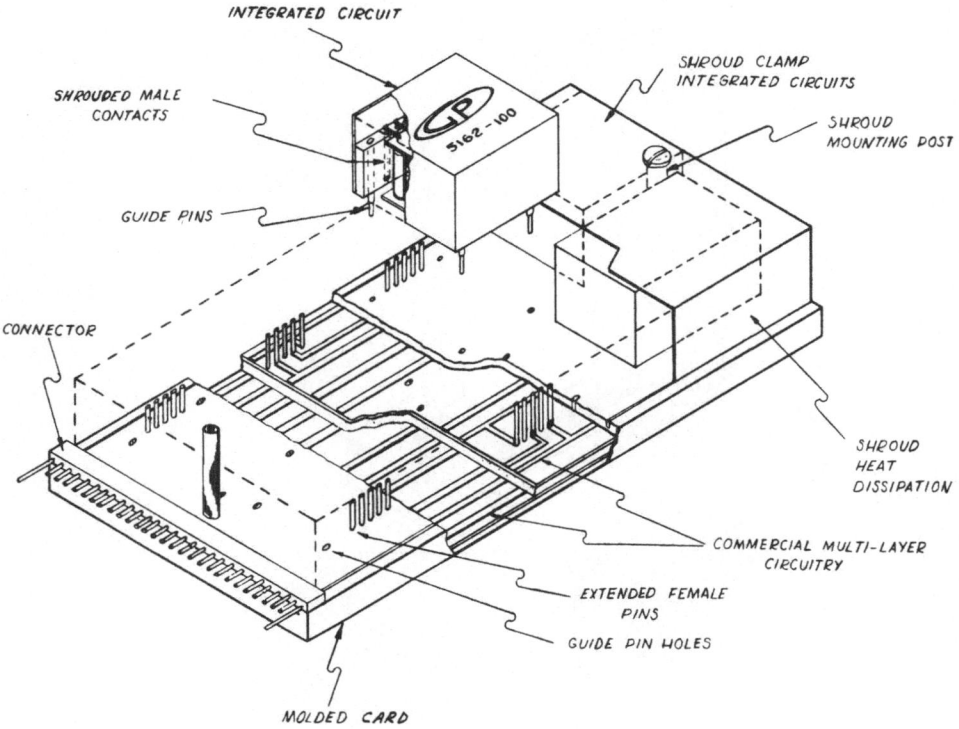

Fig. 6. Card construction.

to conserve volume and physically protect the strip. Note that this design permits the modules to extend completely to the board edge, avoiding the loss of board surface area normally found with connector assemblies.

TEST FEATURES

The design provides for tests to be conducted with minimum inconvenience at the board or module functional level. A complete set of test jacks is molded as an integral part of the end of the circuit board. These are to be color coded for rapid identification and are illustrated in

Fig. 7. Electronic board—modules—shroud.

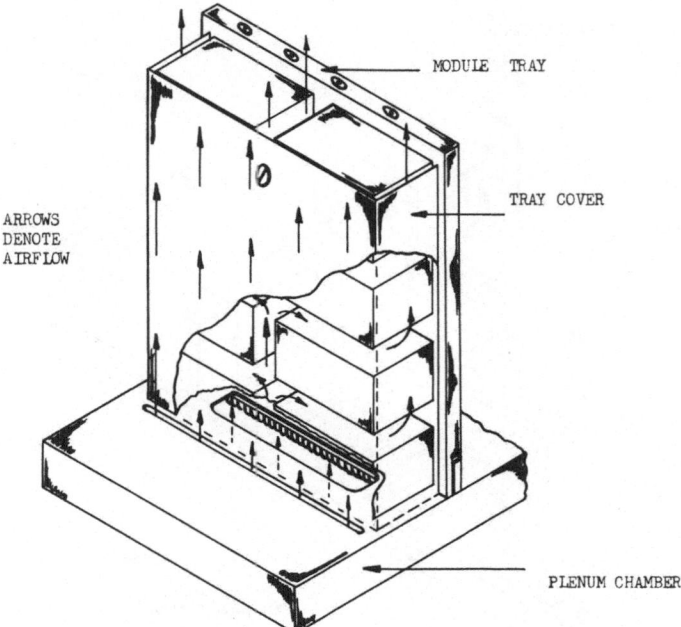

Fig. 8. Cooling airflow pattern.

Fig. 9. Board mounting bracket.

Fig. 2. The jacks selected should be of the fully enclosed type to avoid resin penetration and assure optimum performance.

An additional row of jacks extends above the surface of the boards, between the row of modules as shown in Fig. 2. These are placed to permit direct readout of the functional outputs of modules and may be obtained directly from the card without a separate test fixture or removal of the module from the card. The convenience afforded permits simplified and rapid routine maintenance of the electronics box.

HEAT TRANSFER

The transfer of heat from module surfaces is accomplished through a channel-shaped aluminum shroud slipped over the mounted units as is shown in Fig. 7. The shroud is fastened to tall, threaded posts molded into the molded board. This cover performs a secondary function of securely holding the modules to the board, eliminating the need for individual module clamps. The design of modules with uniform height permits positive clamping action and the intimate contact necessary for effective heat transfer. The internal design of modules orients the position of elements with high heat dissipation adjacent to the top or inside panel of the module. The resin thickness between hot elements and cooling surfaces has been kept to a minimum to reduce the length of the thermal path. Studies are in progress to determine allowable magnitudes of power dissipation per module volume.

The proposed arrangement of an air plenum chamber and an appropriately placed opening permits the flow of cooling air to pass through a channel between the row of modules and over the outer surface of the shroud. This is illustrated in Fig. 8, which also shows the module connector cabling channeled through the plenum chamber.

BOARD MOUNTING RACK

The board is mounted in the electronic box, using a rack of the spring-loaded clip type shown in Fig. 9. Clearance has been left on each edge of the board for the clips to hold the unit securely in position. Extensive tests on radar systems in the field have proven the effectiveness of this mounting technique. A group of ten boards compose the electronics portion of the equipment as shown in Fig. 10.

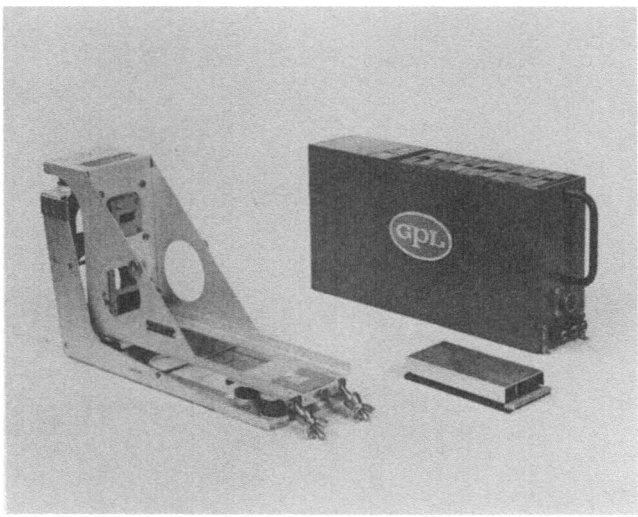

Fig. 10. Electronic box.

SUMMARY

The need for development of a method to assemble compact module functions into the electronics box has resulted in the board structure presented in this paper. Advantages are listed which can be attributed to the packaging technique described.

1. Rapid maintenance feature—plug-in modules.
2. Test jacks provide board and module functional data.
3. Heat shroud and channeled air flow for cooling.
4. Economical standard connector sections used in modules and board.
5. Epoxy molded board structure provides unit rigidity.
6. Circuit interconnections may be welded standard nickel patterns or multilayer printed circuitry in the board.
7. Total surface of board used for module mounting due to recessed connector.
8. Spring-loaded clip holds tray securely.
9. Modules firmly clamped to board by heat shroud.
10. Keying pins prevent errors when inserting modules.

The molded board or tray offers opportunity to vary the form of testing mode. Test jacks may be replaced with bulb-type indicators providing a built-in failure feature. The number of boards required may be reduced as increased use is made of thin-film and integrated circuit devices. The size of modules and boards may remain unchanged, however element densities in the module will of necessity increase sharply.

The packaging method does provide flexibility, opportunity for reliability, and excellent features for maintainability. Weight and volume are factors which are considered in the utilization of total board surface and the thermal analysis of module designs.

The board or tray technique offers many advantages as presented, and can readily be adapted to the electronics box design requirements as applied to the development of a new avionic system.

A High-Frequency Multiple-Signal-Conductor Transmission Line

R. C. Paulsen* and W. K. Springfield†

IBM Components Division
Endicott, New York

The application of a multiple-signal-conductor transmission line for use in modern high-speed miniaturized computers is discussed. Electrical and mechanical requirements are outlined. Design aspects showing analytical and experimental results with analog techniques on round and flat wire configurations are reviewed. Noise coupling, characteristic impedance, loss and propagation delays versus frequency, and termination aspects are considered.

GENERAL

MODERN COMPUTERS make stringent demands on the wiring used for the transmission of signals. As miniaturization reduces the size of data-processing apparatus, conventional single wires, twisted pairs of wires, and coaxial cables become impractical for interconnecting the various elements in the apparatus. Logic elements have been reduced to a fraction of their former size. Logic decisions are made in such short times that the velocity of propagation of the signal through the circuit wiring is significant. The interconnecting media have not kept pace with this technology.

Miniaturization has resulted in less space for wires; high-speed transistor circuits have increased sensitivity to noise and transmission losses. High-speed computers require signal transmission lines having a uniform impedance, low crosstalk, and low line losses; maximum flexibility, reliability, minimum size, and cost are important. Because these are not mutually compatible attributes, efforts to reduce the bulk of transmission lines are complicated. A thorough search of existing transmission media failed to yield a practical solution.

Although coaxial and twisted-pair cables meet the electrical requirements, they are not acceptable for reasons such as size, cost, flexibility, terminations, and routing requirements. Many multiple-conductor ribbon cables that are commercially available satisfy most of the mechanical requirements, but are not acceptable because they fail to meet the electrical needs such as impedance levels, crosstalk, and attenuation.

ELECTRICAL REQUIREMENTS

An analysis of IBM's high-speed system yields the following electrical requirements:

1. Nominal characteristic impedance, 95 ohms $\pm$ 10% up to 180 megacycles.
2. Crosstalk, down 20 decibels between adjacent signal wires.
3. Attenuation, less than 15 decibels per 100 feet, measured at 60 megacycles.
4. Conductor capacity, 9- and 20-signal conductors with the width of the 9-conductor cable one-half of the 20-conductor cable width.
5. Resistance, less than 0.25 ohm per foot at 25°C.

* Present address: IBM Systems Development Division, Poughkeepsie, New York.
† Present address: IBM Systems Development Division, Endicott, New York.

6. Current-carrying capacity, 400 milliamperes in each signal conductor simultaneously, for a 15°C temperature rise.
7. Dielectric strength, 90 volts dc, which requires a hipot test in excess of 900 volts ac.
8. Insulation resistance, must exceed 100 megohms between adjacent conductors or to an outside metal surface at 40°C, 90% relative humidity.

MECHANICAL REQUIREMENTS

The overall package sizes dictated the cable dimensions, flexibility, and foldability. A ribbon-like cable structure, termed a flat cable, was to be manufactured in a continuous length for use in automated termination and cable assembly equipment.

The following specific mechanical properties were imposed:

1. Dimensions, 0.050 × 1.150 inches maximum for 20-conductor cable.
2. Flammability, cable must be flame retardent according to UL standards.
3. Flexibility, cable must withstand 6000 flexures around a two-inch-diameter mandrel.
4. Foldability, cable shall withstand a sharp, creased fold which is used to make 90° changes of direction in a cable channel.
5. Terminations, must be automatable and of low cost.

ELECTRICAL DESIGN

Several configurations of noncoaxial-wire ground structures were evaluated. These included wire over a ground plane, parallel signal wires, twisted pair, alternate ground and signal wires in a single plane, and the flat-cable structure. The structures are compared in Fig. 1.

Consideration was given to round and flat (etched) conductors over a ground plane. Grounding systems tried were solid copper, etched copper, copper screening, ground wires, conductive cloths, and points. All were either too costly, cracked under flexing, or exceeded any reasonable stiffness limits. No satisfactory means of obtaining a continuous, highly conductive, flexible ground plane were found. A single-layer cable was evaluated. It was found to be marginally acceptable electrically, when physically isolated. Variations of coupling and Z_0 were found to be too severe when the cable was brought into proximity with other cables and metallic surfaces.

To facilitate design as well as to provide engineering data, an analog technique was developed. This technique employs a conductive paper* having a high, uniform surface resistivity. A scaled-up cross section of the desired cable design is drawn using a low resistance silver paint.† This provides a good approximation of the inductance, capacitance, characteristic impedance, and a plot of the fields around the conductors. Measurements are made using a dc voltage source and a high impedance voltmeter as a detector. A description of this technique is given in Appendix I.

The field plots for the Fig. 1 cables are shown in Fig. 2. For comparison, each design is adjusted for approximately the same characteristic impedance. In each case, 2-volt dc is impressed across a signal conductor and the common ground. The ground wires at each end of the line are commoned. The 0.5-volt equipotential field line is traced with a high impedance voltmeter and is shown by the dotted lines. A comparison of the field plots for conductors in the cables of Figs. 2a through 2c indicates the relative electromagnetic radiation and line coupling of the signal conductors.

In Fig. 2d the 0.5-volt equipotential line encloses less area. Accordingly, the crosstalk is less for the flat cable than for the other cables. Figures 2a and 2b have crosstalk ratios of about 1/4 (0.5/2.0) while the Fig. 2d cable has a ratio of about 1/12.5 (0.16/2.0). Thus the flat cable has approximately $\frac{1}{3}$ as much crosstalk as the other designs.

Figure 2c shows good isolation between adjacent signal conductors within the same cable. However, coupling is introduced by the common ground shared by two adjacent signal

---

* A facsimile recording paper produced by the Western Union Telegraph Company, New York, New York.

† A conductive silver paint produced by Engelhard Industries, Inc., East Newark, New Jersey.

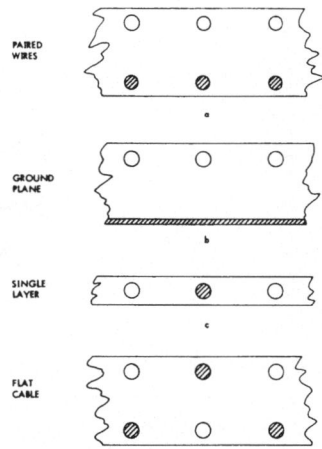

Fig. 1. Various cable designs. All wires   0.007   inch   diameter.
○ Signal wire; ⊘ ground wire.

conductors. There is greater coupling between signal lines in other cables stacked in the same channel in a machine. Greater cable width is required per signal conductor. If the horizontal dimensions are restricted, fewer signal conductors can be used at the higher impedance levels.

A detailed cross section of the flat cable is shown in Fig. 3. The empirical equation given relates the characteristic impedance to the mechanical dimensions and the dielectric constant, and is accurate within ±5%. Derivation of this formula is given in Appendix II. The round-wire equivalent may be converted into a flat-wire version and is shown in Fig. 4.

FABRICATION AND ASSEMBLY

The electrical requirements dictated a low-dielectric-constant, low-loss material. A low-dielectric-constant material results in a thin mechanical structure which decreases the stress in

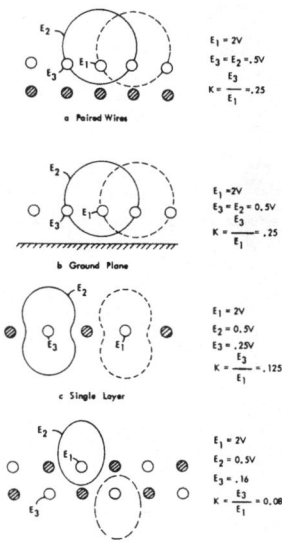

Fig. 2. Cable field plots. ○ Signal wire; ◉ ground wire.

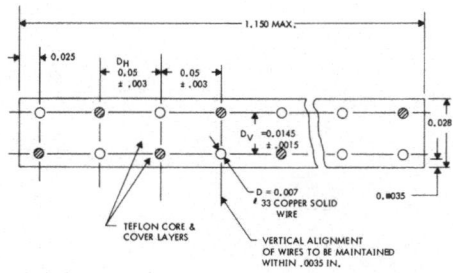

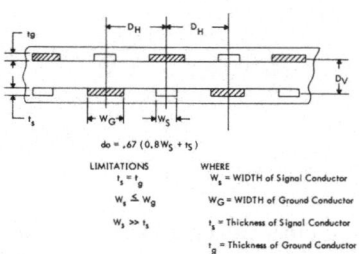

Fig. 3. Final design of flat cable. ◯ Signal wire;
◍ ground wires, commoned at terminated ends.

$$Z_0 = \frac{100}{\sqrt{\varepsilon_r}} \ln \frac{3D_i D_h}{(2D_r + D_h)d_0}$$

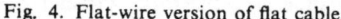

Fig. 4. Flat-wire version of flat cable.

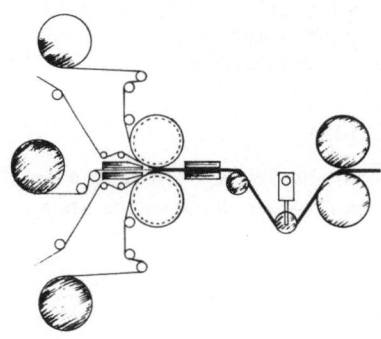

Fig. 5. Hot-roll laminating machine.

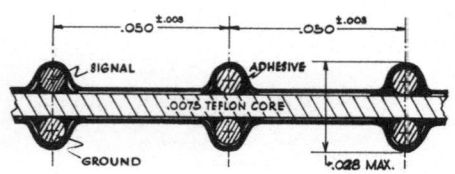

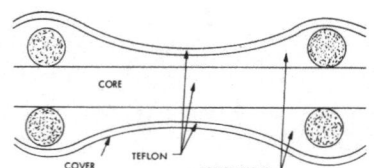

Fig. 6. Cable cross section.

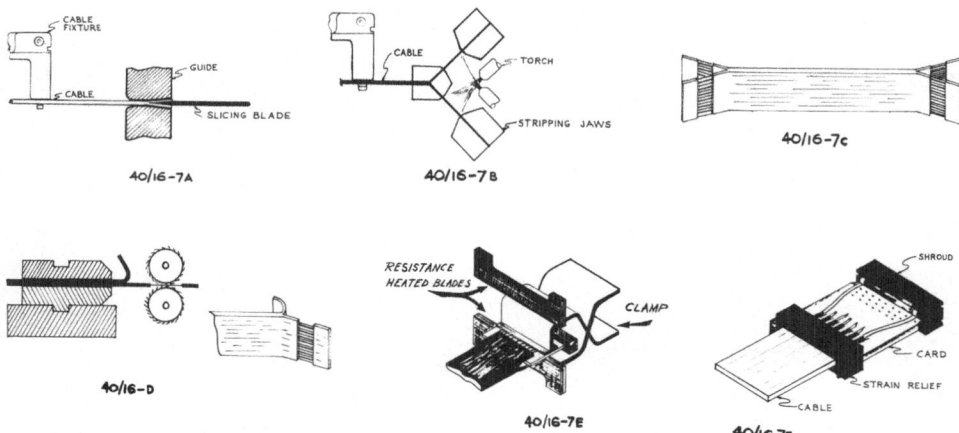

Fig. 7. *a*. Cable splicing. *b*. Cable stripping. *c,d*. Cleaning residue from wires. *e*. Application of strain relief. *f*. Completed cable assembly.

the wires during bending. Because of the flammability requirements, Teflon* is used rather than the lower-cost fluorocarbon compounds. Number 33 AWG silver-plated copper wires are used. Silver is required because of the heat used in the insulation-stripping process.

A thin layer of polyethylene is extruded on the two surfaces of the Teflon core material and one surface of the Teflon overcoat layer. Under elevated temperature and pressure in a hot-roll laminating machine, the polyethylene bonds the Teflon layers and fixes the physical position of the wires. The laminating machine is illustrated in Fig. 5. One advantage in using Teflon is that during the laminating processes temperatures are not high enough to distort the material, and excellent vertical spacing between wires is maintained. A typical cross section of the cable is shown in Fig. 6. This process produces continuous cables in excess of 100 feet in length.

Assembly posed the problems of dividing the cable into two layers, stripping the cable insulation, and simultaneously making 44 terminations. This was accomplished with the following sequence of operations:

1. Cut to length and mark fold locations.
2. Slice ends—this is accomplished by machining the center core layer of Teflon which is only 0.0075 inch thick.
3. Strip insulation—this is accomplished by flame stripping; the ends are held together with a strip of insulation until the wires are terminated.
4. Solder—a technique has been devised to solder all connections simultaneously with a heated blade.
5. Apply strain relief.
6. Apply protective coat to termination area.

These operations are illustrated in Fig. 7.

The cable is routed through the machine in 1.25-in.-wide raceways that form the grid, or framework, which holds large printed circuit panels. There are 12 possible cable entrance points on each panel, 36 possible cable positions on the panel, and over 100 panels in some typical installations. The cables change direction in the raceway by means of right-angle folds. A typical folded assembly is shown in Fig. 8.

The cables are stacked in the raceways as many as 40 deep. Consideration must be given to this factor—significant length can be added to the cable depending upon its relative level

* A product of the E. I. DuPont Company.

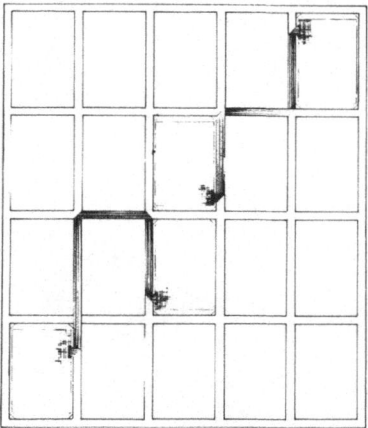

Fig. 8. Typical flat-cable assembly folded for installation.

in the stack. Because there is an almost infinite number of different cable length–folding combinations which can be specified, a computer-controlled method is used to handle the cable-assembly design, billing, and control. A punched-card program has been developed so that cable design is a step-by-step analytical process, and the engineering request to the manufacturing line for cable production is a permanent record in punched-card format.

MEASUREMENTS

Characteristic Impedance, Z_0

Z_0 is measured using a Rhode & Schwartz ZDU diagraph. Figure 9 is a plot of Z_0 versus frequency for a 5-foot length of flat cable. The nominal impedance from 30 to 120 megacycles is observed to be 91 ohms. This method is accurate within $\pm 5\%$. Values of short-circuit impedance, Z_{sc}, and open-circuit impedance, Z_{oc}, which are greater than 500 ohms and less than 5 ohms are eliminated because they are out of the range of accuracy on the Smith chart. Z_0 is determined by the equation $Z_0 = \sqrt{Z_{oc}Z_{sc}}$. Another method which can be used is the pulse-reflection technique. The results are comparable. Figure 10 illustrates this method.

Attenuation, α

Flat-cable attenuation versus frequency for a 5-foot sample is shown in Fig. 11. The short-circuit input impedance, Z_{sc}, is determined at half-wavelength multiples. The short-circuit

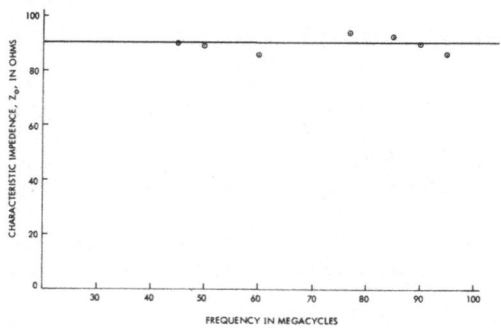

Fig. 9. Z_0 vs frequency. 5-foot sample, flat cable, diagraph measurement, all grounds commoned.

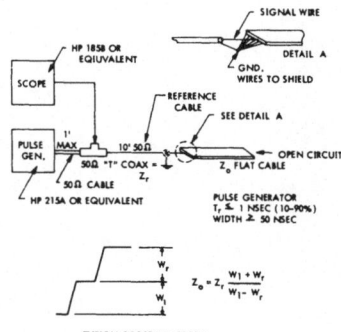

Fig. 10. Characteristic impedance by the pulse-reflection method.

admittance, Y_{sc}, is determined at the quarter-wavelength frequency and odd multiples thereof. The attenuation points are determined by the following relationships:

$$\alpha_Y = \frac{Y_{sc} Z_0 868.6}{\text{length in feet}} \text{ decibels/100 feet}$$

$$\alpha_Z = \frac{Z_{sc} 868.6}{Z_0 \times \text{length in feet}} \text{ decibels/100 feet}$$

$$\alpha = \sqrt{\alpha_Z \cdot \alpha_Y}$$

The value of α at 75 megacycles is used as a reference for determining overall cable quality. The sample length has been chosen so that the $\frac{1}{4}$ wavelength point falls within the range of interest.

Coupling

Open-circuit-line capacitive coupling measurements are made to determine coupling ratios between adjacent signal lines. The coupling ratio is determined by measuring the capacitance between a signal wire and ground, C_{wg}, and the direct capacitance between two adjacent signal lines, C_{ww}. The coupling coefficient, K, is determined by

$$K = \frac{1}{1 + (C_{wg}/C_{ww})}$$

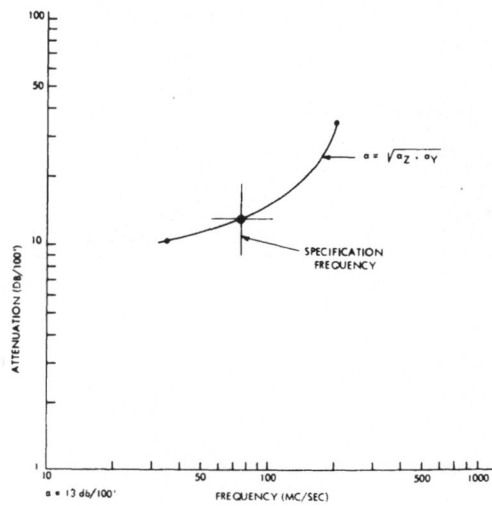

Fig. 11. Attenuation vs frequency. SLT flat cable.

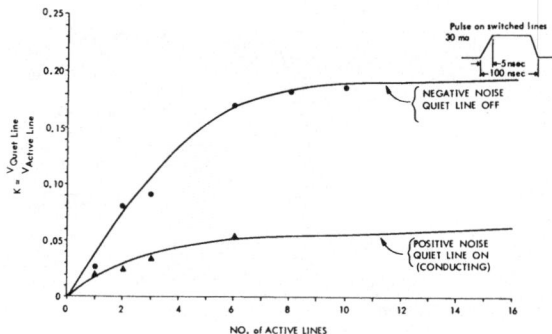

Fig. 12. Voltage coupled on quiet line *vs.* number of lines switching simultaneously.

The capacitance is measured on a 5-foot sample of cable using a Boonton Electronics Model 75B Bridge or equivalent at a test frequency of 1 megacycle.

APPLICATIONS

Coupling

Noise coupling occurs between signal conductors within the same flat cable as well as between signal conductors in stacked cables. Figure 12 shows the coupling versus the number of lines simultaneously switched in the same cable. Figure 13 is a plot of coupling between signal wires of adjacent cables versus cable spacing. Some critical applications may require a physical spacer about 20 mils thick as indicated in Fig. 13.

Figures 12 and 13 are for worst-case conditions, i.e., lines are switched simultaneously, and the spacing between cables maintains the wires in alignment for the entire cable length. In practice, signal switching and cable alignments are of a random nature and considerably less coupling is experienced.

Phase Velocity

The propagation delay is proportional to the square root of the relative dielectric constant of the material. Figure 14 is a field plot of the flat cable. Energy sent down the cable propagates

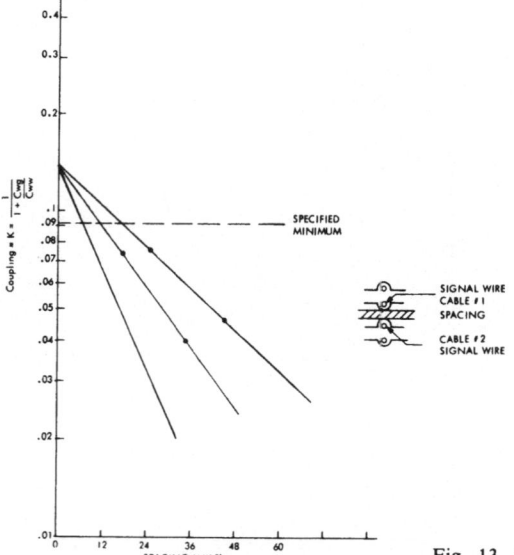

Fig. 13. Capacitance coupling between cables *vs* spacing.

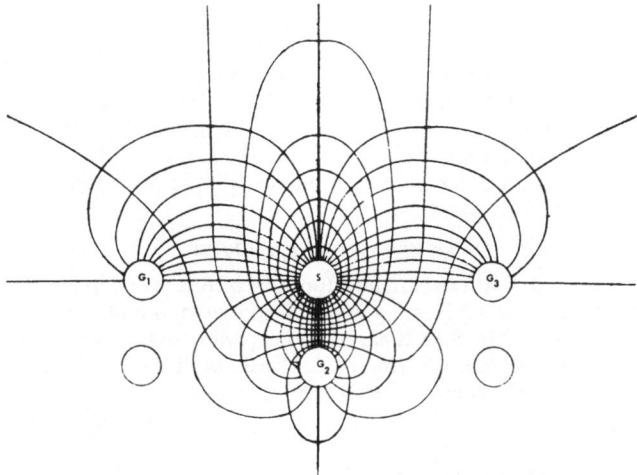

Fig. 14. Analog field plot for flat cable.

with two wave velocities. One, phase velocity, is governed by the dielectric constant between a signal wire and its vertically-adjacent ground wire. The greatest portion of the electric field is contained between these wires; the dielectric constant approaches 2.0. The remaining ground wires and the signal wire form another propagating system, and the dielectric constant approaches 1.5. The delay is 1.39 nanoseconds per foot with an apparent dielectric constant of 1.86 for the cable.

Cable Termination

Careful attention must be given to the grounding system. The connection between the cable ground wires and the external ground must have low inductance. This is of particular importance where simultaneous switching of many lines occurs. Multiple ground paths reduce this exposure.

Open grounds within a cable assembly are very difficult to detect. All grounds are commoned at each end and simple resistance checks will not detect a single open ground. To detect such opens, a dc current of about 0.5 ampere is passed through the common ground system. A Hall probe is used to detect the flux about the individual ground wires. Chart-recorded measurements for a perfect cable, that is, no open grounds, and for a cable having one open ground, are shown in Fig. 15.

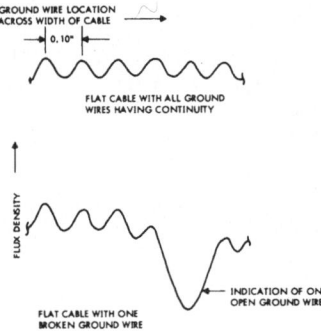

Fig. 15. Test method for detecting open grounds.

SUMMARY

The flat-cable design meets the electrical and mechanical requirements of the IBM System/360. The multiple-signal conductor transmission line having a uniform characteristic impedance with low crosstalk and attenuation was designed for high-speed miniaturized circuits. An automated manufacturing process has been developed which places the wires into a plastic medium to close mechanical tolerances. Wire terminations are made employing automated manufacturing techniques.

ACKNOWLEDGMENTS

The problems associated with the manufacture of the IBM production cable, cable terminations, stripping, flexing, folding, and testing were solved by the Manufacturing Research Department under the direction of Mr. Ken Knapp. Others who contributed to the work reported herein are Messrs. B. Hairabedian, J. S. Harris, K. Hinck, M. R. Marshall, A. Shah, W. J. Walsh, and V. D. Winkler. Their efforts are sincerely appreciated.

APPENDIX I

Resistance-Paper Analog for Determining L, C, and Z_0 of Transmission Lines

A picture of the cross section of the transmission system to be analyzed is painted on paper having a uniform resistance per square (1200 ohms) with a conducting paint which has a very low resistance per square (10 ohms). The conducting paint corresponds to the conductors and the resistance paper corresponds to the material surrounding the conductors of the transmission system. If the same set of voltages is applied to the electrodes on the resistance paper as exists on the conductors of the transmission system (in the TEM mode), then the field pattern on the paper is the same as the field pattern in the cross section of the transmission system.

Using Maxwell's equations it can be shown that the current field on the paper will have the same form as the H field in the transmission system, and that the E field on the paper has the same form as the E field in the transmission system. It can also be shown that $Z_0 = V_{line}/I_{line}$, and $R_{measured} = V_{paper}/I_{paper}$, for corresponding conductors and voltages are related by a simply-calculated constant. That is:

$$Z_0 = \frac{V_{line}}{I_{line}} = \frac{\sqrt{\mu/\varepsilon}}{R_{square}} \frac{V_{paper}}{I_{paper}}$$

where μ is the permeability of transmission line media; ε is the permittivity of transmission line media; and R_{square} is the resistance per square of the resistance paper.

In the same manner it can be shown that

$$L = \mu \frac{V_{paper}/I_{paper}}{R_{square}}$$

$$C = \varepsilon \frac{R_{square}}{V_{paper}/I_{paper}}$$

The advantages of the resistance-analog method are that it gives answers which are accurate to $\pm 10\%$, uses a modest amount of inexpensive equipment, requires only a fraction of the time necessary for an analytic solution of the problem, gives far greater insight into the problem than the analytic method does, and also serves as its own permanent record.

EXAMPLE

Consider a parallel wire line as shown below:

$$Z_0 = \frac{120}{\sqrt{\varepsilon_r}} \ln \frac{2D}{d}$$

$$\varepsilon_r = 2.1$$

For the above case

$$Z_0 = \frac{120}{\sqrt{2.1}} \ln \frac{2.50}{10} = 190 \text{ ohms}$$

The dimensions of the line are scaled up to facilitate drawing. The size of the sheet used for the drawing should be about ten times the size of the drawing so that the field patterns will not be disturbed.

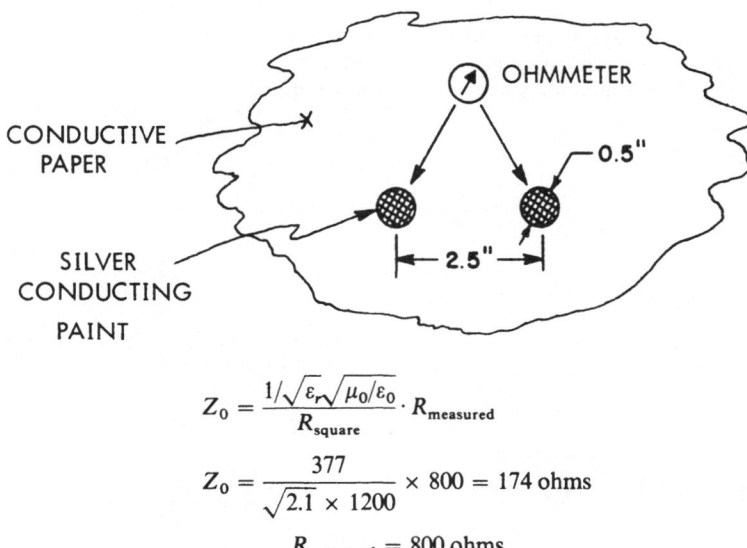

$$Z_0 = \frac{1/\sqrt{\varepsilon_r}\sqrt{\mu_0/\varepsilon_0}}{R_{square}} \cdot R_{measured}$$

$$Z_0 = \frac{377}{\sqrt{2.1} \times 1200} \times 800 = 174 \text{ ohms}$$

$$R_{measured} = 800 \text{ ohms}$$

This is within 10% of the calculated 190 ohms.

APPENDIX II

Flat-Cable Formula for Characteristic Impedance, Z_0

The sketch shows a cross section illustrating the general form of the flat cable.

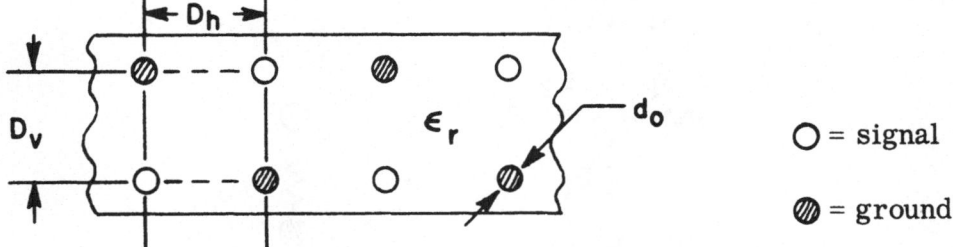

O = signal

⊘ = ground

An empirical formula which has less than 5% error over a wide range is

$$Z_0 = \frac{100}{\sqrt{\varepsilon_r}} \ln \frac{3D_v D_h}{(2D_v + D_h)d_0} \qquad (1)$$

where ε_r is the relative dielectric constant; D_v is the vertical spacing of conductors, center-to-center; D_h is the horizontal spacing of conductors, center-to-center; and d_0 is the diameter of the conductors. If slightly different size conductors are used for ground and signal lines, d_0 is obtained from

$$d_0 = \sqrt{d_s d_g}$$

where d_s is the diameter of the signal conductor and d_g is the diameter of the ground conductor. If D_v or $D_h \leq 2d_0$, the formula

$$Z_0 = \frac{100}{\sqrt{\varepsilon_r}} \cosh^{-1} \frac{3 D_v D_h}{(2D_v + D_h) 2 d_0} \tag{2}$$

yields values which correspond closer to the measured value.

Formula (1) was derived using the following reasoning:

(a) Consider the formula for a five-wire transmission line:

$$Z_0 = 75 \ln \frac{kD}{d_0} \tag{3}$$

$$(k = 1)$$

The flat cable can be considered a special case of the five-wire line. If D_v and D_h are made equal, it should have the same general formula with different constants.

$$Z_0 = k' \ln \frac{k'' D'}{d_0} \tag{4}$$

The constant k' was assumed to vary in proportion to the number of grounds. Thus, k' is assumed to be 100 to compensate for the removal of one ground.

D_v and D_h can vary independently. A function of D_v and D_h is required which includes the combined effects of these spacings and satisfies the physical boundary conditions. That is, $D' = f(D_v, D_h)$ such that $D' = D_v$ when $D_h = \infty$, $D' = D_h/2$ when $D_v = \infty$, and $D' = 0$ when D_v or $D_h = 0$. The expressions for k'' and D' are developed below.

(b) Consider the following model:

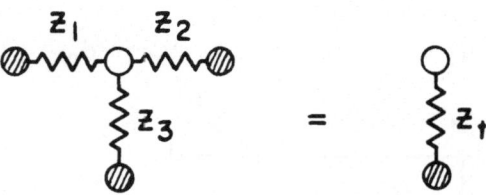

The total impedance Z_t is made up of the three parallel branches, Z_1, Z_2, and Z_3. Therefore

$$Z_t = \frac{Z_1 Z_2 Z_3}{Z_1 Z_2 + Z_1 Z_3 + Z_2 Z_3} \tag{5}$$

Z_1 always equals Z_2, and these values are determined by D_h. Z_3 is likewise determined by D_v. By substitution of D_h for Z_1 and Z_2, and D_v for Z_3 in (5),

$$Z_t = \frac{D_v D_h}{2D_v + D_h} \tag{6}$$

Take the special case where $D_h = D_v = D'$. Equation (6) becomes

$$3Z_t = D' \tag{7}$$

Substituting $3Z_t$ for D' in (4), we obtain

$$Z_0 = 100 \ln \frac{3Z_t}{d_0} \tag{8}$$

or

$$Z_0 = \frac{100}{\sqrt{\varepsilon_r}} \ln \frac{3D_v D_h}{(2D_v + D_h)d_0} \tag{9}$$

A rigorous approach to develop a formula for the flat-cable configuration employs basic field theory and yields the formula

$$Z_0 = \frac{60}{\sqrt{\varepsilon_r}} \left| \ln R - \frac{2 \ln (D_v/R) \ln D_h + \ln (D_h/R) \ln D_v}{\ln (D_h D_v^2/R^3)} \right| \tag{10}$$

where $R = \frac{1}{2}d_0$ and R, D_v, and D_h are in meters.

Formulas (1) and (2) are of simpler form and are within the same range of accuracy as noted when compared with experimental data.

High-Frequency Interconnections

WILLIAM L. THIBODEAU

Raytheon Corporation
Bedford, Massachusetts

A technique for interconnecting high-frequency circuits in which an interconnection path is cut into a grooved aluminum plate and a printed circuit run is centered in this groove is given. The remaining exposed portions of the groove are filled with an encapsulation compound to the height of the aluminum plate. The final step is to plate the encapsulation compound, and in this manner all interconnection runs are completely surrounded by metal, thereby effecting complete shielding of all runs.

HIGH-FREQUENCY INTERCONNECTIONS

THERE ARE MANY TECHNIQUES available to the design engineer to solve high-frequency interconnection problems. Such things as shielded cable, coaxial cable, coaxial connectors, grounding springs, and fancy boxes or bracketry to isolate one high-signal-level stage from a susceptible stage adjacent to it are all solutions to this problem which all of us have used and are familiar with.

The subject of this paper is high-frequency interconnections. It is my intention, however, to report on one technique as applied to an integrated receiver presently being designed for use in an advanced version of a weapons system by the Raytheon Company. The technique described in the following pages accomplishes the shielding requirements and may be applied through frequency ranges to 1000 Mc.

The design concept for the receiver modules is cordwood. This particular technique has been amply discussed by the industry for the past few years and further amplification of the approach is not a part of this presentation. It is the technique for interconnecting these modules and isolating the various stages from one another, that we hope to cover.

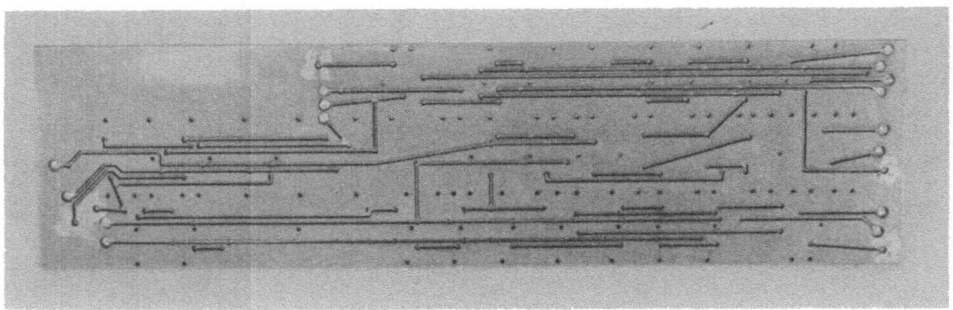

Fig. 1. Rigidax ground plane.

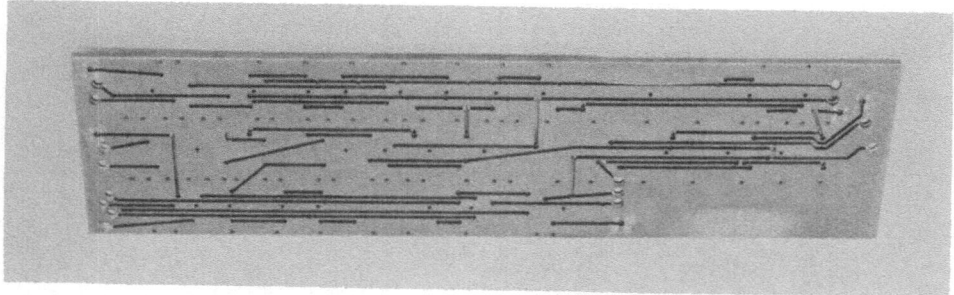

Fig. 2. Plane with signal-voltage runs installed.

The technique is called *Rigidax*. The system is, literally, rigid coaxial cable.

To interconnect any given series of units, we establish the pattern of runs, that is, voltage, signal-in, and signal-out, on paper. This layout is analogous to the layout of a single-sided printed circuit board, the criterion being that no two points cross one another although they may intersect or join.

Line widths for a printed circuit run are established at this time as well as interconnection land areas. The spacing desired between adjacent runs is also established, and module design is reoriented to comply with the interconnection pattern. Up to this point the system does not vary appreciably from the established criteria in the design of any interconnecting scheme using printed circuit patterns.

The variation in the approach now manifests itself. Any printed circuit pattern is made up of a series of individual conductors. If these interconnections were completely surrounded by metal, we would theoretically have perfect shielding, and in essence, this is what we attempt to accomplish.

The individual runs are die cut into separate pieces from the master pattern, after being suitably etched. An epoxy border is left which surrounds the interconnection pattern. These pieces are then placed in a grooved aluminum plate (see Figs. 1 and 2).

The land areas where intraconnections are to be made are then plugged. The remaining exposed portion of the cavity is filled with an encapsulation compound to the height of the aluminum plate (see Fig. 3). The plate then presents a flush surface, top and bottom. The plate surfaces, after all contact areas are masked, are then coated and made conductive so that it may be plated. Each conductor path, signal or voltage, is thereby surrounded by metal.

The initial, grooved metal pattern developed was a milled piece of hardware, suitable for R & D evaluation but too costly for production fabrication. Production design is now in

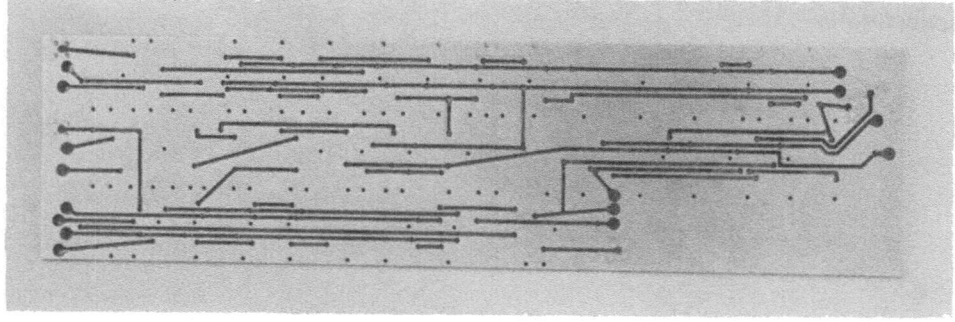

Fig. 3. Plane filled with encapsulation.

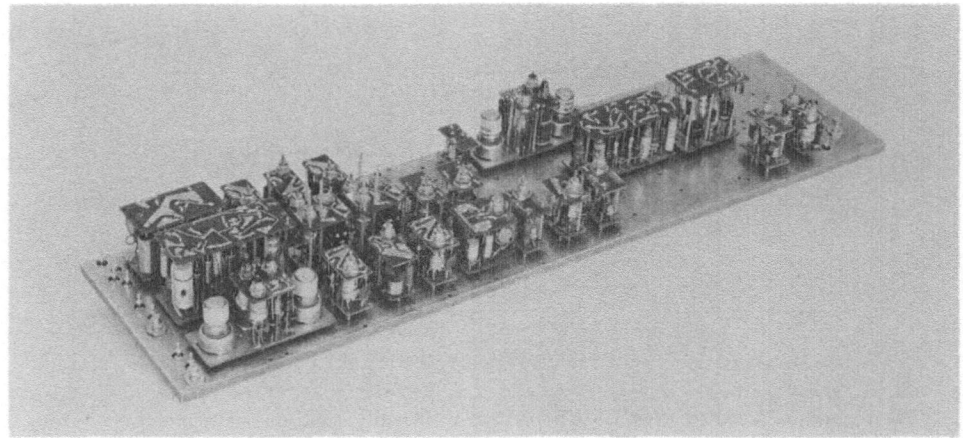

Fig. 4. Modules installed on ground plane.

process to fabricate this as a two-piece assembly; the top plate, which provides holes for module wires to be joined to the printed circuit runs, would be punched or stamped as a separate piece. The second part, which is the carrying agent for the printed circuit, would be cast or die cut. These two pieces would be joined by brazing or soldering to yield functional hardware, reproducible at minimum unit cost.

After the masking has been removed, the modules are added to the system through the pigtail leads affixed to the land areas on the printed circuit runs (see Fig. 4).

The modules are removable from the base plate through the use of a solder vacuum device which frees each lead individually, allowing the removal of the assembly.

Fig. 5. Modules installed with covers.

The masses of metal involved in both the module covers and the Rigidax plate provide a heat-sink capability as an inherent characteristic of the design. It is reasonable to assume that with careful attention to layout detail, fins could be provided as required, to yield any degree of heat transfer (see Fig. 5).

The assembled system is made impervious to moisture after all subassemblies are affixed to the plate (see Fig. 3). The land module interconnection areas up to this point are exposed. The simple expedient of filling these cavities with a silica gel seals the entire assembly. The balance of the conductive paths, due to their construction, are inherently moisture protected.

The spatial relationship of the dielectric medium and its depth and width placement in the base plate have the following determinable characteristics:

1. Capacitance—field between the ground plane and the signal carrier.
2. Inductance—width and thickness of the carrier.
3. Characteristic impedance $= \sqrt{L/C}$.
4. Resistance ac.

The dielectric material, independent of its function in the system, must be selected for the following characteristics (see design example):

1. Attenuation—measure of the losses in the material.
2. Power handling—carrier width and thickness.
    a. Breakdown—dielectric material and its separation from the ground plane.
3. Q—power into the carrier divided by the carrier losses.
4. Resistance—impedance to dc flow in the line.

If the wavelength of the carrier frequency is much greater than the length of line being used, then capacitance, inductance, characteristic impedance, and dc resistance may be ignored. Frequencies whose wavelength approaches the length of line being used should consider the reactive properties of the line as shown in the design example.

Power handling capability of the carrier is treated in the design example at the end of the paper. It is a major determining factor in selection of copper-clad thickness, width of the dielectric, and spacing (see Fig. 9).

The current-carrying capacity of an individual line is dictated by the I^2R losses. The selection of material, carrier thickness, and width may be varied to suit an individual requirement, and is only limited by the cross-sectional area available.

An application where high voltage or high current occurs is constrained by the dielectric material selected, and spacing. The selection of both of these items is a design parameter and is limited only by available space.

Radio-frequency energy fields, either radiated or conducted into a signal path, degrade system performance. The Rigidax technique offers a potential solution to this problem because it surrounds the conducting paths with adequate thickness of metal, assuring little or no penetration of this stray energy.

Attenuation in decibels (db) may be calculated as shown in the design example. This attenuation is the loss due to the conductor and the selected dielectric.

Shield effectiveness is defined as the ratio of energy transmitted into a shielded conductor to the energy leakage to the outer ground plane. The design example will show that the fields that exist outside of the signal-voltage carriers are negligible and may be ignored for a broad range of frequencies. The penetration of these fields is a function of plating thickness and selected base material.

The preferred ground plane for RFI suppression is generally defined as one where the length exceeds the width by approximately five to one. With Rigidax, careful placement of functional modules allows design within these constrictions.

The entire structural electrical member is the system ground plane. Ground loops are virtually eliminated due to the large cross-sectional area of the package and careful consideration of module placement. Individual circuits also obtain ground from this same broad plate therefore assuring the same ground for the entire system.

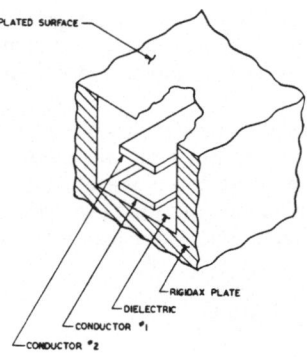

Fig. 6. Twin conductor coaxial line.

The attachment of shielded modules to the interconnection plate establishes a mechanical ground for each assembly, and assures the same ground potential for all system components.

A twin conductor coaxial line (see Fig. 6) can also be incorporated into a channel thereby allowing flexible grounding techniques. One signal carrier would therefore serve as a single point ground and its mate serve as the signal carrier. This technique has its maximum utilization in high-impedance low-level circuitry.

In summary, this paper presents a basic technique having many possible applications. A fixed delay line, for example, could be built into the cover of a piece of equipment, be tuned to the given equipment, and add no volume requirement to the overall package. A second application, one presently being integrated into the receiver presented here, is a local oscillator strip transmission line device at microwave frequencies.

Utilization of this technique through today's ranges of operational frequencies is theoretically possible, and allows the product designer greater flexibility in fabrication and layout. Smaller more reliable packages with maximum heat dissipation per unit volume are the desired objectives and may thus be realized.

DESIGN EXAMPLE

The following example has been selected to aid the design engineer in implementing this technique. A center frequency of 30 Mc has been used as a basis for all calculations, and the following characteristics have been determined: (1) capacity; (2) resistance dc; (3) resistance ac; (4) characteristic impedance; (5) inductance; (6) attenuation—conductor and dielectric; (7) shield attenuation; (8) power handling; (9) quality—unloaded.

w = width of conductor = 0.043 in.
t = thickness of conductor = 0.002 in.
b = spacing between ground planes = 0.122 in.
E_n = property of dielectric = 2.55

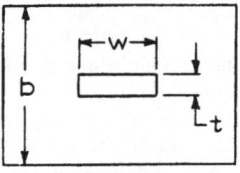

w/b=0.35
t/b=0.016

Summary of Calculations:
1. Capacitance   0.686 $\mu\mu$f/in.
2. Resistance dc 0.79×10^{-2} ohm/in.
3. Resistance ac 6.3×10^{-2} ohm/in.
4. Characteristic impedance   74 ohms
5. Inductance   0.0037 μh/in.
6. Attenuation, dielectric and conductor 13.3×10^{-3} db/in.
7. Shield attenuation (depth of plating required)   0.00236 in.
8. Power handling   520 W
9. Quality   20.8

Capacity

The capacitance of the line is equal to the plate-to-plate capacity (C_p) plus the fringe-field capacitance ($4C_f$). Capacitance for various width-to-spacing ratios (W to b) is given in Fig. 6. The total fringe and centerstrip capacitance is

$$C = C_p + 4C_f$$

or

$$C = \frac{(35.4E_r W/b)}{(1 - t/b)} + 4C_f \; \mu\mu f/m$$

From Fig. 6, the capacitance for a W/b ratio of 0.35 is found to be

$$C = 27 \; \mu\mu f/m \text{ or } 27/39.37$$

$$C = 0.686 \; \mu\mu f/in.$$

This is total line capacitance. To determine plate capacitance or fringe-field capacitance, substitution in the previous formulas is required, or it may be read directly from the graph.

Resistance dc

The dc resistance is determined by

$$R_{dc} = \frac{1}{\sigma^{tw}}$$

where σ is the conductivity of the conductor, 1.47×10^6 mhos/in.,

$$R_{dc} = \frac{1}{(1.47 \times 10^6)(0.002)(0.043)}$$

$$R_{dc} = 0.79 \times 10^{-2} \text{ ohm/in.}$$

Resistance ac

The high-frequency resistance takes into account the distance below the surface of a conductor where the current density is 68% of its value at the surface.

$$R_{ac} = \frac{2R_s}{W} = \frac{(2)(2.52 \times 10^{-7})\sqrt{f}}{W} \text{ ohms/in.}$$

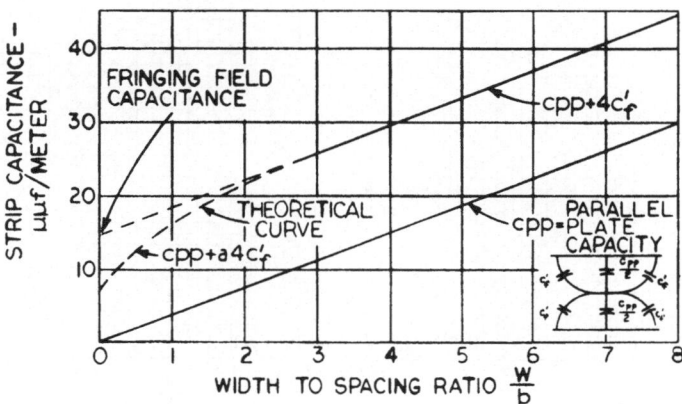

Fig. 7. Graph # 1: Strip capacitance *vs.* width-to-spacing ratio.

where R_s is the skin-effect resistivity of the conductor

$$R_{ac} = \frac{2(2.52 \times 10^{-7})(30 \times 10^6)^{\frac{1}{2}}}{0.043}$$

$$R_{ac} = 6.3 \times 10^{-2} \text{ ohm/in.}$$

Characteristic Impedance

The characteristic impedance of Rigidax is determined by the width and thickness of the signal-voltage carrier, the ground plane separation, and the dielectric constant of the material. Maximum power transfer results when the characteristic impedances of the line and terminations at both input and output are equal.

Characteristic impedance is obtained from Fig. 8.

$$Z_0\sqrt{\varepsilon_r} = 117$$

$$Z_0 = \frac{117}{\sqrt{2.55}}$$

$$Z_0 = 74 \text{ ohms}$$

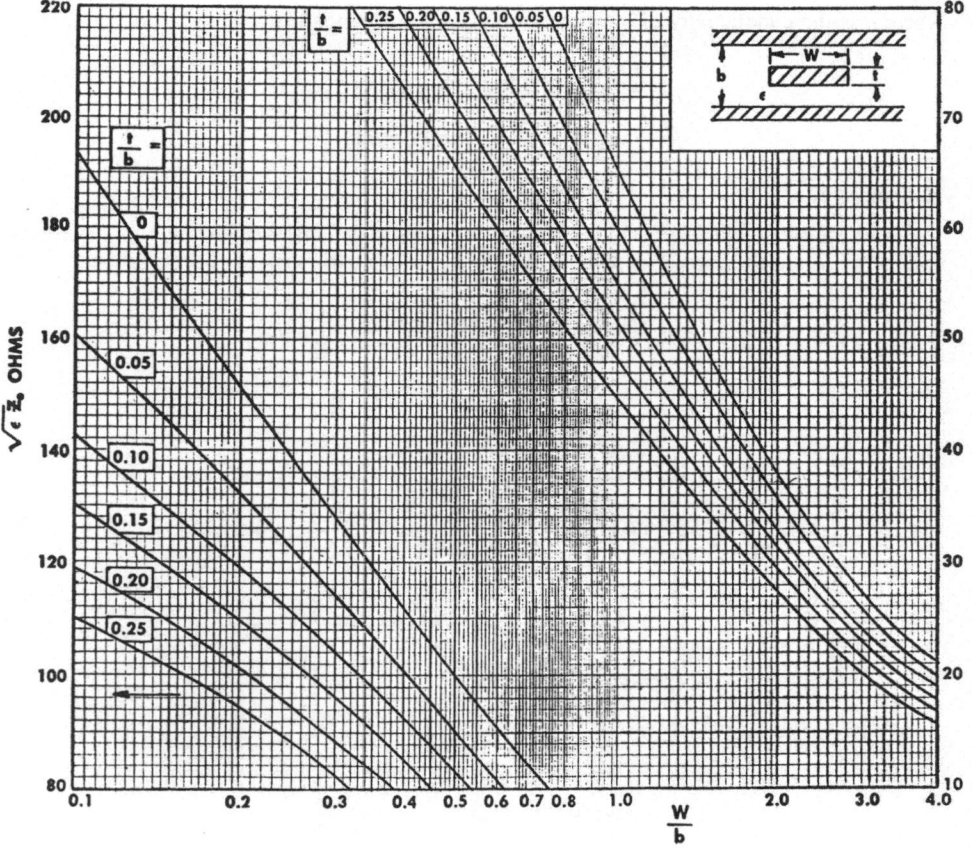

Fig. 8. Graph # 2: Characteristic impedance *vs.* width-to-spacing ratio.

Inductance

Inductance of the line per unit length is determined from the impedance expression:

$$Z_0 = \sqrt{\frac{R + j2\pi f L}{G + j2\pi f C}}$$

where R is the line resistance, G is the conductance loss of the dielectric, L is the inductance per unit length, and C is the capacitance per unit length. For this selected problem, a low-loss dielectric is used, where $G \approx 0$ (due to low-loss tangent of the dielectric) and $R \approx 0$ (as determined by calculations).

The characteristic impedance expression becomes

$$Z_0 \approx \sqrt{\frac{L}{C}}$$

The inductance per unit length is

$$L = C(Z_0)^2$$
$$L = (0.686 \times 10^{-12})(74)^2$$
$$L = 0.0037 \ \mu\text{h/in.}$$

The inductance can also be approximated by

$$L \approx 2\mu\left(\frac{t}{W}\right)$$

where μ is the permeability of copper, $(4\pi \times 10^{-7})/39.37$ h/in.

$$L = 2\frac{(4\pi \times 10^{-7})}{39.37} \times \frac{(0.002)}{(0.043)}$$
$$L = (0.636 \times 10^{-7})(0.465 \times 10^{-1})$$
$$L = 0.003 \ \mu\text{h/in.}$$

Attenuation

Attenuation of this line is the sum of the conductor losses and dielectric losses. Dielectric-loss attenuation at 30 Mc is

$$\alpha = \frac{\pi f}{3 \times 10^8}\sqrt{\varepsilon_r} \text{ (loss-tangent of dielectric) (8.686) db/in.}$$

$$\alpha = \frac{(3.14)(30 \times 10^6)}{30 \times 10^7}\sqrt{2.55}(5 \times 10^{-5}) = (8.686)$$

$$\alpha = 2.18 \times 10^{-3} \text{ db/in.}$$

The attenuation through the dielectric medium can be neglected for this design example because of the selected dielectric and frequency range.

Attenuation due to the conductor losses is given by

$$\alpha = \sqrt{\frac{\mu\pi f/\sigma}{tZ}} \ (8.68) \text{ db/m}$$

for copper, $\sqrt{\mu\pi f/\sigma} = 2.52 \times 10^{-7}\sqrt{f}$

$$\alpha = \frac{2.61 \times 10^{-7}(30 \times 10^6)^{\frac{1}{2}}(8.68)}{(0.002)(120)}$$

$$\alpha = 543 \text{ db/m}$$

$$\alpha = 13.3 \times 10^{-3} \text{ db/in.}$$

Attenuation Through the Plating

The magnitude of current radiated from the conductor into the surrounding media is attenuated. This radiated current density decreases exponentially with penetration into conductor. The path of least attenuation is through the plating material.

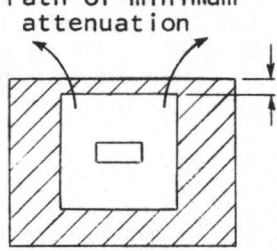

Path of minimum attenuation

The minimum plating depth for zero current density is determined by

$$d_{min} = \frac{5}{\sqrt{\pi f \mu \sigma}} \text{ (meters)}$$

where f is 30 Mc carrier frequency, μ is the permeability of the plating material (for copper it is $4\pi \times 10^{-7}$ h/m), σ is the conductivity of the plating material (for copper it is 5.80×10^7 mhos/m).

$$d_{min} = \frac{5}{\sqrt{(3.14)(30 \times 10^6)(4)(3.14 \times 10^{-7})(5.80 \times 10^7)}} \text{ m}$$

$$d_{min} = 6.0 \times 10^{-3} \text{ m or } 6.0 \times 10^{-5} \times 39.37 \text{ in.}$$

$$d_{min} = 0.00236 \text{ in.}$$

Power Handling Capabilities

Figure 9 gives empirical data for average power versus frequency in megacycles. It is obvious that the Rigidax plate has a power capacity far in excess of the power required for our design example, so that it need not be considered.

Radio-frequency power transmission, however, must consider the width of the conductor, conductor thickness, and spacing between ground planes. Power functions are also inversely

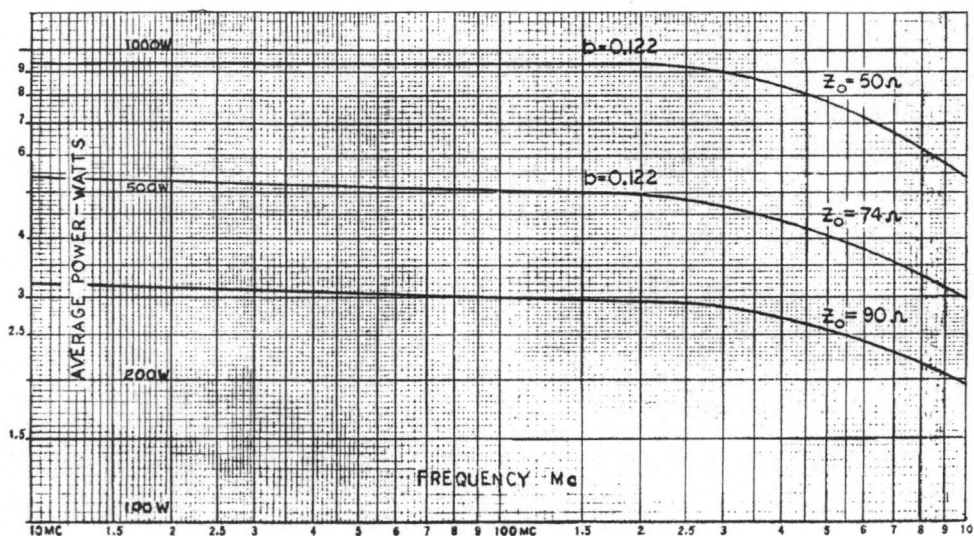

Fig. 9. Graph # 3: Average power *vs.* frequency.

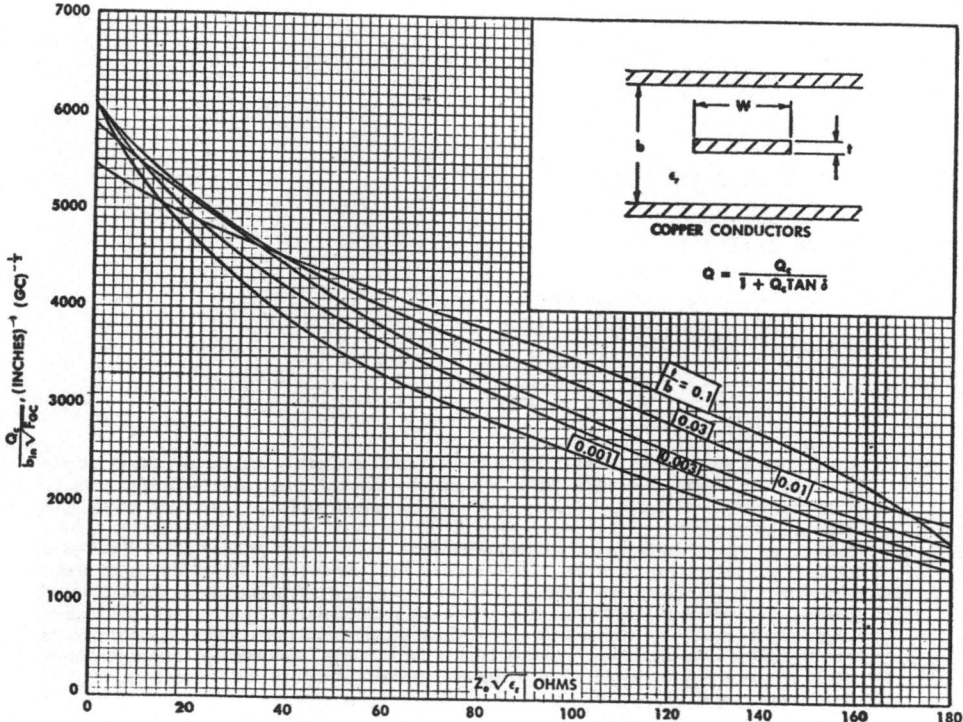

Fig. 10. Graph # 4: Theoretical Q in a copper-shielded strip line in a dielectric medium.

proportional to the line impedance and frequency. Design compromises between these two considerations are necessary as shown in Fig. 9, and may be varied for any given design problem.

Unloaded Q

Theoretical Q may also be calculated as follows: The abscissa of Graph #4 (Fig. 10) yields

$$Z_0\sqrt{\varepsilon_r} = 117$$

$$\frac{Q_c}{b_{in}}\sqrt{f_{gc}} = 3100$$

where Q_c is the unloaded quality factor of the line, f_{gc} is the frequency being sent down the line and is expressed in 1000 Mc units, and b_{in} is the separation of ground plane in inches.

Select a carrier frequency of 30 Mc, then

$$f_{gc} = \frac{30 \text{ Mc}}{1000} = 0.0030$$

If $b_{in} = 0.122$,

$$Q_c = 3100 b_{in}\sqrt{f_{gc}}$$

$$= (3100)(0.122)\sqrt{0.0030}$$

$$Q_c = 20.8$$

REFERENCES

1. S. B. Cohn, *Characteristic Impedance of the Shielded Strip Transmission Line*, Stanford Research Institute, Stanford, California.
2. *Fields and Waves in Modern Radio*, John Wiley & Sons, Inc., New York.
3. *Reference Data for Radio Engineers*, Fourth Edition, ITT Corp.
4. *The Microwave Engineers Handbook and Buyers Guide 1964*, Horizon House, Inc.
5. *Handbook of Tri Plate*, Sanders Associates, Inc.

Interconnection Systems

STEVE ZELENCIK

Systems-Packaging, Amphenol Division
Amphenol-Borg Electronics Corporation
Broadview, Illinois

The continued advancement of microminiature electronic equipment and the significant increase in the use of integrated circuits and thin films have created interconnection nightmares. In many cases the development of these new devices has outstripped interconnection technologies. With proper recognition and consideration of the six basic levels of interconnection—intramodule, module-to-motherboard, intramotherboard, motherboard-to-back-panel, back-panel wiring, and input–output—inherent interconnection and system design difficulties can be greatly minimized. By discussing new approaches and new ideas, Amphenol–Systems Packaging will present fresh concepts for the interconnection of microminiature systems and components. Subjects to be reviewed are interconnection and packaging of thin films and integrated circuits, new motherboarding techniques which minimize back wiring problems yet retain flexibility, and a new concept of simplified prepositioned weldable circuitry for microminiature interconnection applications.

BASIC INTERCONNECTION PHILOSOPHY

ELECTRON EQUIPMENT, like the human body, depends upon a finely developed nerve system to perform its functions. Each subsystem relies on its own intraconnected nerve circuit which in turn connects to the master system.

Consider the nerve, or interconnection, system of a miniature computer: components, cables, connectors, circuit cards, terminations, and wire. In essence, the wire lead on a single resistor is an interconnection system.

Placing emphasis on specific portions of an interconnection system has produced specialists who can provide recommendations on such entities as wire, connectors, terminations, and circuit cards. Once the performance capabilities of a system are determined, lead time considerations usually become so critical that a hurried marriage of these separate interconnection entities must be made by a coordinator.

Viewing the interconnection requirements of electronic equipment in separate entities is no longer acceptable. Like the nerve system of the body, today's electronic equipment requires a balanced, wholly-integrated nerve system. We can no longer "piecemeal" together the interconnection system of our equipment after all other design parameters have been established. Interconnection systems must be thought of in parallel with other basic design requirements, lest we pay overwhelming consequences in size, reliability, and, particularly, economics.

There are six levels of interconnection to be considered in any electronic system or equipment (Fig. 1): (1) intramodule; (2) module-to-motherboard; (3) intramotherboard; (4) motherboard-to-back-panel; (5) back-panel wiring; (6) input–output. Each level requires specific interconnections, but the entire system must be mutually compatible when considering performance, reliability, and economics. This can be accomplished only by a total systems approach.

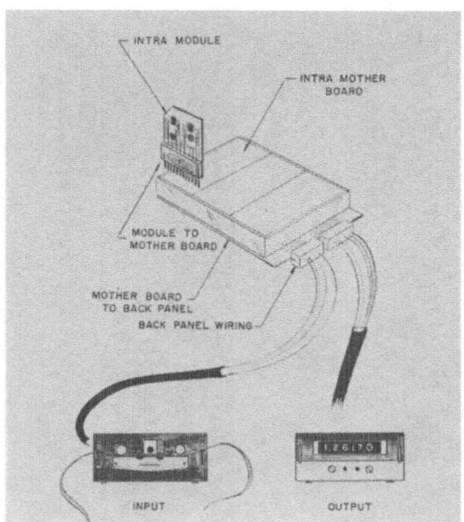

Fig. 1. Six levels of interconnection.

Techniques developed at one level of interconnection do not necessarily restrict themselves to that level, as was so often seen in the past. By being fully cognizant of the circuit functions and circuit requirements, the designer can employ those techniques which best fulfill the requirements of the level and the system without regard to obsolete conventions.

It is the purpose of this paper to discuss specific examples of the microminiature-interconnection-systems approach and show how insight into the total system permits marked product improvement from engineering, reliability, and economic aspects.

A NEW APPROACH TO STANDARDIZATION OF DESIGN AND LAYOUT OF WELDED INTERCONNECTING CIRCUITRY

Amphenol Intercon circuitry, a weldable prepositioned circuitry with raised tabs, was first introduced in 1960. At that time Amphenol had little interest in designing or fabricating anything beyond interconnecting circuitry.

When we expanded our capabilities to include the design and production of modules, motherboards, and other items related to interconnection systems, we recognized the need for standardization of techniques—particularly those which would permit the design of welded-circuit-board requirements more compatible with the time and economic demands of the system.

To satisfy our requirements in systems packaging, the following criteria were established: Any new circuit technique must (1) provide standardization; (2) package latest component configurations; (3) adapt to all levels of interconnection; (4) reduce cost and time.

In essence, then, our criteria simply reflected the total design consideration rather than the specifics of a particular module. We had the basic product in the form of Intercon; we needed a new technique of using it.

A thorough review of the past and present Intercon circuitry and packaging requirements indicated that 75 % of these applications could be satisfied by a grid system.

At this point, the course of action became well defined.

A unique grid pattern was developed placing interconnection tabs on both 0.100- and 0.050-in. centers (Fig. 2 and Table I).

The grid, called Intercon Production Grid, established positive parameters for conductor and component placement. By employing the grid with its associated termination techniques, layout time, reaction time, and costs can be reduced by a factor as great as 50 %.

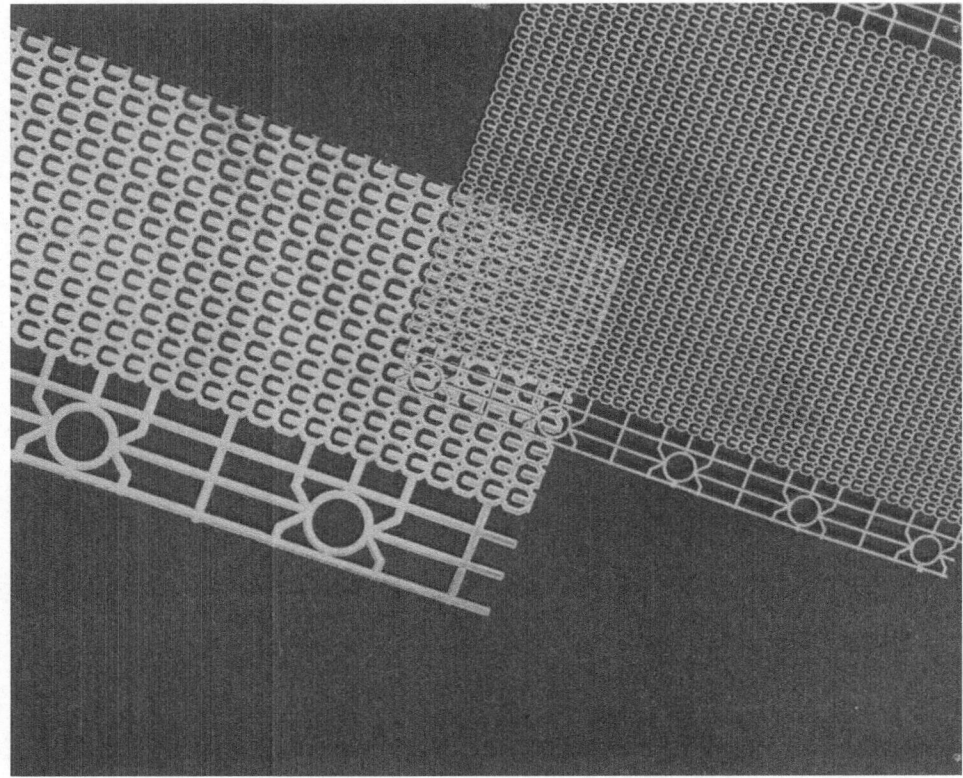

Fig. 2. Intercon grid, 0.100- and 0.050-in. tab centers.

The following is an outline for design and layout techniques when applying the grid principle (Fig. 3):

Ten-to-one reproductions of the grid are used for layout. Because the component and conductor placement is controlled by the grid parameters, direct transposition from the circuit schematic to the layout pattern is possible without fear of violating the mechanical

TABLE I
Intercon Production Grid Dimensions and Tolerances

Element	0.100-in. Grid	0.050-in. Grid
Conductor:		
width	0.020	0.010
thickness	$0.007 {+0.001 \atop -0.0005}$	0.0035 ± 0.0005
Tab:		
width	0.030	0.015
thickness	$0.007 {+0.001 \atop -0.0005}$	0.0035 ± 0.0005
length, straight or "T"	0.049 ± 0.005	0.027 ± 0.002

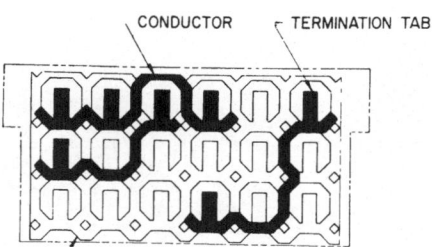

Fig. 3. Layout using grid principle.

or electrical limits of the circuitry. The designer can immediately begin to place components and begin circuit routing over the pattern to form discrete circuits.

A 10:1 photographic positive of the grid artwork is altered to represent the discrete circuitry pattern (Fig. 4). This is accomplished by removing or adding portions of the pattern required in the circuitry composite. (The artwork lines are narrower than the finished conductor because the circuitry is plated rather than etched.) From this point the discrete positive is photographically reduced to a 1:1 size, and it becomes circuit artwork.

The nickel is deposited, the circuitry laminated to a substrate, and the tabs either raised or left flat depending upon the types of component devices to be terminated (Fig. 5).

The grid approach met all the parameters set forth for it:

1. Standardization was achieved while maintaining flexibility.
2. The 0.050-in. configuration will accommodate the newest integrated-circuit flat-pack devices.
3. Interconnections can be achieved at all levels by employing various substrate types and multilayer circuitry constructions.
4. Circuitry designs that have taken up to six weeks in lead time are now being satisfied in three weeks or less with corresponding dollar savings in design and fabrication.

One added yet very significant feature of the grid technique is its adaptability for prototyping. Circuitry can be prototyped using a piece of sheet grid to form the exact circuitry configuration required for an operational module. We placed the grid circuitry on a prepunched substrate with the holes corresponding to the tab locations (Fig. 6). By physically interrupting the conductors, a pattern can be made which duplicates the circuit layout of a specific module. The tabs are easily raised and a discrete board is made. Thus, we have been able to make in prototype an exact representation of a production unit. The advantages of this technique require little discussion.

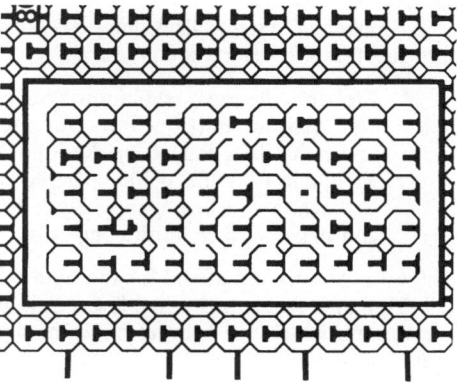

Fig. 4. Artwork for production grid.

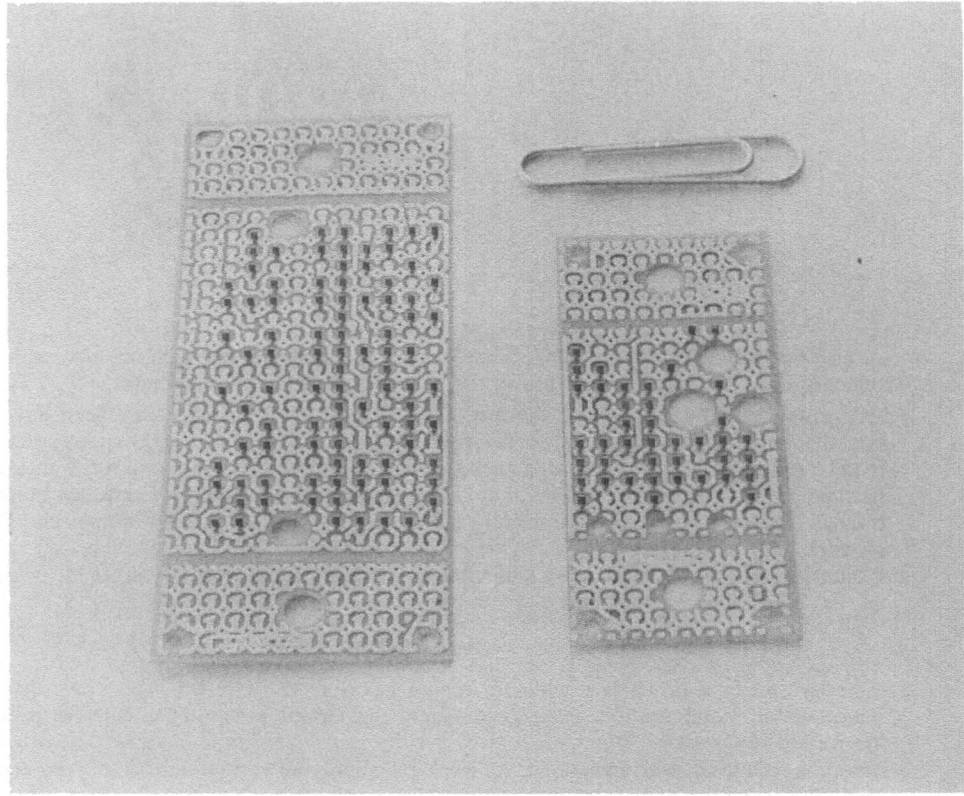

Fig. 5. Finished production grid circuit boards.

PACKAGING TECHNIQUES FOR HYBRID THIN-FILM CIRCUITS AND OTHER PLANAR-TYPE MODULES

The advent of thin-film circuitry and silicone integrated circuits has reestablished the need for an effective method of packaging planar-type modules and subsystems.

The problems of packaging are especially evident when considering thin-film networks because thin-film depositions are usually made on materials which do not lend themselves to direct plug-in designs.

Amphenol–Systems Packaging has solved the packaging problems concerned with thin-film substrates in several unique ways.

A. Probably the most revolutionary approach to thin-film packaging involves the use of a new product developed by Amphenol-Borg Corporate Research and Engineering under the direction of Dr. Rodolfo M. Soria. This product development is Precision Moldable Ceramic #100. The ceramic can be molded to a tolerance of ±0.001 in. per inch, is vitrified, completely gastight, withstands temperatures in excess of 2000°F, and has a surface finish compatible with thin-film deposition requirements (Table II).

The value of such a material for thin-film substrate applications is very evident.

Using the Precision Moldable Ceramic as a base, Amphenol–Systems Packaging developed two designs which, we believe, represent significant steps forward in the packaging of hybrid thin-film modules.

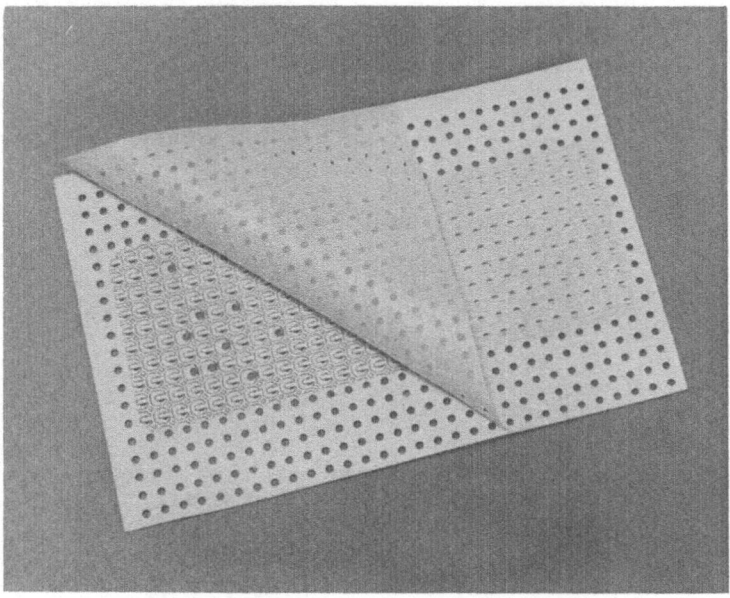

Fig. 6. Intercon 0.100-in. grid on prepunched substrate.

1. The first design is an integral substrate header which is positioned perpendicular to the motherboard and can be placed on 0.100-in. centers (Fig. 7). The substrate is molded with integral contact pockets. The thin film is made with conductors deposited directly into the contact pockets. The pockets accept Amphenol wire-form contacts which can be spaced as closely as 0.075 in. The contacts are soldered or welded directly to the film conductors. A band of epoxy is placed across the contacts for added mechanical strength (Fig. 8).

TABLE II
Property Chart
Precision Moldable Ceramic #100

Tensile strength (psi)	6000
Compressive strength (psi)	65,000
Flexural strength (psi)	15,000
Impact strength (izod, ft-lb/in.^{2})	1.0
Water absorption (%)	0.006
Gastightness	$<2 \times 10^{-10}$ helium/sec*
Specific gravity	2.72
Maximum temperature endurance (°F)	2000
Thermal expansion (10^{-6}/°C)	8.0
Thermal conductivity (cal/sec/0°C/cm/cm^2)	0.015
T_e Value (°C)	650
Volume resistivity (ohms/cm)	$>6 \times 10^{14}$
Dissipation factor (10^5 cycles/sec)	0.0007
Dielectric constant (10^5 cycles/sec)	5.5
Loss factor (10^5 cycles/sec)	0.0038
Dielectric strength ($\frac{1}{8}$-in.-thick v/mil)	300
Arc resistance (sec)	>420
Surface finish (μin. rms) (unglazed)	19 to 25
Surface finish (μin. rms) (glazed)	2 to 3
Tolerance (in./in.) (unglazed)	±0.001
Flatness (at 0.030 in. thick over 2 in.)	±0.004 to ±0.006

* Helium mass-spectrometer test on 0.010-in.-thick section.

Fig. 7. Ceramic # 100 molded integral header.

2. The second design is an integral substrate header positioned in the same plane as the motherboard, which permits high-density stacked constructions. The substrate is molded with deposition cavity and contact slots as integral parts (Figs. 9 and 10). The contact employed is a modified Amphenol tuning fork. The contact is retained in the slot by mechanical means. A special finger on the contact makes a pressure interconnection with the thin-film conductor. The contact is then welded or soldered to the thin-film conductor. The contact tails are designed for board plug-in, wire-wrapping, welding, soldering, or plugging into mating hermaphroditic contacts, to permit horizontal stacking of the modules (Fig. 10). The entire unit is functional to 400°F.

B. By employing relatively standard-line connectors, some very effective planar packages have been produced.

 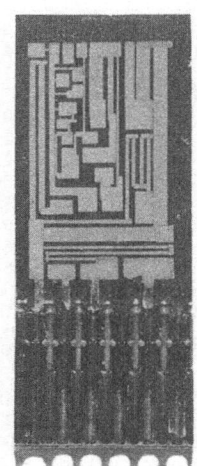 

Fig. 8. Header with hybrid thin-film circuit and wire-form contacts.

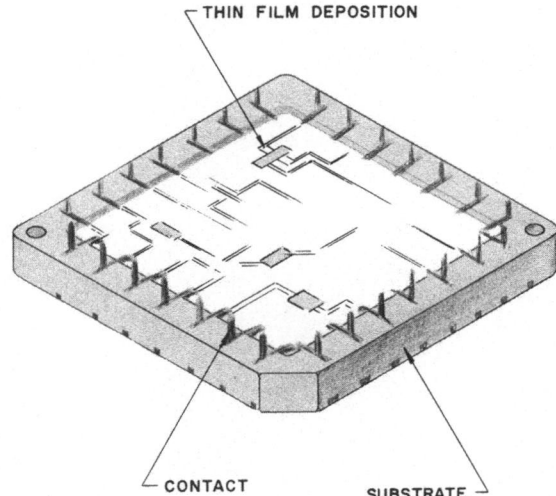

Fig. 9. Header with hybrid thin-film circuit and tuning fork contact.

The standard wire-form strip connector is a dielectric strip accommodating a single row of contacts on either 0.075- or 0.100-in. centers. The dielectric is cut to the size of the substrate. The contacts have leads crimped-in which reach the termination pads of the thin-film deposition. The substrate is attached to the header and the leads are soldered to the deposition (Fig. 11).

The variations of the above technique are unlimited. For one particular application, we made a molded unit that functions as a connector header and thin-film substrate frame (Fig. 12). Another requirement was satisfied by employing an existing Amphenol micro-min connector assembly, which is a double-sided 38-contact connector with 0.050-in. contact spacings (Fig. 13). The substrate was cradled between the leads, and attached to them at the conductor pads. This particular design provided a high number of contact leads for a relatively small package size.

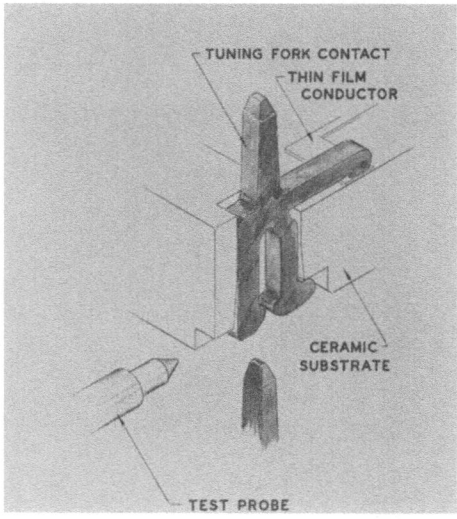

Fig. 10. Detailed view of contact in header.

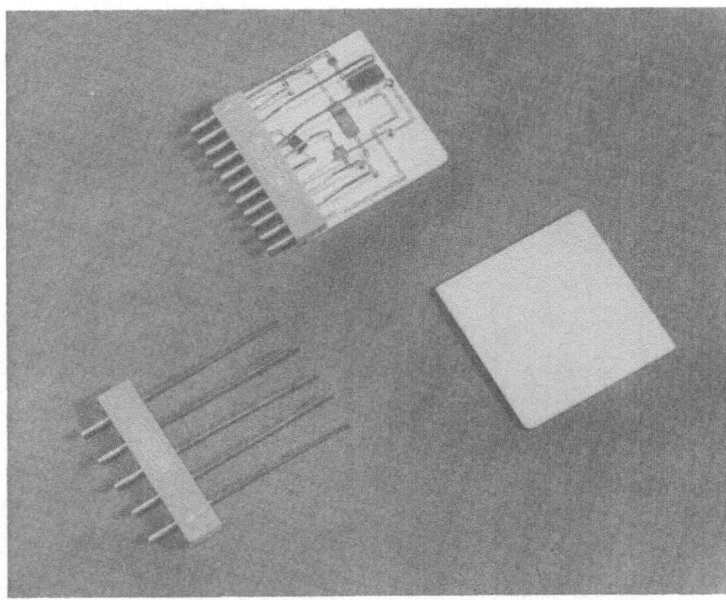

Fig. 11. Wire-form strip connector used as substrate header.

A HIGH-DENSITY MOTHERBOARD

Employment of the wire-form strip connector as a basic interconnection device for plug-in modules has enabled us to design and produce motherboards with extremely high-density capabilities (Fig. 14). The strip connectors or special 450 contact blocks are assembled to form the required motherboard (Fig. 15).

Bussing can be accomplished by using strip connectors machined out of copper or by plating the dielectric part. This technique reduces the backwiring necessary on the motherboard and enhances overall system reliability.

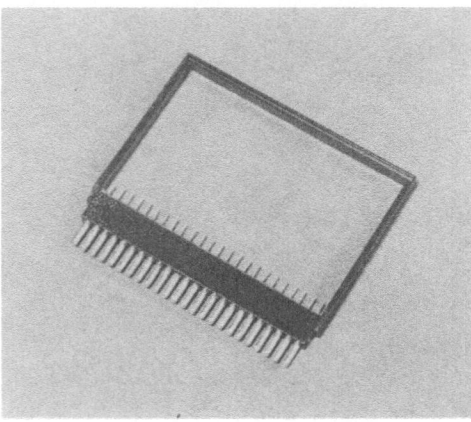

Fig. 12. Special molded thin-film header frame.

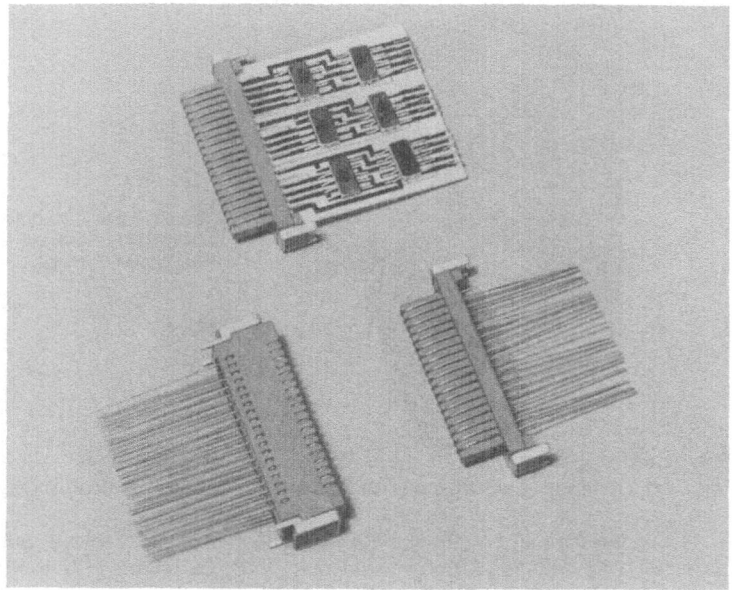

Fig. 13. Micro-min connector employed as a substrate header.

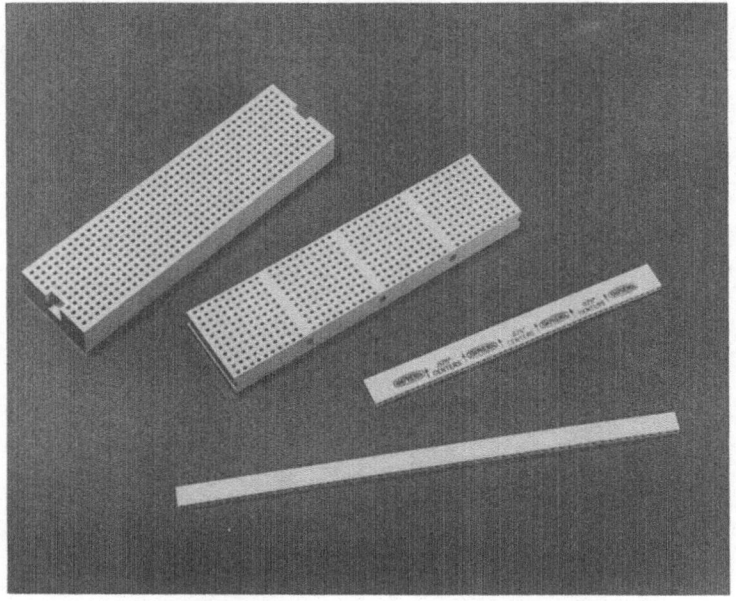

Fig. 14. Wire-form strip connectors and 450 contact blocks.

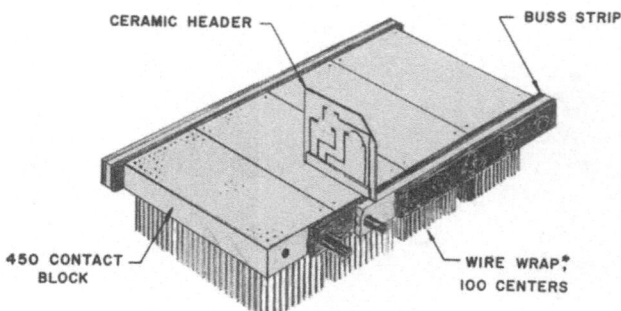

CERAMIC HEADER ⌐   ⌐ BUSS STRIP

450 CONTACT
BLOCK
                              WIRE WRAP*
                              100 CENTERS

Fig. 15. Typical motherboard assembly with bussing strips and Wire-Wrap* contacts.

For one particular application a one-piece Wire-Wrap* female contact was developed for 0.100-in. centers. The entire backwiring is then accomplished automatically (Figs. 16, 17, and 18).

Backwiring of this motherboard technique has been accomplished by several other means (Fig. 19): (1) crimped wire-routing; (2) wave-soldering of P. C. boards; (3) welding using 0.100-in. Intercon grid.

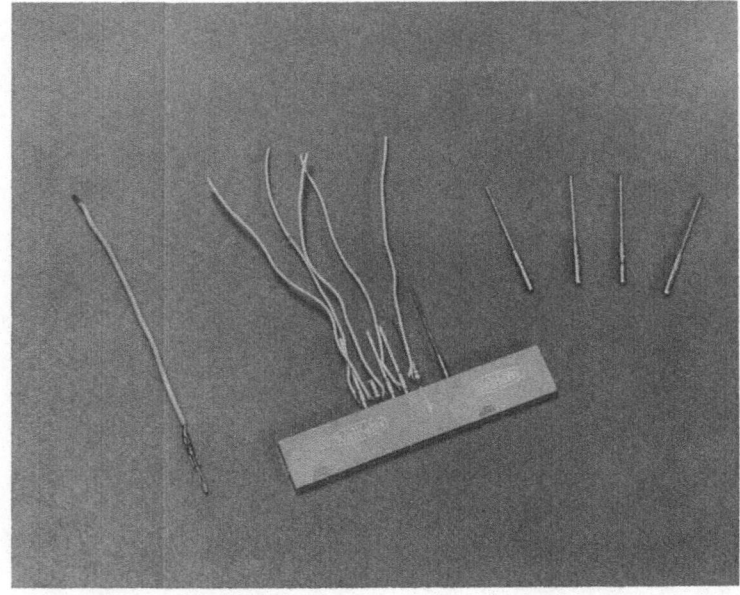

Fig. 16. Wire-Wrap* contact.

* Trademark Gardner-Denver Company.

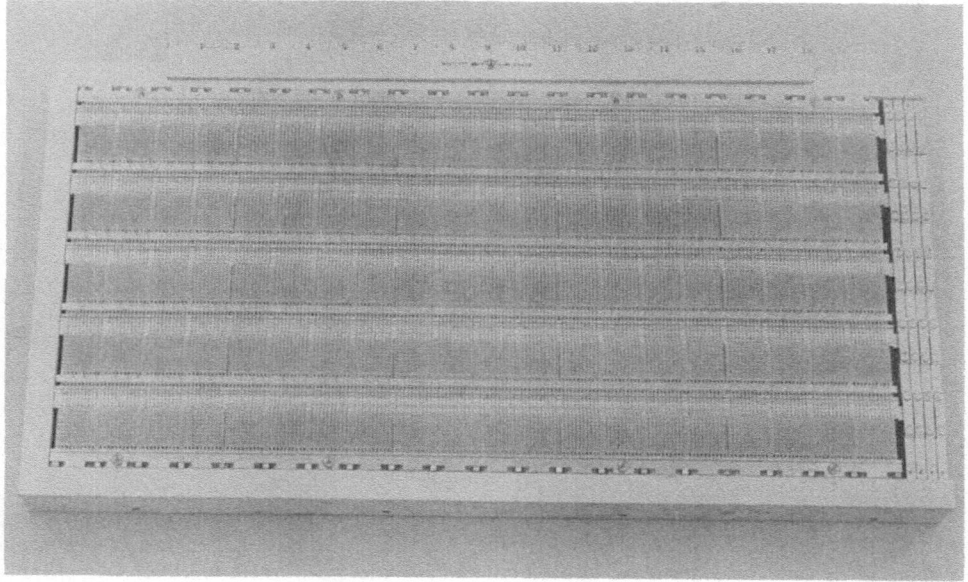

Fig. 17. Operational high-density motherboard assembly with bussing strips, front.

Fig. 18. Operational high-density motherboard assembly showing Wire-Wrap tails, back.

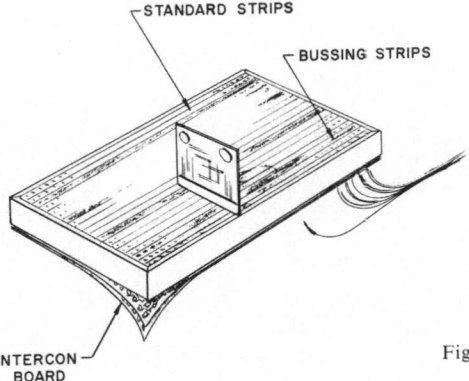

Fig. 19. Variation of high-density motherboard using other backwiring techniques.

CONCLUSION

The concepts presented satisfy specific design requirements of interconnection systems; yet, no one concept is restricted to the particular application shown. By viewing total system requirements with recognition and application of the six levels of interconnection, Amphenol–Systems Packaging is convinced that marked product improvements can be achieved.

Substitution of a Printed Conductor Ring Harness for a Conventional Cable Harness on the Mariner Series Spacecraft*

EARLE R. BUNKER, JR.

*Jet Propulsion Laboratory, California Institute of Technology
Pasadena, California*

This paper covers problems and design considerations encountered in the substitution of a printed conductor cable harness for a conventional wire harness in a Mars-type spacecraft. This harness provides over 700 electrical connections for interconnecting electronic subsystems located in eight bays around the periphery of the spacecraft. Conductor ratings in this harness include dc and square-wave ac power busses, low-level transducer signals, and pulse-command signals. In addition to the details of the harness, the paper contains: (1) A simple mathematical approach (based on the impedance, voltage level, and estimated run length) used to quantitize the susceptibility of each signal circuit to noise. The printed conductors were then located so that the more sensitive circuits were separated from the noisy ones. (2) A discussion of the interface problem between the printed conductor ring harness and the eight bay cable harnesses, including the criteria for selection of an acceptable connector. (3) Specification requirements to improve reliability.

INTRODUCTION

IN THE NEVER-ENDING QUEST for weight reduction of spacecraft components, the use of an etched printed conductor board to replace conventional wire cable harnesses appeared to be a logical step. Normal weight reduction procedures in spacecraft systems are usually measured in ounces—in this case a rough computation showed that in a Mariner-class spacecraft a printed-board substitution could result in a weight saving of the order of 10 lb for one cable harness assembly alone. It is recognized that printed conductor boards, especially of the flexible type, have been substituted for cable harnesses many times. However, as far as can be determined, no substitution has been made in as complex a system with such a variety of voltages, currents, signal levels, etc.

Several other advantages are obtained by the use of a printed conductor board for interconnection of electronic circuitry. It is extremely controllable and reproducible. Once the artwork, photography, and initial tooling are completed, procurement of additional boards is possible in a matter of days at a relatively low cost. In the process of test and system checkout, rapid changes and rerouting of circuits on a printed conductor board are possible by substituting regular wire for the printed conductors. Since circuit changes are easy to accomplish on the artwork or negatives, they can be kept up to date with these changes and modifications. When system checkout is completed, a new board can be fabricated from the artwork, and the old one discarded.

* This paper presents the results of one phase of research carried out at the Jet Propulsion Laboratory, California Institute of Technology, under Contract No. NAS 7-100, sponsored by the National Aeronautics and Space Administration.

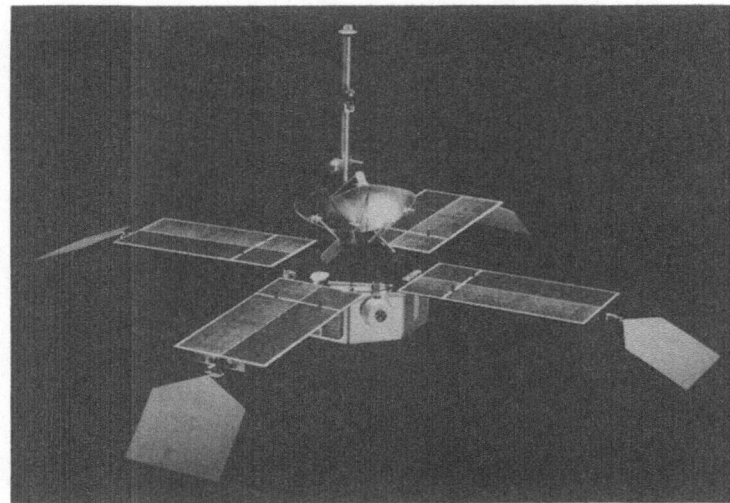

Fig. 1. Side shot of a Mars-type spacecraft with solar panels extended.

This report covers the progress to date of the development of a printed conductor ring harness (PCRH) which could take the place of the upper ring harness assembly of a Mars spacecraft.

DESCRIPTION OF PRESENT RING HARNESS

The primary objective of the early Mariner program is to achieve close-up (flyby) scientific observations of the near planets Mars and Venus and to transmit the results of these observations back to the Earth. With the success obtained by the 1962 flyby of Venus, attention has been turned to Mars. In addition to the usual fields and particle experiments, future Mars vehicles will carry TV equipment to permit pictures of the Martian surface to be transmitted back to Earth. An attempt to identify organics in the atmosphere and on the surface will also be made.

The inflight configuration of a Mars-type spacecraft is shown in Fig. 1. The electronic systems are mounted in eight cases, or bays, arranged in octagonal form about the center of the spacecraft. Individual electronic subsystems in each bay are interconnected by means of a wire-cable bay harness which also provides the input and output connections to each bay. Lower and upper wire-cable ring-harness assemblies provide the interconnections between bays. The upper ring-harness assembly actually consists of two cable harnesses, one for power and the other for signal. Figure 2 shows the ring-harness assembly, including the cable trough which provides a hard mounting for the interface connectors for the bay case harnesses.

The ranges of voltages and currents and some of the complexity can be seen from Table I. Electrostatic and electromagnetic noise pickup reduction by twisting and/or shielding is somewhat difficult to achieve on a printed conductor board in the quantities employed in the present ring harness. By proper design, a few critical circuits can be transposed from side to side of the PC board to simulate twisting, and shielding can be obtained by a suitable sandwich of grounded conductors.

Table II shows the magnitude of the termination problem in the present ring harness. There are a total of 653 connections to the signal harness and 57 to the power harness, making a grand total of 710 connections around the eight bays. Of course, these are not evenly distributed. Bays 1 and 4 contain nearly half of the connections, which adds an additional complication.

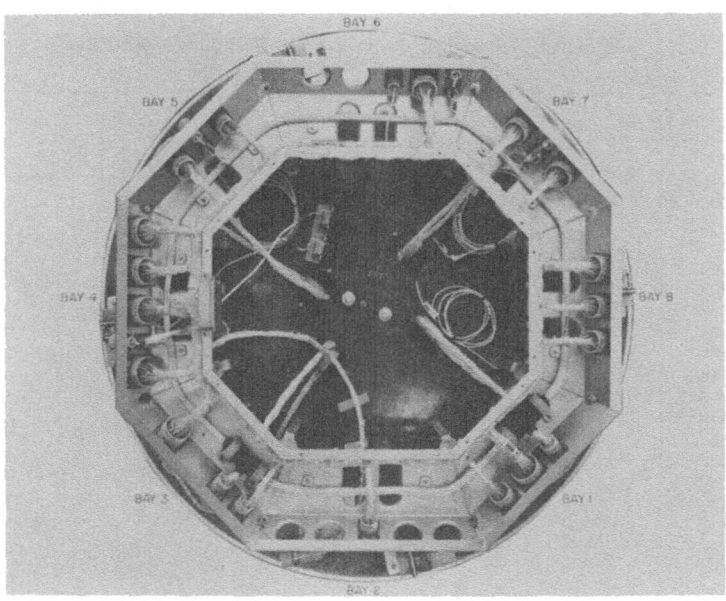

Fig. 2. Top view of upper ring harness assembly including trough.

Table I
Ranges of Voltages, Currents, Frequencies, and Waveforms in the Ring Harness Assembly

Rating	Power				
	AC			DC	
	2400-cps square wave	400-cps sine wave 3 phase	400-cps sine wave 1 phase	Solar panels	Battery
Voltage	50 V peak	26 V rms	32 V rms	0–45 V	23–28 V
Current	1.0 A	0.75 A	0.14 A	1.25 A	9.1 A
Watts	52	19.41	4.38	—	255
Circuit impedance	—	—	—	—	—

Rating	Signal						
	AC				DC		
	38.4 kc	20 to 100 msec pulse	3 to 10 msec pulse	5 μsec pulse	Telemetering		
Voltage	31.5 V	9 V peak	27 V peak	6 V peak	0–3 V peak	0 $\pm$ 1.5 V peak	0–100 mV
Current	—	—	—	—	—	—	—
Watts	—	—	—	—	—	—	—
Circuit impedance	39 K	27 Ω to 27 K	550 Ω to 2.9 K	1.5 K	6 Ω to 2.5 K	10 K	90 Ω to 1.5 K

Table II

Single-Shielded and/or Twisted-Wire Breakdown per Bay

Wire terminations	Bays							
	1	2	3	4	5	6	7	8
No. 22 not shielded:								
Single	11	—	4	—	4	—	4	5
Twisted pair	—	—	2	—	—	2	—	2
Twisted trio	12	—	12	—	12	—	12	48
Twisted quad	8	—	—	—	—	—	—	8
No. 22 shielded:								
Single	—	—	—	—	—	—	—	—
Twisted pair	20	—	16	8	8	4	6	—
Twisted trio	3	—	—	—	—	—	3	—
Shield connection	4	—	—	—	—	—	1	—
No. 24 not shielded:								
Single	35	3	3	143	16	19	49	25
Twisted pair	20	16	18	16	12	6	14	8
Twisted trio	—	3	9	6	6	—	—	—
Twisted quad	—	—	4	8	4	—	—	—
Twisted 5	5	—	—	5	—	—	—	—
Twisted 6	6	—	—	—	—	—	6	—
No. 24 shielded:								
Single	—	—	—	6	2	4	—	—
Twisted pair	2	—	2	—	4	—	2	—
Twisted trio	—	—	—	—	—	—	—	—
Twisted quad	—	—	—	—	—	—	—	—
Shield connection	—	—	—	1	1	1	1	—
Total connections in each bay	126	22	70	193	69	36	98	96

At this point there would be a doubt that such a substitution of a PC board for this cable harness would be practical from a geometry standpoint. As a matter of interest this program was initiated and a board designed for the Mariner-1962 Venus spacecraft. This was hexagonal rather than octagonal, and had only 625 connections. The PCRH was designed and fabricated as shown in Fig. 3. Although intended for test in the Mariner Venus 1962 system, this board also served to prove that a complicated harness system could be laid out on a two-sided PC board without resorting to multilayers, or using jumpers. Low-level, sensitive-signal circuits were separated as far apart as practical from noisy power circuits.

Unfortunately, this board was fabricated after the completion of the Venus flight. The Proof Test Model (PTM) test setup was no longer available for checking of the PCRH.

DESIGN REQUIREMENTS

The present PCRH is intended for test in a Mars-type PTM which will be available for a complete checkout. The PTM is identical to a flight model and serves as a backup available for tests if trouble is encountered during checkout at the launch site, or after launch, later in flight. This condition establishes a requirement which would not be applicable in a flight model. The experimental PCRH must be able to be substituted for the upper ring harness assembly without any modification of the spacecraft. Thus, jumper cables are required between the PCRH and the present interface connectors of the case harnesses. Thus constraint is necessary so that, in case of trouble at the launch site or in flight, the PTM can be returned to the initial configuration as soon as possible, with its original ring harness assembly. This design constraint of additional jumper cables requires a compromise in the configuration which may cause some degradation of the results.

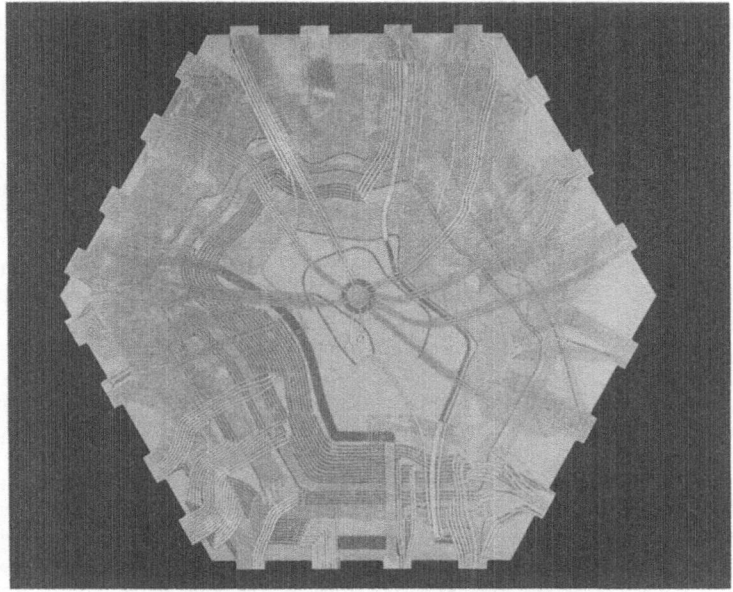

Fig. 3. Mariner Venus 1962 PCRH.

Other overall design constraints and requirements are as follows:

1. Signal circuits must be isolated from noisy circuits whenever possible by separation rather than twisting and shielding.
2. The PCRH shall be of two-sided, rather than multilayer, construction.
3. Techniques shall be provided for changing wiring or substituting shielded wire for circuits found to be critical with regard to noise pickup.
4. All measures must be taken to insure adequate reliability.

PROBLEM AREAS

Before the PCRH could become feasible, two major problem areas had to be overcome. These were (1) circuit-noise consideration and (2) selection of interface connectors.

Circuit-Noise Consideration

As can be seen from Table II, the large number of shielded and twisted wires must be reviewed and drastically reduced if the PCRH is to become a practical reality. Good practice normally dictates that twisting and/or shielding should be used for sensitive signal circuits whenever a possibility of noise pickup exists. Power circuits are also shielded to reduce electrostatic coupling to other circuits. Although system operation could be checked to see if twisting and/or shielding of low-level circuits was necessary, usually time and cost considerations require that these measures be taken when in doubt "just to be on the safe side."

A short experiment was run on two bay harnesses in bay 3 of the Mariner Venus 1962 spacecraft. One set of case harnesses was fabricated identical to the specification control drawing, but with all shielding left out except for the 2400-cps square-wave power lines. Comparative tests with specification case harnesses showed that the unshielded version had equal or less noise pickup in sensitive circuits. Other data indicated that the shielding requirements could be materially reduced without degradation of spacecraft system operation.

Many circuits employ twisted wire without shielding to minimize electromagnetic (and also capacitive or electrostatic) coupling between lines. While twisting may be approximated to a certain degree by transposition of conductors from side to side of the etched board, it is obvious that these should be kept to a minimum. The 2400-cps square-wave and 400-cps sine-wave power leads would appear to have the most need for transposing.

The tightness of twisted wire has come under consideration and analysis. For a long while twisted wire for the Mariner spacecraft has been a so-called tight twist of 3 or 4 lays per inch, rather than the standard $1\frac{1}{2}$ in. per lay. This was based on the philosophy that the tighter the twist the more transposition of conductors per unit length and hence a greater reduction in pickup. Analytical studies not completed as yet seem to indicate that the electromagnetic field very close to a transposed wire system on a PC board might be significantly higher than that due to parallel wires. Whether this applies to a smoothly twisted pair or not awaits completion of the analysis.

Shielding in general is only effective against electrostatic (or capacitive) rather than electromagnetic noise pickup. The relatively low ac currents flowing in the power circuits, and the impedance levels of the signal circuits make it apparent that the electrostatic-noise pickup is much more than the electromagnetic; consequently, the main effort will be with regard to electrostatic-noise reduction.

Based on the above findings and the consideration that sensitive circuits may be spaced much farther away from the noisy circuits on a PC board than is possible with a tightly bundled cable harness, it appears that at least a trade-off of shielding and/or twisting versus separation may be possible.

Since it is desired to separate as far as possible the low-level signal circuits from the noisy power circuits, with those less affected in between, it becomes necessary to establish a quantitative comparison of noise sensitivity. Assuming capacitive coupling between a 2400-cps square-wave power line and a low-level circuit parallel to it, a mathematical model was set up and the following equation obtained:

$$S_N = 0.52 \frac{LR}{V_s} \, \mu\text{V/signal V}$$

where L is a maximum length in inches between bays as a function of the intervening bays as follows: adjacent bays, 20 in.; one bay between, 30 in.; two bays between, 40 in.; three bays between, 50 in.; R is the estimated impedance in Ω; and V_s is the voltage in V for 100% output signal.

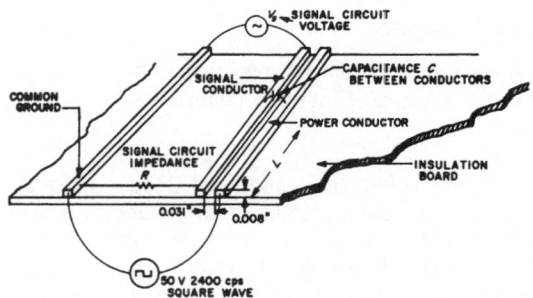

Fig. 4a. Assumed geometrical model for computation of noise susceptibility number S_N.

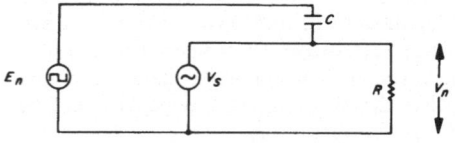

Fig. 4b. Equivalent schematic of geometrical model.

Derivation of this equation is shown in the appendix. If it is assumed for the time being that all circuits with equal S_N are affected equally, then the various circuits could be arranged with increasing S_N placed farther from the 2400-cps power lines. This assumption is not very valid because it does not take into consideration the inherent noise rejection capability of different types of circuits. For example, a pulse circuit with a very long or very short time constant with respect to the 2400-cps power would be much less affected than a sine-wave circuit operating at frequencies in the range of 2400 cps, even though both had equal S_N. It can be seen that the assumption errs on the conservative side.

Computations of S_N range from a low of zero to 156×10^3 μV per volt of signal.

Selection of Connectors

The interface between the PCRH and the bay harnesses requires the use of suitable connectors providing the high density of connections required, yet a design to provide a reliable, failure-proof cable termination. A rather complete survey of the manufacturers showed that no connectors which could be used on the PCRH had undergone JPL proof tests for the space environment. The connector finally selected was on an interim basis only and had not been flight accepted. It is felt that the connector is a reasonable compromise because, first of all, the PCRH must operate satisfactorily during a complete spacecraft system checkout. Only when the system is shown to be operating properly will environmental tests be considered.

In view of the above, the choice of a connector for use with a Mars spacecraft printed circuit ring harness is based primarily on four parameters. In the order of importance, these items are reliability, weight, small size, and availability. The ideal connector configuration would be as follows: It would have about 0.050-in. spacing between pins, and would have solder pot terminations for the jumper cable, and potting-cup provisions. It would not depend upon solder joints for mechanical attachment to the PCRH, and would have a documented reliability test history. Since there is a multitude of manufacturers, and the variety of connectors is extremely large, the connectors were divided into two general classifications for the investigation; these were: (1) edge connectors, and (2) permanently-mounted connectors.

Edge Connectors. This classification included those types which plug directly onto the printed circuit board and make contact with the printed circuit conductors on the board through mechanical pressure. As can be seen in Fig. 3, this type was used as the interface connector on the Mariner Venus 1962 PCRH.

The main advantage of the edge connector is that the mating connector—the PC board itself—involves no additional weight. A major disadvantage is that misalignment of the edge connector during insertion may damage the contacts. Close control of the circuit-board thickness is required to obtain a reliable contact, because thickness variations cause unequal contact pressure and contact resistance.

Because of the large number of connections required in the Mariner 1962 PCRH, together with the basic limitations that only the periphery of the hexagon was available for contacts, spacing selected between conductors was 0.050 in. The investigation of edge connectors with 0.050-in. center-to-center spacing disclosed that only flat tabs are provided for cable connections. The tabs are generally 0.023 in. wide by 0.008 in. thick. It would be extremely difficult to fasten a stranded wire to this tab, and the resulting connection would be of doubtful reliability. The fabrication of test cabling would therefore become more difficult. Edge connectors with wider spacings would not allow a sufficient number of connections to be obtained. All edge connectors investigated were found to have one or more of the above listed defects and, hence, the general classifications of edge connectors was considered to be undesirable for use on the printed conductor ring harness.

Permanently Mounted Connectors. The second classification of connectors considered those types which are permanently attached to the printed circuit board by such methods as soldering or welding. Typical of this type is a connector with pins that protrude through the circuit board and are soldered to the circuitry. The mating half of this connector, which serves as a termination for the cable, plugs into the permanently mounted connector.

The main advantage of this type of connector is that it is inherently more rugged and less susceptible to damage. Guide pins prevent misalignment during insertion. In case of damage to the connector mounted on the PCRH, replacement is possible rather than scrapping the board. Another advantage is that the pin-and-socket contact design can be optimized for current-carrying capability and reliability.

The major disadvantage, of course, is the increase in weight over the edge connectors. Also, connector design must allow for inspection of the joints between the contacts and the PCRH conductors. Typical connectors suitable for the voltage and current ratings have pin diameters of approximately 0.030 in. and center-to-center distances of 0.100 in., supported rigidly in an insulating connector body. The variations include configurations with pins that bend at a right angle inside the connector-body material, pins that protrude straight through, and pins that bend at a right angle after passing through the connector-body material.

Since hand soldering instead of dip soldering will be used to attach the pins to the PC pads, it will be necessary to have access to both sides of the PCRH at the connector–board interface to assure a reliable connection. Thus, all the straight–pin type and internal-right-angle-type connectors were eliminated because the connector body would prevent this accessibility and visual inspection of the solder joint on the connector side of the PCRH.

One problem common to all permanently mounted connectors with 0.100-in. pin spacing, or less, is the difficulty of running etched circuitry from the row of outer pins between the row of pins toward the center of the circuit board. The dimensions are such that no space is available for solder pads around the pins, even when using the minimum specified dimensions.

Results of Investigation

After a thorough literature search and contacts with suppliers, it became apparent that only two connectors were available that would meet the requirements of a Mars PCRH. One connector, the Becon-Type 2700–011, is similar to the edge connectors previously described, in that it employs pressure contact with an etched circuit pattern on the printed board, but does not depend on the circuit-board thickness to maintain the correct pressure. It does not have to be mounted near the edge, but can be placed anywhere on the circuit board. This type would have to be used in pairs on opposite sides of the circuit board to prevent bowing of the board due to the high contact pressure. The disadvantage of using this connector on the printed circuit ring harness is that the contacts have eye tabs for fastening the wires rather than solder cups. There is no easy way to provide strain relief for the cable.

The second connector that was considered is the Continental-Type 600–1–45. This connector is a part of the second classification, that is, permanently mounted on the circuit board. The connector has three rows of pins with a center-to-center spacing of 0.150 in. The three rows are offset such that the equivalent of 0.050-in. pin spacing is obtained. Since the pins go through the connector before making the right-angle bend, both sides of the circuit board are available for inspection. This connector was designed to meet the requirements of MIL-C-8384, but since it has been sold in relatively small quantities, there were no reliability-test data available.

RELIABILITY CONSIDERATIONS

While the present tests planned are only for the purpose of comparing the electrical functional operation of the PCRH with the wire ring harness, several requirements were included to improve reliability.

For example, a major source of trouble with PC boards is the cracking of the printed conductor due to insulation-board stresses. In the PCRH the normal conductor thickness is 0.008 in., rather than the standard 0.0028 in. The thickness of the insulation board is correspondingly reduced so that the overall thickness of the conductors and insulation board would be the same as the conventional PC board, to accommodate standard edge-type connectors. Since a permanent-mounting type had been decided upon instead, this requirement is not as important. The thinner insulation material does achieve a reduction in weight and also a lessening of tensile stresses on the conductors during deflection.

Another possible source of trouble is the plated-through holes connecting conductors on opposite sides of the board. Since the power lines are intentionally transposed to minimize coupling with sensitive circuits, the additional plated-through holes should not decrease the reliability. Accordingly, all plated-through holes in both signal and power circuits are used in pairs to achieve redundancy.

Adverse vibration effects on the thinner insulation board will be overcome by suitable redesign for flight, if the PCRH successfully passes the system tests.

Conductor width is limited to 0.025 in. mininum, which is required to allow all of the runs to be made to the various bays. This width and the thickness of 0.008 in. correspond in cross-sectional area to a wire gauge of No. 26, which is adequate for signal circuits. Power circuits with appreciable current flowing have an increased conductor width so as to obtain a current density of not less than 250 square mils per A. This will keep conductor heating and voltage drops within tolerable limits. This is contrasted to the handbook rating of 3.5 A through a 0.0625-in.-wide by 0.0027-in.-thick conductor for a 10°C rise in temperature, which is a current density of 48 square mils per A.

LAYOUT OF PCRH

A fundamental problem in the layout of the printed circuit ring harness is the minimization of stray pickup by sensitive low-level lines from adjacent noisy or high-voltage lines, such as the 2400-cps square-wave circuits. This is accomplished in the conventional upper ring harness

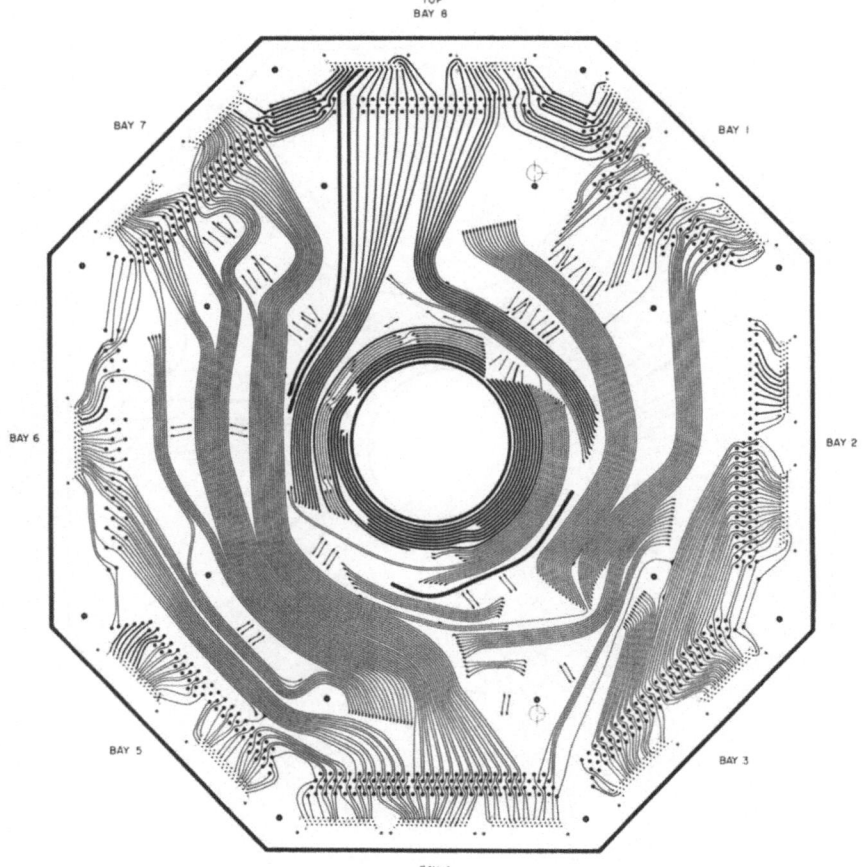

Fig. 5a. Circuit diagram of top of PCRH.

by using two harnesses to separate the 2400-cps power sources from other circuits. In a similar manner, the printed conductor ring harness has the 2400-cps power circuits at the center of the board with the low-level circuits near the outer edges as can be seen from Fig. 5. This enables the power to be supplied to the different bay connectors by radial conductors running from the center straight out to the appropriate connector. Signal circuits farther out thus cross the power circuits at right angles, minimizing electrostatic and electromagnetic coupling. To further reduce pick-up, ground conductors are placed adjacent to the 2400-cps power conductors.

The 400-cps three-phase sine-wave power circuits were routed adjacent to the 2400-cps circuits at the center of the PCRH. The 400-cps circuits also have conductors routed radially to the connectors. The remaining circuits were located according to the type of circuit (quiet or noisy), and the noise susceptibility number S_N. Each signal circuit was evaluated using this equation, and the resulting S_N numbers were used as a relative indication of susceptibility to stray pickup and noise. Another factor taken into consideration was the circuit function and how it varied with respect to time. For example, the use of the above equation on temperature transducer circuits indicated a high noise susceptibility. However, additional investigation indicated that because of their long time constant, noise would not be detrimental to the operation. Therefore, the transducer circuits were grouped together without individual shielding.

Based on the required specifications, the nominal line widths and nominal spacings would both be 0.031 in. A plated-through hole diameter of 0.031 in. was selected as a small, reliable,

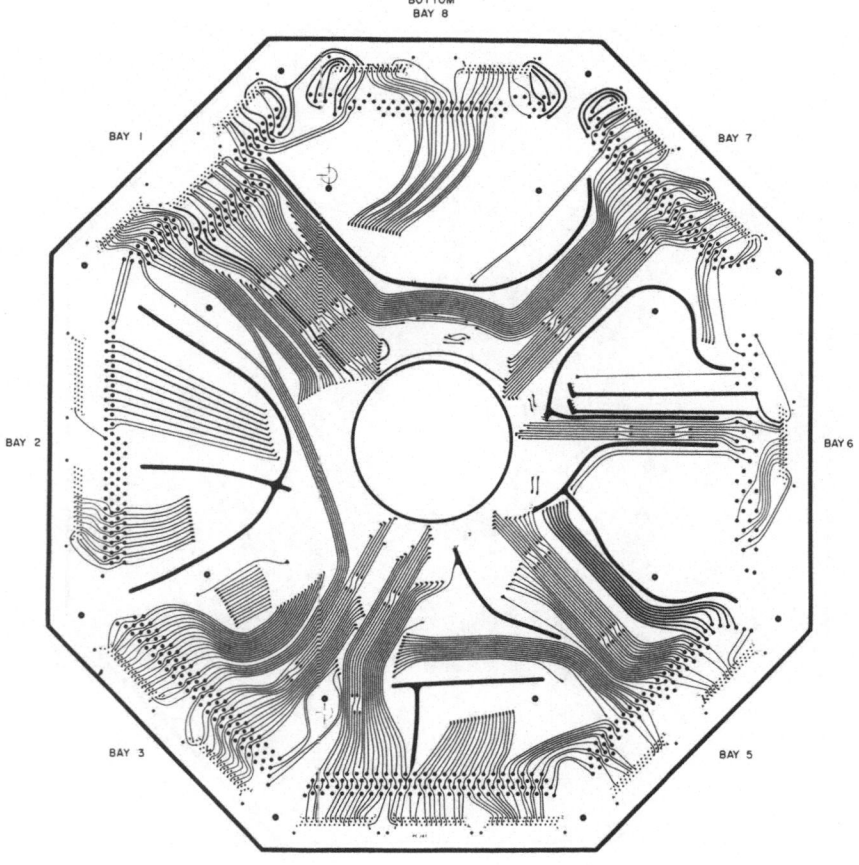

Fig. 5b. Circuit diagram of bottom of PCRH.

readily-fabricated size. Smaller diameters can be used, but are difficult to fabricate and, therefore, tend to be less reliable. A larger size would use an excessive amount of space. Terminal pads are 0.125 in. in diameter, and the holes in the connector pattern are 0.036 in. in diameter. The exception to the above is the line width between the terminals and the connectors, which would be 0.025 in. This exception is required because of the close spacing in the connector circuit pattern. The routing of 0.025-in.-wide circuitry in between connector pins permits a spacing of 0.027 in.

Referring to Table II, it was observed that 193 connections were required by the bay 4 harness. Three connectors on the bay 4 edge of the PCRH permit a maximum of 135 connections. To resolve this, the remaining connections were moved to bay 3. Hence, the jumper cable that connects to bay 4 harness in the spacecraft will have connectors that mate with both of the bay 3 and bay 4 connectors of the PCRH. This is the only jumper cable having connectors that mate with more than one edge of the PCRH.

In the design of this experimental PCRH, it was necessary to make provisions for routing changes, corrections, and addition of shielded wire, if required. This is accomplished by having every circuit connected to a solder terminal mounted adjacent to the connector. Thus, conventional twisted and/or shielded wire can be substituted for runs found to be critical with regard to noise pickup during testing. It is a simple procedure to solder in the substituted wire to the appropriate terminals, and interrupt the corresponding printed conductors. These solder terminals also would make it possible to eliminate the connectors completely by soldering the eight bay cable harnesses directly to the PCRH. An additional investigation and analysis would be required to prove the advisability of this approach.

CONCLUSION

In conclusion, it has been shown that a double-sided printed conductor board can be substituted for a complex conventional cable-ring-harness assembly in the planetary spacecraft from a conductor-routing standpoint. A method of quantitatively estimating the susceptibility of signal circuits to capacitively-induced noise from a 50-V 2400-cps square-wave power system was also described and used to determine relative routings of various circuits. The printed conductor ring harness described is undergoing fabrication and tests to determine its flight capability as an integral part of the future planetary spacecraft systems.

APPENDIX

Derivation of Noise Susceptibility Number

The model used to compute the noise susceptibility of various signal circuits is shown in Fig. 4a. From the schematic shown in Fig. 4b it can be seen that the noise voltage V_n induced capacitively into the signal circuit is a function of the signal circuit impedance (assumed resistive for all cases), the capacitance between signal and power connectors, and the frequency and amplitude of the noise voltage. The 50-V 2400-cps square-wave power used in the spacecraft is assumed to be the noise source. To retain the impedance concept, the Fourier harmonics of this square wave will be employed in the analysis.

While the induced noise voltage for any frequency can be calculated from Fig. 4b, the actual detrimental effect on the signal circuit is also a function of the signal voltage; obviously a signal circuit operating at a higher voltage level will be affected less adversely by a given induced-noise voltage magnitude than one operating at a lower voltage level, other parameters being equal.

Let

E_n = rms noise voltage of the nth harmonic
C = total capacitance between signal and power conductor
R = impedance of signal source, assumed resistive
V_n = induced rms noise voltage in signal circuit
$\omega = 2\pi f$, where f = frequency of nth harmonic
E_{max} = amplitude of square wave = 50 V

then by the voltage-divider equation, taking magnitudes only,

$$V_n = \frac{R}{\sqrt{R^2 + (1/\omega C)^2}} E_n$$

Rearranging and substituting for ω

$$V_n = \frac{1}{\sqrt{1 + (1/2\pi f C R)^2}} E_n$$

The capacitance between two parallel plates of area A separated by a distance d using inch units is given by

$$C = 0.225 \, e_r \frac{A}{d}$$

since for air $e_r = 1$ and $A = Lt$, where L is length and t is thickness of printed conductor. Let

$$c = \frac{C}{L} = \text{capacitance per unit length}$$

then

$$c = \frac{C}{L} = \frac{0.225 \, t}{d}$$

In this case

$$t = 0.008 \text{ in.}$$

$$d = 0.032 \text{ in.}$$

$$c = 0.56 \text{ pf/in.}$$

To simplify the analysis, and since $C = cL$, assume

$$\left(\frac{1}{2\pi f c L R}\right)^2 \gg 1$$

then

$$V_n = 2\pi f c L R E_n$$

To check this assumption, from the constraints of the present geometry, let

$$\frac{1}{2\pi f c L R} = 10$$

which is a good assumption for the above inequality. Substituting in

$$f = 2400 \text{ cps}$$

$$L = 50 \text{ in. (maximum run, 3 bays between)}$$

then the maximum value of R for this assumption to hold is

$$R = \frac{1}{2\pi c L \times 10} = \frac{1}{2\pi \times 2400 \times 0.56 \times 10^{-12} \times 50 \times 10}$$

$$R = 240 \text{ K}$$

Most signal circuits in the PCRH have a much lower R or have shorter runs, so this limitation is not critical in the analysis.

The Fourier components for the nth harmonic (where n is odd) of a symmetrical square wave of amplitude E_{max} about the origin can be expressed as follows:

$$e_n = \frac{4E_{max}}{\pi}(-1)^{n-1}\frac{\cos n\omega t}{n}$$

The rms amplitude of the nth harmonic is

$$E_n = \frac{4E_{max}}{\sqrt{2}n\pi}$$

Also for a 2400-cps square wave

$$n = \frac{f}{2400}$$

where f is the frequency of the odd harmonic

$$E_n = \frac{4E_{max} \times 2400}{\sqrt{2}f\pi}$$

Since

$$E_{max} = 50 \text{ V}$$

then

$$fE_n = 1.08 \times 10^5$$

Substituting in

$$V_n = 2 \times 1.08 \times 10^5 \times 0.5 \times 10^{-12}LR = 3.8 \times 10^{-7}LR \text{ volts}$$

Thus it can be seen that V_n is a function only of the length of run and the resistance of the signal circuit. We define the noise susceptibility number S_N in terms of V_s, the maximum signal voltage (dc or peak ac),

$$S_N = \frac{\sqrt{2}V_n}{V_s} = 2 \times 3.8 \times 10^{-7}\frac{LR}{V_s} = 0.54\frac{LR}{V_s} \ \mu\text{V noise/signal V}$$

where L is a function of the number of intervening bays as shown in the body of the paper.

Design of Connectors for Electronic Packaging

Homer E. Henschen

Connector Products Division, AMP Incorporated
Harrisburg, Pennsylvania

The role of connectors in electronic packaging along with connector design philosophies and parameters are discussed in this paper. Topics considered are reliability, performance, versatility, utility, economy, and specifications. Five connector contact designs are presented and analyzed for their applicability to interconnecting the five basic modules—cards, cubes, substrates, flat-packaged integrated circuits, and transistor-canned integrated circuits—with the basic parent wiring systems.

INTRODUCTION

CONNECTORS ARE CONTROVERSIAL. Their role in electronic packaging is difficult to assess, but it can be said that they are indispensable. Let's look at some elements of the controversy.

1. Connectors save time and money when building and servicing equipment, but are an expense in themselves.
2. Connectors increase the reliability of transitions from component groupings to wiring systems, or from one type of wiring system to another, but decrease reliability in themselves.
3. Connectors enable quick replacement of malfunctioning units, but contribute new failure modes to a system.
4. Connectors enable closer stacking of units, but they occupy part of the total volume.
5. Connectors contribute to transfer of heat from source to sink, but they generate heat.

It should be evident that such simple statements as "We'll eliminate connectors entirely" or "We'll make everything pluggable" are not realistic.

Normal requirements which must be met by electronic packaging engineers require complex decisions regarding connector usage. Electronic components, whether they be discrete, integrated, deposited, hybrids, or mixtures, must be (1) grouped together in various combinations; (2) electrically interconnected and insulated to perform functions; (3) mechanically supported and protected from physical damage; (4) thermally controlled; (5) protected from environmental damage. Packaging engineers must make many compromises to achieve the desired results within the allowable cost limitations and minimum required reliability.

The first determination is that of the smallest permanent, nonrepairable grouping of components. With highly reliable components, the smallest grouping can be quite large. Conversely, it may be necessary to package relatively unreliable components individually or in very small groups. The minimum and maximum number of components required to perform various functions may dictate the sizes of the groupings. Thermal control may offer problems with large groupings. The maximum permissible cost for the nonrepairable function generally bears heavily on the size. These are general considerations and in addition, all equipment has its own specialized requirements which must be analyzed.

CONNECTOR DEFINITION

All equipment has levels at which a group of connections must be made or opened without the use of highly skilled labor, special tools, or equipment. The connecting device is the connector. It is defined as a device in which a multiplicity of connections can be made or broken simultaneously and repeatedly by physically separating two portions of the device without degradation, which would affect the performance of the equipment in which the connector is used.

A lower level of disconnection is frequently accomplished by the breaking and remaking of permanent connections. The methods are usually destructive, requiring replacement of at least one-half of the connection. They require the same level of skill as required at original manufacture, and must be subjected to the same inspection and testing procedures used at the manufacturing level. Devices to assist in this procedure are available and, although they are sometimes called connectors, they fall outside of the definition of connectors as the term is used in this paper.

ADVANTAGES

The connector offers very specific advantages. Component subassemblies, with plug or receptacle halves of connectors as part of them, may be assembled in facilities designed for that purpose. Wiring systems, including the mating halves of the connectors, can be assembled in other facilities suitable for that purpose. Where some operations require highly specialized facilities such as white rooms while other operations can be done in ordinary facilities, the savings in manufacturing costs achieved through segregation of the operations are considerable.

The various subassemblies can then be tested, adjusted, burned-in, and assembled into the next higher level of assembly by mating connectors with a minimum of skill and special tooling. Test equipment terminated in appropriate connector halves can be used with great savings in connection time and with more confidence that the test hook-up has been made correctly.

Malfunctions can be identified and replacements made during check-out and field servicing of the equipment by means of the same connectors. Field-service personnel do not need to have equipment comparable to that used in manufacturing for making and testing connections.

CONNECTOR TYPES

The component grouping is generally called a module, and may consist of one of five basic types: cards, cubes, substrates, flat integrated circuit packages, or transistor-canned integrated circuit packages. These modules are interconnected by any of several parent wiring types: single-layer or multilayer printed circuit boards, welded-wire matrices, soldered harnesses, crimp-on snap-in harnesses, or machine-programmed point-to-point systems using Termi-Point* clips or Wire-Wrap† terminations. Thus it is conceivable that 25 connector types could be required to provide for each possible module-to-parent wiring combination.

We will concentrate on concepts applicable to the sophisticated high-density packaging systems requiring contact spacings of 0.100 in. or less, and which offer the greatest challenge to the packaging engineer. For these systems, more connections must be made within a given volume, higher reliability is required, means for dissipating heat are generally mandatory, and time delays or attenuation caused by poor interconnection designs may prevent the proper functioning of the unit.

This challenge extends to the connector designer. Even within these limitations, variations of approximately 20 connector application types must be available to satisfy the individual

* Trademark of AMP Incorporated.
† Trademark of Gardner Denver Co.

requirements for innumerable package designs. Because of the vital importance and intimate relationship between connectors and package design, the packaging engineer must wisely choose the connector.

CONNECTOR DESIGN CRITERIA

Connector concepts are abundant. Proven designs are less plentiful while production items for which extensive reliability and performance data are available are few in number. The selection then must be made with available data and a large measure of good judgment. Five criteria—reliability, performance, versatility and utility, economy, and specifications— can be used to evaluate the applicability of a connector design to a given packaging design.

Attributes which we use in evaluating our connector concepts with respect to these criteria are discussed in the next section. Then, five proven contact designs are presented and analyzed in relation to these criteria and their applicability to the 20 connector types established earlier.

Our understanding of why connectors perform reliably, or fail, is constantly improving. Much research, performance testing, and reliability testing are constantly in progress to extend our knowledge as well as to provide specific data for our products.

Reliability

Reliability has become a prime requisite for all phases of electronic packaging. Even severe cost limitations do not eliminate the need for high reliability. A number of physical features have prime importance in determining the inherent reliability of a connector design. These are, the introduction of a minimum number of additional joints or interfaces into the circuit, the use of familiar metals such as copper or nickel alloys, the use of suitable plastics and dielectrics, and the use of established finishes such as gold plating over nickel plating [1,2]. The use of untested and unproven materials requires that their properties be established in addition to the design itself. This is desirable and necessary, of course, but time must be allotted to do it.

There are many facets to good contact design. Contacts should wipe during engagement to insure cleaning of foreign materials and breaking of oxides or other chemical films. Contact pressures should be controlled by spring designs, which are not critically related to minor variations in tolerances of the parts during manufacture. Similarly, contact pressures should be virtually independent of the relative positions, within reasonable tolerances, of mated connector blocks.

The area of the interface between mated connector contacts is directly related to the contact forces and must be sufficient to pass the desired current. High contact forces insure large contact areas but result in high wear rates and high mating and separation forces. Excessive wear rates will destroy the protective plating, permitting oxidation or corrosion which may make the connector inoperative. High mating and separation forces are no problem with a small number of contacts or where mechanical assistance in the form of jackscrews or levers can be used. But with large numbers of contacts, other damage may result, as will be discussed later. Low contact forces may not provide sufficient wiping action and may allow the contacts to open during vibration or shock.

For most packaging applications, connectors need a wear life of many hundreds of matings without severe degradation. To accomplish this the contact interface should be a relatively large area subjected to moderate forces. Two or more contact areas which are sprung independently will increase the numerical value for the inherent reliability if each area individually has the same reliability as one. Too frequently, however, redundancy, in its true definition of superfluity, is accomplished instead of the desired contact multiplicity; or multiplicity is accomplished at the expense of more series connections in the circuit resulting in a reduced overall reliability.

The connector design should offer protection against damage during the life of the connector. It is not enough that a mounted and assembled connector protect the contacts. The design must also offer the maximum protection against damage during shipment, assembly, and check-out. Removable contacts should be resistant to damage when outside the connector blocks. Unusual stresses that occur accidentally or intentionally during check-out phases should not degrade the connector.

Perhaps a look at reliability's antonym would be helpful. Connectors can fail in many ways. One of the most common is failure of the external connections to the contacts. Unless external connections can be made easily, thoroughly checked out, and then properly protected against subsequent damage, this is certain to be a troublesome area.

Failures of connectors result frequently from destructive contact mismating. Contacts become bent or splayed, are pushed out of the plastic blocks, or suffer other damage which results in open circuits or short circuits. Contacts that are easily bent or deformed or inadequately protected against mechanical damage are the usual causes. Inadequate control of alignment during engagement and separation may result in contact damage. Lack of contact float to permit proper mating of contacts misaligned because of normal manufacturing tolerances will result in excessive wear and high mating and separation forces.

The presence of foreign materials such as dust and dirt, and oxides and other chemical films frequently result in excessive resistance, intermittent open circuits, or excessive wear. Defective or inadequate plating can cause similar difficulties.

Mechanical and electrical failures of plastic connector blocks are very common. Mechanical overstressing through careless handling accounts for most breakage. Overheating caused by soldering and encapsulating operations can cause deterioration that may result in subsequent failures. Chemicals used in the industry frequently attack certain plastics. The resulting deterioration may cause mechanical or electrical failures.

Excessive heating of connectors due to overcurrent or lack of thermal control may cause deterioration of plastic materials and, if severe enough, deterioration of the electrical contacts, causing eventual malfunctions. Similarly, overvoltage applications may cause flashovers, short circuits, or reduced insulation resistance.

So far, we have discussed only those connector failures which would result in obvious equipment malfunction. A failure category that is not well understood is the inability of a connector to meet certain performance test requirements, although it is still functional. Perhaps, we could define these as out-of-specification failures. Where the performance requirement and the circumstances under which the tests are performed can be related to actual operating requirements there is no serious problem. But too frequently, arbitrary values not related to either an actual requirement or a statistical basis for determining reliability are selected as minimum criteria for reliability testing.

The principal reason for lack of knowledge on out-of-specification failures is that they may occur very frequently in actual usage, but go unnoticed because they do not cause equipment malfunctions. However, if the performance is degrading with time or usage, equipment malfunctions may occur at some future time. Much effort and money are being expended to relate the results of laboratory testing with reliability under actual working conditions.

Performance

The performance of connectors is not unrelated to reliability but is more a quantitative set of information about the electrical and mechanical properties. Testing which has customarily been required for compliance to various industrial and military specifications gives us performance data rather than reliability data. Units which pass can be said to meet the requirements visualized by the specification writer. But statistical assurance that the units will perform this way indefinitely can be obtained only through additional reliability testing and experience in actual usage.

The prime requisite for connectors to be used in electronic packages is continuity at low-level signal voltages or dry circuit conditions. This test is generally applied to connectors after they have been exposed to a certain amount of testing such as vibration, shock, durability cycling, temperature cycling, and possibly corrosive atmospheres.

Many values for the open-circuit test voltage have been suggested. We have used values from 0.1 μV to 50 mV with no differences in results. Whitley believes that 50 mV is the minimum necessary to determine whether contamination or chemical films are interfering with proper conductivity [3].

The increase in circuit resistance by inserting connector contacts is next in importance. Generally speaking, the total resistance, including the external connections, should be 25 mΩ or less since this compares approximately with the resistance of a similar length of the conductors used in packaging. Uniformity and resistance to change are more important than the absolute values for most applications. Resistance measurements are usually made at a low-level current between 1 and 20 mA, and then again at a current near the maximum rating, which will usually be between 0.5 and 3 A.

Insulation resistance is generally important. The minimum values required depend on the application but, again, the insulation resistance after exposure to environments such as temperature or humidity cycles is more important than the original value. A typical minimum required value is 500 MΩ.

Dielectric-withstanding voltages can generally be low because of the very small voltages used in sophisticated systems. However, accidental application of high voltage during testing or other unusual circumstances should not cause destructive flashovers. A test voltage of 900 V ac RMS is adequate for most purposes.

Mating forces of connectors become very important when large numbers of contacts are required. High insertion and withdrawal loads cannot be tolerated in most applications because of the mechanical stresses on the various parts attached to the connector blocks. Also, there is not room for jackscrews or other hardware useful for mating or unmating the connector in high-density packages. Reasonable changes in mating forces can be tolerated during exposure to environments as long as the connector can still be successfully operated without degradation of its other properties. Maximum average values of 2 to 4 oz. per contact are desirable. Maximum individual values must generally be limited to 8 oz.

Durability cycling, or the number of times a connector can be mated without degradation, has become of prime importance. We have been using 250 as the minimum number for all designs, while most are subjected to 500 for test purposes. It is not unusual to receive requests requiring connectors to be mated thousands of times without degradation of their other critical properties. Incidentally, these large numbers of insertions frequently occur only at the manufacturing level. The concern of some packaging engineers is that the connectors will be worn out before the equipment has been checked-out, adjusted, and shipped to the customer.

So far, we have discussed the performance of connectors primarily for use in ground-based equipments. Aircraft and space vehicles demand additional performance requirements [4]. Resistance to radiation damage [5] and resistance to sublimation in hard vacuums [6] at various temperatures are primary requirements. Temperature extremes are greater than for most ground-based equipment. Exposures to a variety of fluids and atmospheres without degradation are necessary. In addition each application usually has its own unusual aspects that require special attention.

Versatility and Utility

Versatility and utility are mandatory so that connector designs may be effectively used in several types of package designs. This is especially true for contacts. Performance and reliability data are accumulated only through repeated testing and experience gained in actual usage. Therefore, to be able to supply specialized connectors without long lead times, connector designers must use contacts for which they have accumulated data and experience.

Variability in size, form factor, and applicability to various means of external connection are some of the prime considerations. The maximum possible latitude in materials and finishes for contacts and connector blocks will enable the best design necessary for the specified purpose.

The maximum utility is realized when connectors have true contact center-to-center spacings which are multiples of 0.025 in. There seems to be little advantage in spacing capabilities between the 0.025-in. multiples because standardization attempts are now made at these intervals instead of the fractional intervals, as was common in previous connector designs. Row-to-row spacing is less well standardized. Staggered patterns and row spacings different from contact spacings are frequent on printed circuit board constructions. Square grid patterns are virtually universal on systems wired with automatic equipment.

Contact spacings and corresponding row spacings of 0.100 in. and 0.075 in. are used commonly, while 0.050-in. contact spacings with somewhat larger row spacings are now being required in the most advanced packages. We have not seen legitimate requirements for 0.025-in. contact spacings. Wiring systems and interconnection methods preclude their proper use today. As wiring and interconnection systems become practical at 0.025-in. spacings, the connector industry hopefully will have available the necessary connectors.

Economy

Closely allied to versatility are cost considerations. To make fair comparisons the total applied cost including inspection and testing after installation must be used as the basis. It becomes evident that easy and reliable installation means are important since they are such a very large part of the total applicable cost. Segregation of manufacturing operations, easier assembly, check-out, and maintenance were pointed out earlier as some of the areas which must be considered in the economic evaluation of connectors.

Specifications

Few military or industrial specifications apply directly to the connectors used in electronic packages. Some standard connector specifications apply in a general way and attempts are made to specify connectors for packaging through their use. This has led to difficulties in that numerical values often are not realistic for packaging connectors. Also, the general requirements of these specifications do not equally well describe good packaging and standard connectors.

Good descriptive product specifications can be written only through the combined efforts of packaging and connector engineers. Results of performance and reliability testing in functioning packages must be fed back to the connector people for assimilation, correlation, and integration with other data. This enables product improvement in several ways. Failure modes can be eliminated or MTBF values increased through better control of materials and manufacturing processes. Marginal performance traits can be improved. Or, in other words, capabilities can be matched to the actual requirements.

This is not to say that the packaging engineer must compromise his requirements to less than he actually needs to use a particular connector. Instead, the ultimate specification should describe a product or products which will fulfill those requirements and which have been adequately related to test procedures and conditions that will ensure compliance with the actual requirements.

CONTACT DESIGNS

Five contact designs are required to fulfill the necessary connector requirements of the twenty possible types established earlier. Table I lists their applicability. Several blocks are vacant but I doubt that connectors for those applications are seriously needed. On the other hand, some blocks have several entries. These represent choices that are available to the packaging engineer. We developed these contacts for the electronic packaging industry and have accumulated performance data for them.

Active Pin Contacts

Active pin contacts (Fig. 1) are one-piece stampings in which the contact spring force is supplied by a split pin design. One leg is hooded by an extension of the other leg to prevent accidental splaying. The pin, when sprung closed, is 0.028 in. in diameter.

The mating contacts may be suitable cylindrical holes in a printed circuit board such as plated-through holes, or eyelets, or tubular sockets. The pins are projected beyond the mating face of the connector block when plugging directly into a printed circuit board. They are generally recessed within the cavities when mating with extended tubular sockets.

The contacts are firmly retained in resilient polyurethane cavities to prevent mismating due to uncontrolled contact float. Contacts misaligned due to normal manufacturing tolerances

TABLE I

Connector Contact Application Chart*

Parent wiring type	Module type				
	Cards	Cubes	Flat I.C. packages	Substrates	Transistor-canned I.C., transistors, diodes, resistors, etc.
Single-layer and multilayer boards					
Plug to plated holes	Active pin	Active pin	Active pin		
Plug and receptacle type	Helical MECA Active pin	Helical Active pin	Helical	Helical	Component lead socket
Welded wire matrix	Helical MECA Active pin	Helical Active pin	Helical	Helical	Component lead socket
Soldered harness	Helical MECA Active pin	Helical Active pin	Helical	Helical	Component lead socket
Crimp-on, snap-in harness	Channel Active pin	Channel Active pin			

* Only preferred styles shown.

on spacings will align themselves during mating through deflection of the resilient cavity walls. The contacts are retained in the connector block by the engagement of bottlenecks in the contact bodies and the cavities and can be removed.

When connectors are too large for the polyurethane to provide sufficient rigidity by itself, a combination molding technique is employed. Polyurethane is injection-molded into a skeleton of a more rigid plastic or metal. The contact is completely enclosed within the polyurethane, permitting the normal advantages of its resilience while the rigid skeleton provides dimensional and structural stability.

Spacings of 0.075 in. can be accomplished with polyurethane connector blocks, while spacings of 0.100 in. are usually recommended when the combination molding technique is used. Maximum prolonged operating temperature is 105°C.

The contacts may be terminated by welding or soldering to the integral tab extending from the rear of the contact. Another version has wire barrels to provide crimp-on snap-in ability.

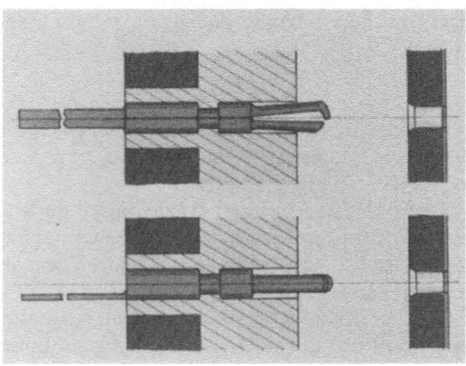

Fig. 1. Active pin contact.

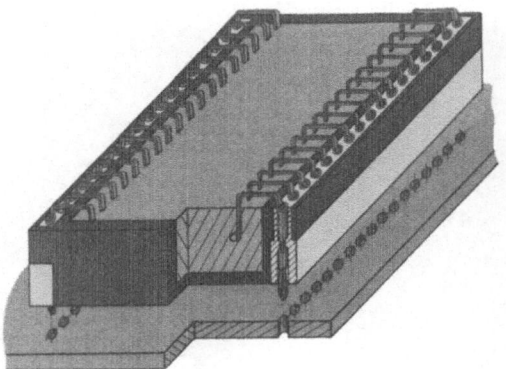

Fig. 2. Active pin module.

Figure 2 shows a typical module for plugging directly to a printed circuit board. There are 36 contacts on 0.100-in. centers. Guide pins are used in the corner cavities.

The contact tab extensions are bent over for direct connection to the module wiring. Leaving the tabs exposed when encapsulating provides convenient test probe points that are accessible when the module is plugged into its normal operating location.

The connector body is the potting form. External dimensions of 2 in. by 1 in. by 0.3 in. allow internal dimensions of 1.75 in. by 0.75 in. by 0.25 in.

Helical Contact

The helical contact (Fig. 3) is made up of a spring-and-tab assembly mounted in an appropriate housing. The spring has been deformed so that its axis is a helix. When a pin is inserted, deflection tending to straighten the axis occurs, resulting in forces pushing the pin and tab together, providing the primary current path. Contact multiplicity is provided by the turns of the spring being in contact with the tab and pin.

Physically, this construction is very small, ranging from $\frac{1}{16}$ in. to $\frac{3}{32}$ in. in length, and capable of being spaced on 0.050-in. centers. The tab portion of the contact can be made of any suitable material and finish that may be required for connection to the packaging system. Thus copper alloys can be used where external connections are made by soldering, while nickel alloys can be used for welding, etc. The tab can be of a low-temper or malleable material so that it can be bent to shape while being connected into the system. This feature is especially important in substrate terminations in that direct connection is possible, eliminating the extra connections of the usual jumper wire.

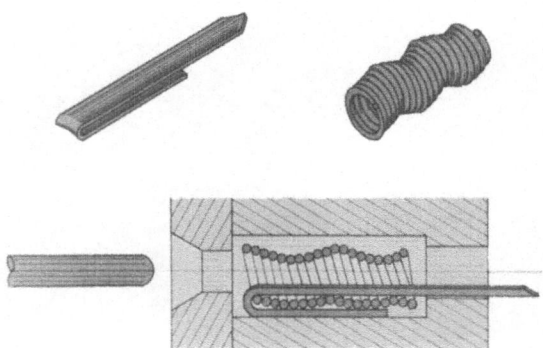

Fig. 3. Helical spring contact.

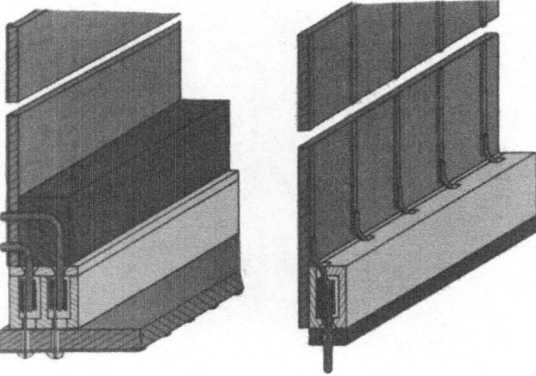

Fig. 4. Helical right-angle connectors.

The contact assembly is independent of its housing except for protection and to provide lead-in for the mating pins. It may float in its cavity to allow for normal manufacturing tolerances on spacings. The housing material may be any suitable dielectric or individual metal eyelets.

The tab extension may emerge from the contact cavity parallel to or at right angles to the contact axis. Pins can be plugged from either end or through the contact. Where extended life of the contact is important, the pin contact must be provided with a tapered or spherical end.

Figure 4 shows typical right-angle card connectors using helical contacts. The arrangement shown in Fig. 4a is used for card-to-motherboard applications. Figure 4b shows an arrangement in which the pin header can be attached to the parent wiring system which may be a welded-wire matrix, a soldered harness, or a printed circuit board. Cards or substrates can be plugged on 0.150-in. centers with this connector.

Figure 5a shows substrates or cards framed in a connector with contacts at right-angles to the cards. The frames are stacked as shown in Fig. 5b. The interconnecting wiring is provided by pins plugged into the contacts. Approximately 25% of the contacts in a frame are connected to the cards in that frame. The others provide through-wiring ability. Forty 2.75 in. by 2.75 in. cards are mounted and completely wired within a volume 3.25 in. by 3.25 in. by 7 in. in height.

Figure 6 shows a connector for a flat integrated circuit package. The lead wires of the integrated circuit are inserted in the helical springs instead of the usual tabs. External connector dimensions for a ten lead, 0.250 in. square IC package are 0.440 in. by 0.340 in. by 0.187 in.

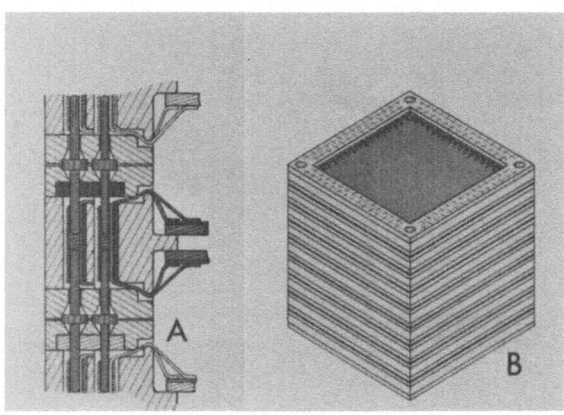

Fig. 5. Helical wiring system.

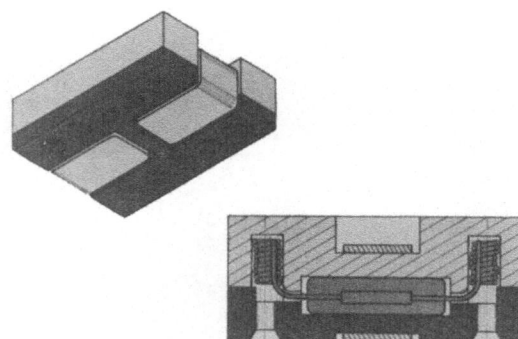

Fig. 6. Helical I.C. connector.

Component Lead Socket

The component lead socket (Fig. 7) is a one-piece stamping which is designed to accept component leads or transistor can leads. This contact permits the use of lead wires without special preparation as pins on modules. Individual conventional components can be plugged as desired.

The contacts are preassembled into a snug-fitting housing of any suitable plastic to prevent damage by overstressing with oversized pins or test probes. Four contact areas are accomplished by the four spring beams. The contact depends upon contact deflection and float of the mating pin to accommodate manufacturing tolerances on spacings. Where the distance between the extreme pin locations on mating pin headers is small enough to avoid large tolerance build-ups, the contact deflection suffices. For large pin header dimensions, the pin contacts are mounted in resilient polyurethane to provide the necessary float.

Contact center spacings of 0.050 in. are achieved with this design.

External connections are made to the integral tab extension by means of welding or soldering.

Figure 8 shows a diode bussing system using these sockets. The commoning bus bar between contacts is an integral part of the strip from which the contacts were stamped. The other leads of the diodes plug into individual contacts to which the external circuitry is connected.

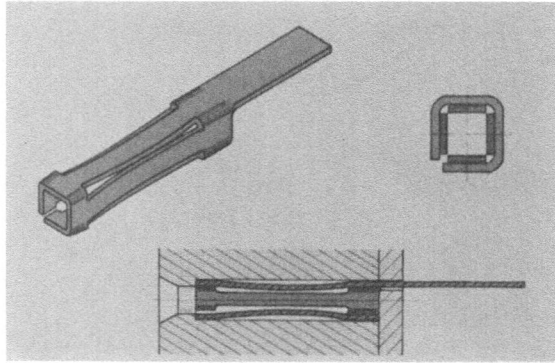

Fig. 7. Component lead socket.

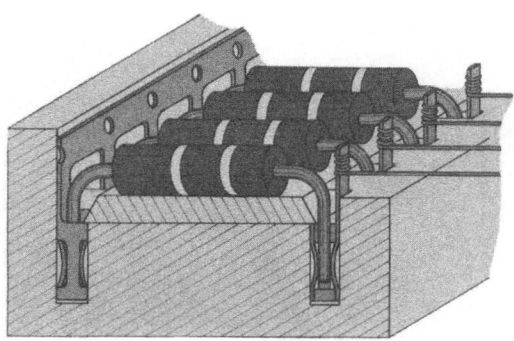

Fig. 8. Diode bussing strip.

Channel Contact Connector

The channel contact connector (Fig. 9) is a right-angled card-to-crimp-on, snap-in wiring system connector. The crimp-on contact is a one-piece stamping with a latching spring protected within the body of the contact for minimum vulnerability during handling and assembly. Contacts are crimped to wires in high-speed semiautomatic applicators. The wire size range is #28 to #32 AWG solid or stranded.

Mounted on the edge of the card are molded-in, right-angle contact elements with the same channel cross section. The spring elements are preassembled in a separate housing, and are exceptionally well protected against accidental damage. This spring block can be semi-permanently attached to either the card or harness side of the connector.

Contact center spacings of 0.050 in. are achieved with this design. Because of the mechanical latching devices, this design can be constructed of compression molded plastics which are necessary for high-temperature and high-altitude applications.

A wire-to-wire connector using the same contacts and springs is also available.

MECA* Connector

The receptacle contacts developed for the AMP-MECA* (Fig. 10) packaging system described by Evans and Swengel [7], Wasiele [8], and Shue [9] have been used in right-angle card-to-motherboard connectors. The contacts are one-piece stampings preassembled into a suitable plastic housing. The mating parts are blades which have been insert molded. The semielliptic spring beams of the receptacle are slotted, providing four areas of contact with the blade.

* Trademark of AMP Incorporated.

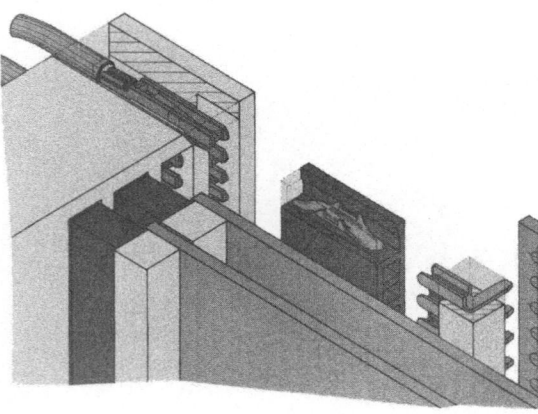

Fig. 9. Channel contact edge connector.

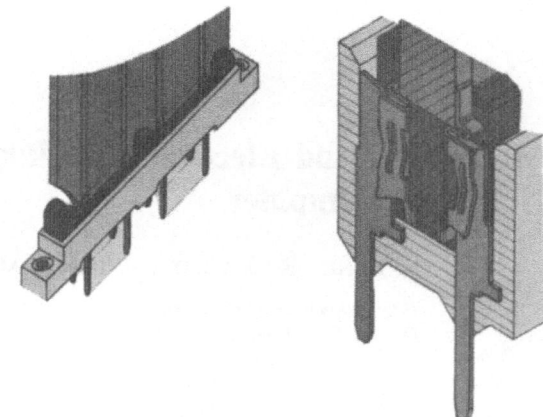

Fig. 10. MECA contact connector.

This connector is uniquely suited to double-sided card modules. The contact extensions are terminated on both surfaces of the card by means of welding or soldering to pads. The blades are soldered into a motherboard. Contacts are spaced on 0.050-in. centers.

Contacts misaligned due to normal manufacturing tolerances on spacings will align themselves through receptacle contact deflection and bending of the blade contacts. There is clearance between the blade and the molded housing to allow the necessary movement.

Construction of this connector offers protection to the receptacle contacts by virtue of their being recessed in appropriate slots. Similarly, the blades are protected by the molding which completely encloses them.

REFERENCES

1. M. S. Frant, "Plated Metals as Electrical Contact Materials," Ballistics Systems Division—Space Technology Laboratories Symposium on Connectors, Los Angeles, California, November 29–30, 1962.
2. R. H. Zimmerman, "Plating, Diffusion, and Oxide Films on Electrical Contacts," Invitational Symposium in Montreal, Ottawa, and Toronto, Aircraft-Marine Products of Canada, Ltd., October 1, 2, and 3, 1963.
3. J. H. Whitley, "Contacts and Dry Circuits," Invitational Symposium in Montreal, Ottawa, and Toronto, Aircraft-Marine Products of Canada, Ltd., October 1, 2, and 3, 1963.
4. L. D. Jaffe and J. B. Rittenhouse, "Behavior of Materials in Space Environments," *ARS J.*, March 1962.
5. R. W. King, N. J. Broadway, and S. Palinchock, "The Effect of Nuclear Radiation on Elastomeric and Plastic Components and Materials," Radiation Effects Information Center, Battelle Memorial Institute, Columbus, Ohio, September 1, 1961.
6. M. M. Fulk and K. S. Horr, "Sublimation of Materials, Problem in Electronic Packaging for Spacecraft," Fourth International Electronic Packaging Symposium, University of Colorado, Boulder, Colorado, August 14–16, 1963.
7. R. C. Swengel and W. R. Evans, "A New Mechanical Approach for the Construction of Modular Electronic Equipments," Aviation Conference of the American Society of Mechanical Engineers, Los Angeles, California, March 12–16, 1961.
8. H. Wasiele, Jr., "Maintainable Electronic Component Assemblies," Second International Electronic Circuit Packaging Symposium, University of Colorado, Boulder, Colorado, August 16–18, 1961.
9. J. I. Shue, "Interconnections, A Problem in Electronic Packaging," Invitational Symposium in Montreal, Ottawa, and Toronto, Aircraft-Marine Products of Canada, Ltd., October 1, 2, and 3, 1963.

Mechanical and Electronic Packaging for a Launch-Vehicle Guidance Computer

S. Bonis, R. Jackson, and B. Pagnani

International Business Machines Corporation, Space Guidance Center
Owego, New York

The packaging design of a launch-vehicle guidance computer is described. The computer uses microminiature technology, of which the basic module is a ceramic substrate with deposited resistors and uncased semiconductor devices. The memory section contains up to eight modules of 4096 words; the modules are self-contained by incorporating toroidal core arrays, memory, drivers, and sense amplifiers as a unit. To minimize structure weight, a magnesium–lithium material is used. This material has the inherent properties of high damping capacity with a high strength-to-weight ratio. Also described is the development program that led to these packaging concepts, along with the in-process testing used to ensure design adequacy prior to final assembly.

INTRODUCTION

THIS PAPER DESCRIBES the packaging design of a launch-vehicle guidance computer, currently in development at IBM's Space Guidance Center, for use in the United States space program. The work reported here grew out of research performed for NASA.

The computer is characterized by high computing speed, large expandable memory capacity, flexible input–output capability, and high reliability for the planned missions. In view of the importance of these missions, the need to protect the lives of astronauts involved, the expense of the vehicles, and the safety of launch equipment, reliability is of paramount importance.

For maximum reliability, triple modular redundancy techniques (TMR) are used in the central electronics. TMR subdivides the computer logic into modules; the modules are triplicated and a voter establishes an output based on a majority decision. One of the modules in each trio may fail (in some circumstances, more than one), yet the computer will continue to function.

The computer uses microminiature technology; the basic module is a ceramic substrate with deposited resistors and uncased semiconductor devices. The memory section contains up to eight modules of 4096 words; the modules are self-contained by incorporating toroidal core arrays, memory, drivers, and sense amplifiers as a unit.

Next to reliability, the second most important design factor is the weight of the computer. To minimize structure weight, a magnesium–lithium material is used. This material has the inherent properties of high damping capacity with a high strength-to-weight ratio. The weight, volume, and power characteristics of the computer described are: weight, 77 lb; swept volume, 2.2 ft^3; power dissipation, 137 W. These figures represent a TMR machine. For a simplex version of the same machine, the overall characteristics would be altered to the following: weight, 21 lb; swept volume, 0.60 ft^3; power dissipation, 37 W.

Included within this paper are descriptions of the mechanical design, the electronic packaging techniques, and the evaluation program pursued during the development phase of the computer.

MECHANICAL PACKAGING AND ENVIRONMENTAL CONTROL

Basic Considerations

Environmental requirements generally defined for aerospace computers include pressure, temperature, cooling media, vibration, shock, steady-state acceleration levels, humidity, and RFI noise levels. Where a vacuum pressure environment is specified, as would be the case for equipment located outside manned areas of the spacecraft, packaging materials are selected considering vacuum stability.

The thermal conditions of the structure and/or ambient temperature surrounding the computer, together with the cooling media provided in the spacecraft, will affect the overall thermal design of the computer package. For extended periods of operation, the major mode of heat transfer is conduction. Judicious use of all available conduction paths is necessary to minimize the overall weight of the computer.

The vibration, shock, and acceleration levels affect the computer structure design. The vibration and shock environments in particular have a major affect on the required component fragility levels, especially at the structure resonant frequency, causing maximum transmissibility to the component parts. These factors in turn affect the weight of the structure required to package the computer.

The humidity environment influences the requirements for overall package moisture sealing as well as the individual circuit module moisture protection. The RFI noise levels specified dictate the RFI filtering and gasketing requirements between the computer core and access cover.

General Configuration

Basic Structure. The structure of the launch-vehicle digital computer (LVDC) is rectangular, and has machined cellular cutouts, as shown in Fig. 1. The computer contains 73 basic electronic subassemblies and a memory section expandable to 8 memory modules. To obtain forced

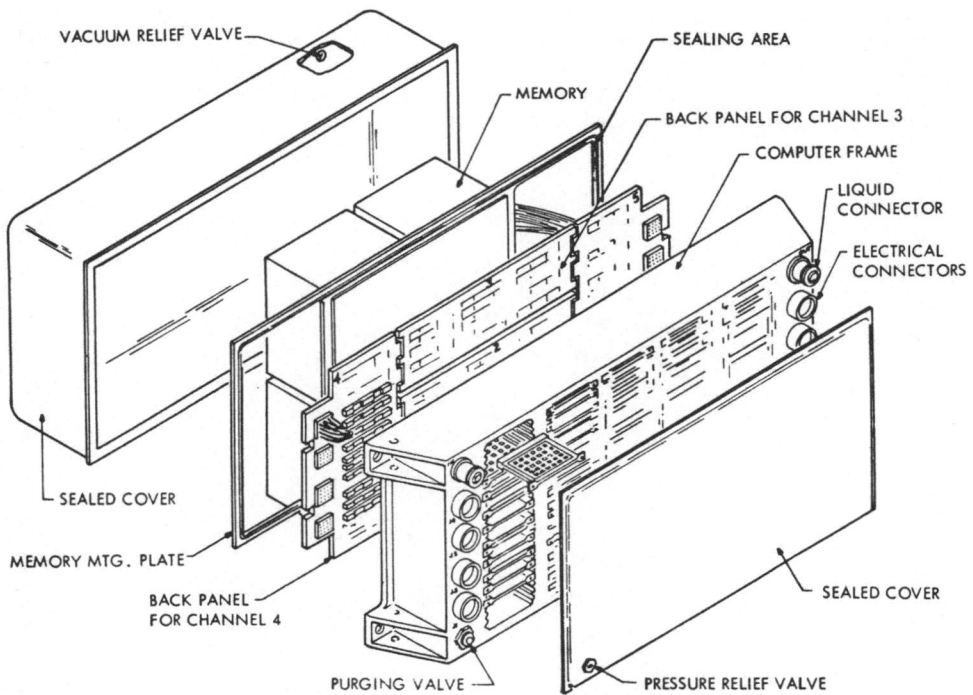

Fig. 1. Launch-vehicle digital computer (LVDC).

convection cooling, circular holes are machined within the frame structure, allowing the passage of vehicle coolant. This passageway allows serial cooling of the unit for all modes of bench and environmental testing, and during flight.

The structure material (magnesium–lithium) serves two purposes: it is (1) a heat-transfer medium (conduction); and (2) a structural housing with inherent damping.

Magnesium–lithium has been tested to verify its usage as a structural material, and in addition for usage in a vacuum environment. Most of this material's mechanical parameters were determined by Battelle Institute testing under contract to NASA, MFSC, Huntsville, Alabama. Additional tests have been conducted at IBM's SGC for evaluation as applied to specific space programs. Some of the advantages and disadvantages of magnesium–lithium alloys are as follows:

Advantages:
1. Lightweight: 25% less than conventional magnesium and 50% less than aluminum.
2. High damping capacity.
3. High stiffness-to-weight ratio (exceeds all others except beryllium).
4. Excellent weldability.
5. Excellent formability at ambient temperatures.
Disadvantages:
1. Low creep strength.
2. Relatively low thermal conductivity ($\frac{1}{3}$ of aluminum).
3. Requires corrosion protection under normal ambient and humidity conditions.

After considering the above, magnesium–lithium was chosen since its advantages outweigh its disadvantages, especially in the consideration of weight. Furthermore, none of the disadvantages proved to be detrimental for the environmental extremes dictated for this application. The structure weight (as shown in Fig. 1) is 15.5 lb.

Installation. Within the instrument-unit section of the parent vehicle, the LVDC structure is bolted to vertical channels on four corner mounting pads shown in Fig. 2. Eight main electrical connectors are mounted on the computer flange areas (four on each end), representing the electrical interface to an adjacent unit.

Also located on the unit flanged areas are two liquid disconnect receptacles for connection to inlet and exhaust loops provided in the spacecraft.

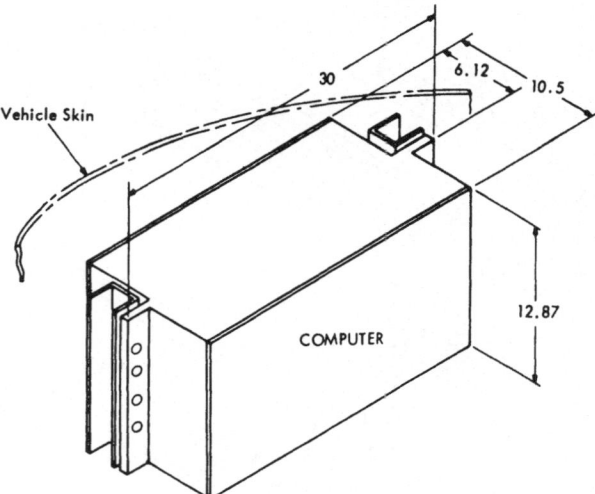

Fig. 2. Computer installation in the vehicle.

Covers. Two covers are fastened to the unit to provide an air seal and the necessary radio-frequency attenuation required for reliable operation. The covers are dished to encompass and clear the protruding memory modules and electronic connectors. The cover gaskets are designed to incorporate both a pressure and moisture seal. The flanged areas at the structure and cover interfaces are finished, and bolt space dimensioned to provide the necessary RFI attenuation.

Two one-way relief valves are installed on the unit case to ensure that a maximum pressure differential of 1.25 lb/in.^{2} is not exceeded between the unit interior and its immediate environment.

Thermal Control

Coolant Parameters. The vehicle provides methanol–water coolant at a flow rate of approximately 2 lb/min at an inlet temperature of 60°F. A maximum allowable pressure drop of 3.5 psi must be maintained from unit inlet to exhaust at the design flow rate. To optimize the coolant passage with respect to heat-transfer coefficient and pressure drop a number of development tests were conducted, resulting in the use of a six-hole configuration as shown in Fig. 3. With this configuration, a heat-transfer coefficient of 55 BTU/hr · ft^2 · °F is realized at a total unit pressure differential of 0.3 psi at the design flow rate. These values are consistent with the overall thermal design goals of the LVDC.

Component Parameters. The overall component design temperatures for the LVDC are defined as follows: maximum allowable semiconductor-junction temperature is 100°C; maximum allowable memory-array temperature is 70°C; maximum allowable memory-array temperature differential is 5°C.

Page Heat Transfer Analysis. Although a 100°C maximum semiconductor-junction temperature is allowable, a thermal design goal of 60–70°C was established to optimize the failure rate criteria from a reliability viewpoint. Since the computer is designed to function under vacuum conditions, in which conduction is the major mode of heat transfer, all analyses were performed considering this mode of operation. With reference to Fig. 4, micromodule

Fig. 3. LVDC mechanical mock-up—vibration setup.

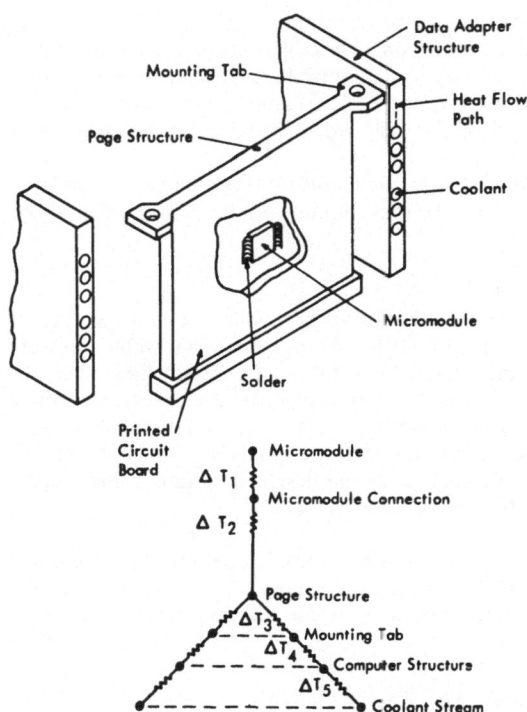

Fig. 4. General heat transfer process for pages.

page temperature differentials that require evaluation are: ΔT_1, micromodule to micromodule solder connection; ΔT_2, solder connection to page structure; ΔT_3, along page structure; ΔT_4, page to structure interface; ΔT_5, structure to fluid path.

The summation of these differentials for a worst case page dissipation is 38°C. Superimposed on this total is the K factor (junction-to-case) of the semiconductor devices.

This resistance has a value of approximately 0.1°C/mW and therefore for a dissipation of 60 mW, an additional temperature differential of 6°C is included. The maximum semiconductor-junction temperature then becomes 59°C.

Vibration Control

As stated previously, one of the major reasons for choosing magnesium–lithium as the frame material was to take advantage of its inherent damping capability. Throughout the development phase of this program, experienced personnel monitored the vibration control of the computer design to ensure compatibility with the environmental specifications. Vibration tests were conducted at various subassembly levels, at increased g levels, to simulate the expected transmissibility from computer mount to subassembly location.

A computer mock-up vibration test was conducted at 1, 3, 4, and 5 g rms to obtain some preliminary response characteristics. Accelerometers were mounted to the structure and various subassembly mock-ups as shown in Fig. 5. The general response characteristics of the accelerometers is shown in Fig. 6.

ELECTRONIC PACKAGING

Hybrid Circuit Modules

The module consists of a ceramic substrate upon which conductor and resistor patterns are applied. The conductor pattern is then tinned with solder, active devices and capacitors are mounted on the substrate, and the assembly is encapsulated. The completed module is then mounted to a circuit board.

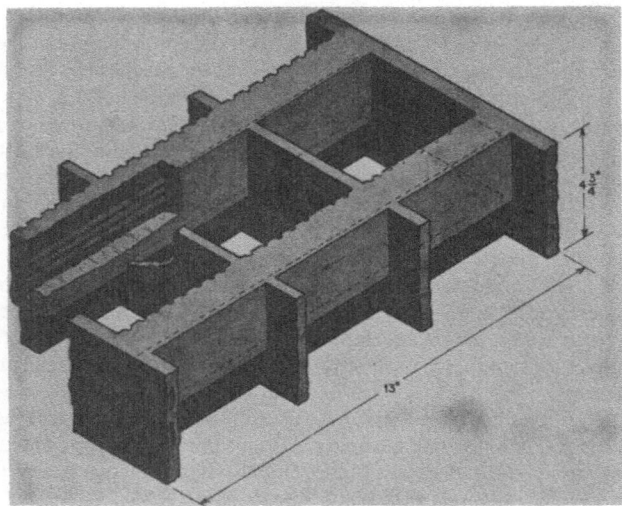

Fig. 5. Section showing cooling paths for LVDC.

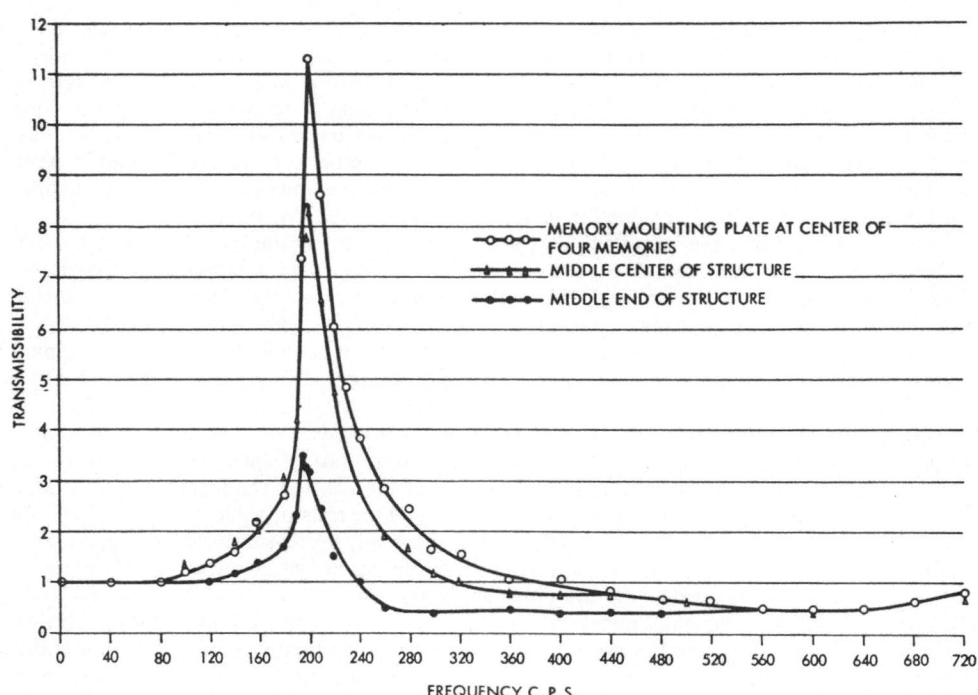

Fig. 6. Computer sinusoidal vibration perpendicular to circumference of missile direction.

The external size of this micromodule is 0.300 in. × 0.300 in. × 0.080 in., with 14 exposed conductor lands for making external connections to an interconnection board. When mounted on these boards in a computer, the average component density is approximately 50 components per cu in. including all related hardware, the mechanical structure, and the external connectors.

The substrate material is aluminum oxide (96%) with trace impurities of magnesium oxide, calcium oxide, and silicon dioxide. The substrate is a flat plate 0.300 in. × 0.300 in. × 0.025 in., with a 0.004 in. radius on all corners and edges. It is produced by pressing a single cavity tungsten–carbide mold in the green state, followed by firing at around 1800°C for 48 hr.

The conductor pattern applied to the substrate is a platinum–gold alloy containing approximately 8% of bismuth trioxide as a bonding agent. It is obtained as a viscous paste of the powdered metals suspended in an organic binder.

The conductor pattern is applied to the substrate by means of four separate screen printing operations. In each of these operations a hard rubber squeegee is drawn across a wire-mesh screen which has been blanked out where no conductor pattern is to appear. The squeegee forces the conductor paste through the openings of the screen and onto the substrate immediately below.

A distinct advantage of this conductor-forming technique is its adaptability to circuit changes. Completely changing a circuit pattern requires only the time to obtain a photographic negative of the desired pattern and the half-hour necessary to apply and expose an emulsion on the bottom side of a blank screen.

After the conductor screen printing and firing operations, but before solder tinning of the conductor pattern, resistors are applied to the bottom side of the substrate by a screen printing process identical with that for conductors. In fact, the same holding and screen printing fixtures are used with a different screen pattern and paste composition. The paste used for resistors is approximately 24% palladium, 52% lead oxide, and 24% silver, all of which are suspended as metallic particles in an organic binder.

Resistance values are dependent upon resistor length, width, height, and on the type of resistor paste. The thickness of all resistors is controlled by the screen printing operation at a constant value. The unit of measurement for resistor pastes is therefore ohms per square. As an example, for a paste of 10 kilohms per square, a resistor one square long (length of resistor = width of resistor) has a value of 10 kilohms, and a resistor three squares long (length = 3 times width) of a 500 ohms per square resistive paste has a value of 1500 ohms. For demonstrating resistor values, the physical size of the respective squares for a given paste is of no significance. However, the total area of the resistor determines the total power dissipation of the resistor. A typical value for this is 10 W/in.^{2} (or 10^{-5} W/mil^2).

After the resistors have been fired, they are coated with an organic overcoat. This coating protects the resistors from contamination and cleaning solvents used in subsequent operations. The coating further serves as an electrical insulator and concomitantly as a solder stop during the dip-soldering operation to tin the conductor pattern.

At this point, the solder tinning of the conductor pattern actually takes place. If tinning were done prior to this operation, the heat of the resistor firing cycle would cause the tin-bearing solder to dissolve the platinum–gold conductor pattern completely.

After the conductor patterns have been tinned with solder, the resistors are trimmed, using a miniature sand-blasting nozzle which erodes away part of each resistor through the resistor overcoat (Fig. 7). By cutting into the width of the resistor, the length-to-width ratio, i.e., the number of squares, of the resistor is increased, and the resistor value increases. The total area of the resistor available for power dissipation, however, decreases during this operation; therefore, this must be taken into consideration in the initial design layout of the substrate.

Active components used in the modules are deposited lead devices produced by IBM. The devices contain three small metal balls suitably attached to the silicon material. During the mounting operation, the devices are positioned on a conductor pattern, and the entire substrate is placed in an oven to melt the solder and form three electrical connections to each of the devices simultaneously. The time required is usually a few minutes. A neutral rosin flux is brushed on the substrate prior to placing the semiconductor on the conductor pattern, and is ultrasonically cleaned off immediately following the mounting operation. During the mounting operation, the surface tension of the molten solder shifts the semiconductor slightly, if necessary, to position it exactly over the center of the conductor mounting pattern.

This form of device is available in various types of NPN transistors and common-anode double diodes. It is a planar, silicon device with a protective glass coating applied as a final processing step during manufacture.

Deposited lead devices possess several distinct advantages over wire lead devices which compensate for the lack of PNP transistors. They are more adaptable to high production processes, since wire lead devices require three separate soldering operations for mounting. Deposited lead devices have higher vibration resistance after mounting. The handling of substrates containing these devices prior to encapsulation is also distinctly simpler than the handling of substrates containing wire lead devices.

After all of the electrical elements on a particular substrate have been formed or mounted, the unit is ready for encapsulation into a module. There are several approaches to accomplishing this operation, the choice depending upon the external environments to which the completed module will be subjected.

The hybrid circuit module was selected for use in the computer because of the following reasons:

The component density obtained with this packaging technique is far greater than that obtainable by the use of standard components.

The circuit design of a module is easily changed for small production quantities. This is reflected in the fact that there are 46 different types of modules in the computer, some used only a few times.

The inherent reliability of both the deposited lead device mounting technique and the infrared solder reflow technique, used to attach completed modules to a printed circuit board, is superior to other interconnection techniques. The vibration resistance is also superior.

The infrared solder reflow technique also requires access to only one surface of a printed circuit board, thus allowing the board to be permanently bonded to a structural member of the equipment for heat transfer and vibration resistance purposes.

The similarity of this module to the solid logic technology used in the new IBM 360 computer system permitted the great amount of reliability and production experience gained in a commercial venture to be directly applied to this project.

The form of a partially completed module is such that quality is easily scrutinized directly in the manufacturing area, resulting in a more uniform and reliable product.

The advantages of this module over the use of single-chip integrated circuits lie in greater circuit flexibility, greater element value ranges accompanied by lower tolerances and TCR's, greater choice of active devices, and increased power dissipation capabilities.

The geometry of this module permitted the application of a 500-hr burn-in period for all production modules, to ensure reliability prior to the final assembly of the computer.

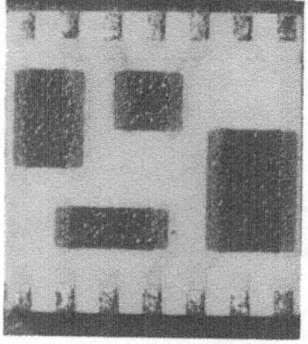 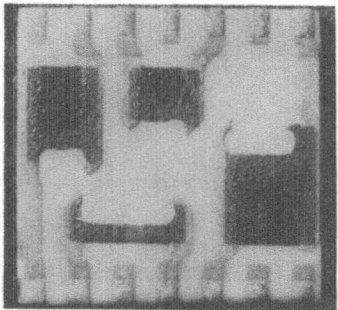

Fig. 7. Resistor before (left) and after trimming (right).

Memory

The computer memory storage requirement is dictated by the functions to be performed by the computer. A magnetic core memory is used. The configuration of the memory package must consider the overall computer package size and shape and make optimum usage of the space available in the computer package.

Once the general memory package configuration is established, packaging details, including support of the magnetic cores and their sense, inhibit, and address wiring, are designed to withstand the mechanical environments to which the computer will be exposed without adversely affecting the cores' response or function. The cores are mounted on their sense, inhibit, and address wires in planes. In addition to the cores and their interconnections, the plane consists of a plastic frame and electrical contacts for the termination of the core wiring. The planes are stacked together and joined by bolts with metal end caps at each end of the stack for structural integrity of the memory package or array. The core terminations on the planes are interconnected, using flat cable jumper strips soldered to the plane terminal pins. An array consists of 14 planes with each plane containing 8192 cores arranged in a 64 × 128 pattern. A continuous thin coating of a silicone material spread over the cores and wires within each plane provides environmental protection with minimum weight.

Interconnections

The general arrangement of the LVDC interconnection system is shown in Fig. 8. Connections between the microminiature circuit modules are made through printed circuit boards. The circuit modules and printed circuit boards are combined in a pluggable assembly called a page. The page (shown in Fig. 9) includes two printed circuit boards bonded to both faces of a magnesium–lithium alloy frame. A 98-pin connector is fastened to the lower edge of the frame, and wire leads from the connector are soldered to terminations in the edges of both boards. Feed-through connections between the boards are provided by insulated pins fastened in holes in the frame. Pin ends are soldered in terminations on both boards. Test point terminals are provided along the upper edges of the boards.

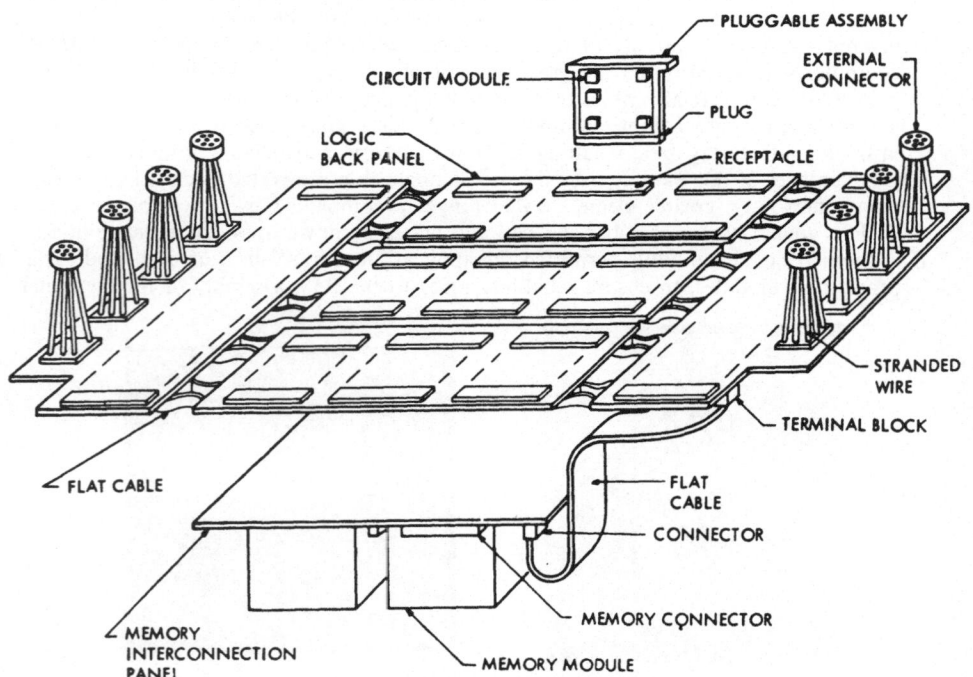

Fig. 8. LVDC interconnection system.

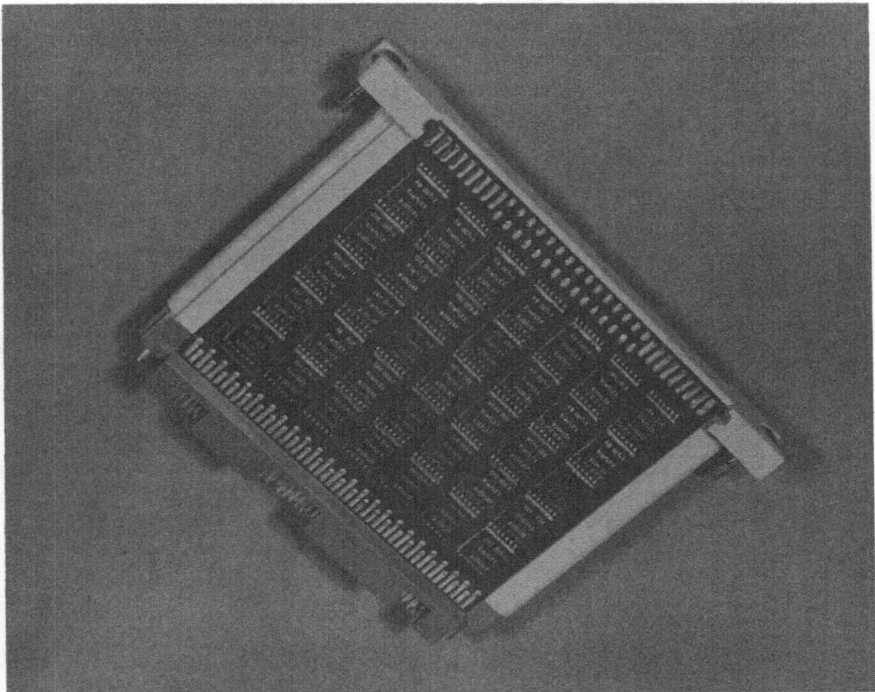

Fig. 9. Pluggable circuit assembly.

Mechanical support for the page is provided on the lower end by the connector shell and guide pins, and on the upper end by two flanges which are fastened to the structure by captive screws. The interface between flange and computer structure forms the main heat conduction path from the page.

Electrical and mechanical attachment of modules to circuit boards is by solder reflow connections. Both board and module attachment surfaces are precoated with solder, and the solder is reflowed by infrared heating. After page assembly and test, a coating is applied for protection in a high-humidity environment. Modules may be removed and replaced using the same infrared attachment device.

The 98-pin connector is shown in Fig. 10. The plug contact is a knife blade which engages a double-leaf spring contact in the receptacle. Solder terminals extend from both sides of the plug for attachment to the page circuit boards. Solder terminals extend from the bottom of the receptacle for attachment to the computer back panels.

Pluggable Assembly Design Criteria. To define the general configuration and size of the basic electronic subassembly, such factors as the following were considered: number of electrical connections to a particular logic group size; type of circuit module; type of interconnections between modules; thermal control; maintenance philosophy; reliable performance in specified environments; ease of manufacture; ease of electrical test.

For the LVDC a pluggable electronic subassembly (page) was chosen because:

The maintenance philosophy for the launch site established the page as the smallest replaceable unit. This could be accomplished best by incorporating a pluggable connector in the page.

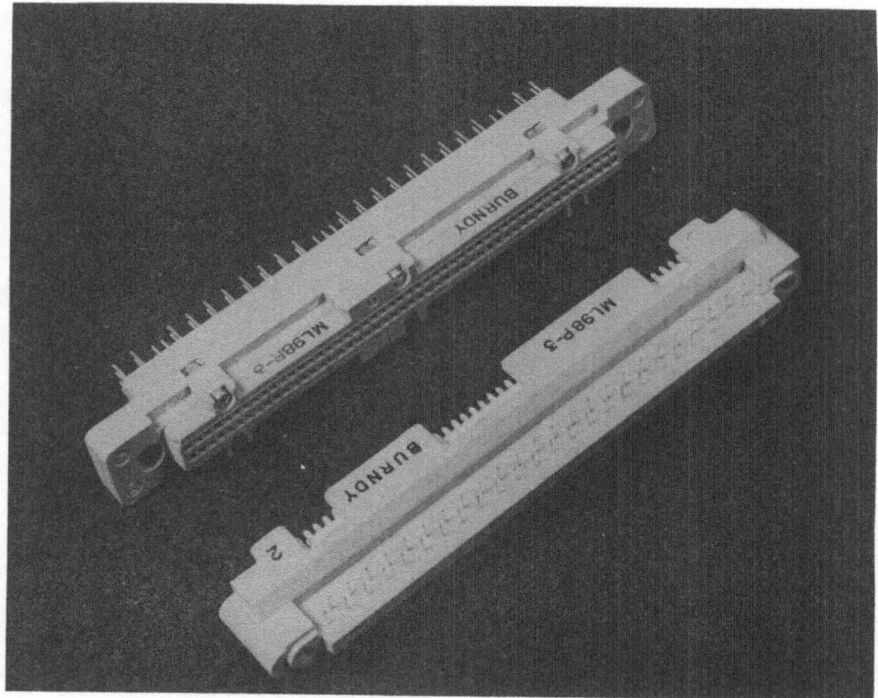

Fig. 10. Page connector.

Page testing during manufacture or after repair would be facilitated by using a pluggable connector.

Circuit modules could be replaced more easily with improved access to the pages when they are removed from the computer.

In order to establish optimum page size a survey was made of typical logic in several existing computers to determine the number of external connections required for various sizes of logic groups. Survey results are given in Table I.

Seventy modules per page was selected as the best compromise because any smaller number would have resulted in a sharp increase in the total number of pluggable connections in the computer while a larger number would complicate automatic test equipment and would add to the problem of developing a practical pluggable connector.

Theoretical analysis indicated that the 70-module page would have a natural vibration frequency well above any resonance of the main structure, and would have thermal characteristics adequate to maintain component temperatures within specified limits.

TABLE I

Circuit modules per page	External page connections per module	External page connections	Logic pages per computer	Total pluggable connections in logic
30	2.46	74	141	10,400
50	1.60	80	81	6500
70	1.40	98	57	5600
100	1.38	138	39	5400

Backpanels. Interconnections between pages of the logic section are made through backpanels. Each backpanel consists of a circuit board, receptacles for pages, terminal blocks for external connections, and a metal plate for mechanical support. The circuit board is bonded to the plate, and the leads of the receptacles are soldered to terminations on the circuit board. The LVDC contains five logic backpanel assemblies. Three of these are identical, each being one channel of the TMR logic section. The other two assemblies contain voter, oscillator buffer, and decoupling capacitor pages.

A sixth backpanel assembly acts as a distribution board for connecting the memory modules to the logic section.

Flat Cables. Connections between backpanels and from logic to memory distribution board are made through etched flat cables. The cables are attached to terminal blocks. The blocks have double ended terminals which accept cable connections on the upper end and are soldered into the backpanel on the lower end. The flat cables are etched from a two-sided copper–teflon laminate. In areas not subject to repeated flexing, signal wiring is etched in one copper layer. A ground plane to provide electrical isolation is etched in the other copper layer. Teflon insulating layers are then applied to both exterior surfaces. In areas where repeated flexing is required, a single copper layer is used. Circuit isolation is provided by alternating ground lines with signal lines in the single copper layer. The transition from two copper layers to a single layer is accomplished through "brazed" eyelets. Figure 11 shows typical cables for interconnecting backpanels. Figure 12 shows a cable assembly for connecting the logic to the memory distribution panel. One end is terminated in a pluggable connector to facilitate assembly and maintenance. Clamps are installed at all flat cable terminations to support the cable and provide strain relief for the soldered connections.

Flat cables were chosen to interconnect backpanels instead of a conventional wiring harness primarily because of the reduced weight and volume. Other advantages of the flat cables include uniform wiring characteristics in successive computers, elimination of wiring errors, and ease of assembly.

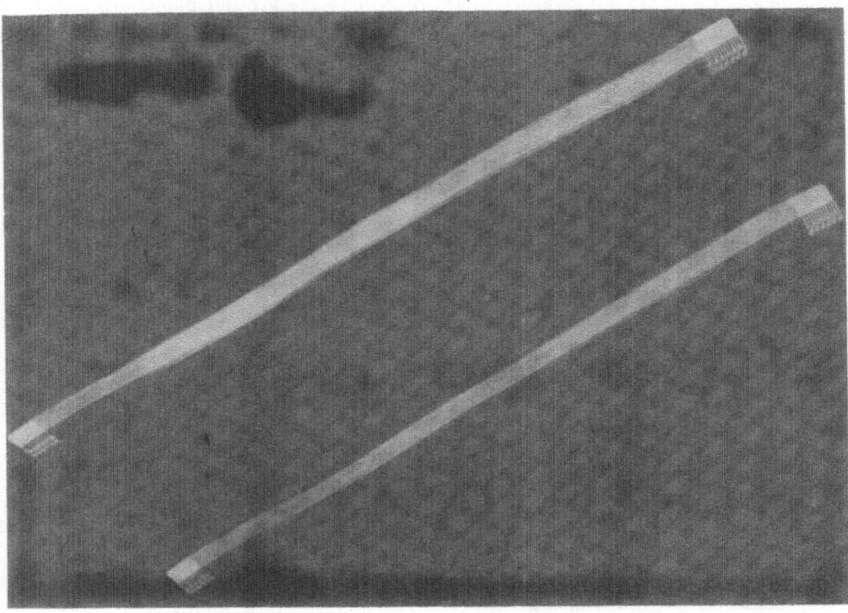

Fig. 11. Flat cables for logic backpanels.

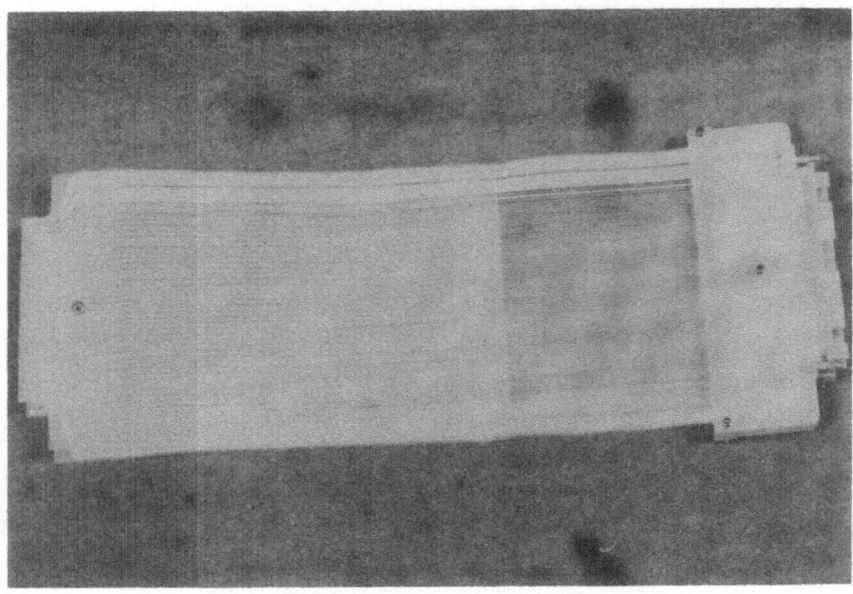

Fig. 12. Flat cables—logic-to-memory connections.

In order to achieve good electrical performance in flat cables where the available space was limited, two wiring configurations evolved. The first configuration provided optimum electrical performance in a cable which was not subjected to a significant flexing after installation. A cross section is shown in Fig. 13.

The signal conductors are staggered with respect to the ground plane conductors in each cable in order to maintain low capacitive noise pick-up between adjacent signal conductors, while minimizing total capacity between signal wiring and ground. This configuration also provides adequate noise rejection for signal conductors in adjacent cables. A cable of this type cannot be flexed repeatedly because the copper conductors are far enough from the neutral axis of the cable to be stressed beyond their yield strength even when the cable is bent over a fairly large radius.

The flat cables between logic and the memory-distribution panel are subject to flexing during installation and maintenance when the connectors are engaged or disengaged. The cable configuration in areas requiring flexibility is shown in Fig. 14.

The single copper layer is located close to the neutral axis, permitting the cable to bend without overstressing the copper. Signal isolation within the cable is provided by alternating ground and signal conductors. A lightweight foam-rubber pad separates adjacent cables to provide electrical isolation.

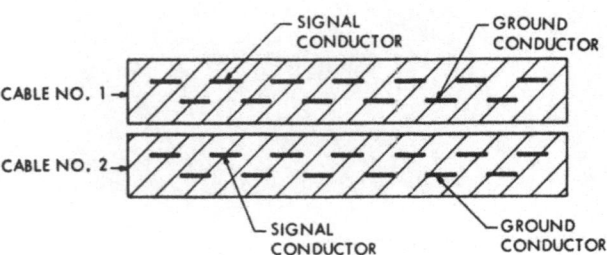

Fig. 13. Flat cable cross section.

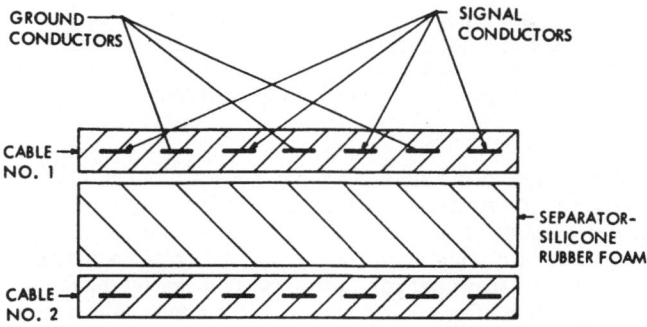

Fig. 14. Flexible flat cable cross section.

Input–Output Connections. Input–output computer connections are made through conventional round connectors fastened to the structure. Crimped standard wire leads from pokehome contacts in the connectors are soldered directly to terminations in two of the computer backpanel circuit boards. The leads are short, and they are separated sufficiently to make shielding unnecessary. Guide blocks fastened to the backpanels support the wires and provide strain relief for the soldered connections.

DEVELOPMENT ANALYSIS AND TESTING

Analysis

Throughout the development phase of the LVDC program, mechanical analysis was performed on a continuous basis to verify packaging compatibility with respect to the environmental requirements. These analyses have been conducted in four basic areas: thermal, vibration, weight, and materials.

Thermal Analysis. Thermal design analyses are conducted on a subassembly and overall unit basis to ensure that operating component temperatures are consistent with the reliability and circuit design requirements. In addition, these analyses are updated, as the computer development progresses, so that all circuit and dissipation changes are incorporated as they occur. This has been facilitated through the use of an IBM 7090 data-processing system heat-transfer program which utilizes an iteration technique to complete the desired temperature profiles. To use this program, a mathematical model (model representation) of the hardware is constructed and thermal paths are simulated as they appear in the actual design. Factors which affect the thermal design, such as conductivity, heat dissipation, contact resistance, and physical dimension, are assigned to those models as required. The 7090 computer then performs the necessary iterations until a predetermined stabilization criterion has been achieved.

With an established set of desirable operating component temperatures, a reliable thermal design can be effected with a minimum weight penalty.

Vibration Analysis. As in the thermal design, the design of each computer package subassembly and unit structure has been analyzed as thoroughly as possible to determine its response characteristics to the expected vibration environment. For a complex structure, such as the computer frame, a 7090 computer program has been used to determine natural frequencies and expected stress levels.

Weight Analysis. The design of each subassembly is also checked to assure that the weight of the part is consistent with that required for thermal and structural design integrity of the package. The purpose of this analysis is to give maximum assurance that the design weight requirement of the computer is not exceeded, and to provide the spacecraft contractor with accurate weight estimates of the computer on a continuing basis. As a part of the weight analysis, computer center-of-gravity and moment-of-inertia data may be determined and updated as required.

Materials Analysis. The materials selected for construction of various computer parts are checked by materials personnel to assure their applicability to the intended function. This analysis includes checking all nonmetallic materials for behavior in vacuum environment, encapsulants and conformal coating materials for their effect on the component parts which they will encase, nonmetallic structural parts for integrity under mechanical environments and handling, and the preparation of process specifications for nonmetallic materials.

In addition, metallic materials are reviewed: for proper application of surface treatments to provide protection of the base material in the operating environments and during handling; to assure that dissimilar materials are not connected; to assure that proper materials and processes are provided for mechanical connections, and to assure that proper materials, from an application mechanical-wear standpoint, are selected for electrical connections. The review and analysis of materials were conducted in parallel with the packaging-design activity so that a design was not committed to hardware without a thorough analysis of the materials of construction.

Testing

In addition to the design analysis, representative parts are tested as the design progresses, to give maximum assurance that the design will meet the computer environmental specifications. This includes fabrication and environmental testing of the package subassemblies as their designs are completed, and eventually, fabrication and environmental testing of the complete computer package.

Considering all aspects of the computer packaging design and the environmental specifications, this testing has been sequenced to provide meaningful test results in time to enable necessary design changes to be incorporated in the computer packaging. This plan carefully considers each development test to be conducted, schedules dates for tests, provides procedures for conducting each test, including instrumentation required to provide the desired information, and provides for the fabrication of the development test hardware consistent with a test-plan schedule.

As a minimum, mock-up parts reflecting a typical design configuration of each major subassembly part comprising the total computer package should be fabricated for development testing.

Some of the basic development tests performed are as follows: heat-transfer and pressure-drop tests on basic flow passage configurations; a mock-up subassembly and a memory-module heat-transfer test; vibration tests on subassemblies and memory modules; a frame vibration test.

A complete computer package will be fabricated and assembled using operational and/or dummy parts as required. Subassembly parts already fabricated (as covered above) and tested may be used in the computer package mock-up. To achieve meaningful results from development tests, particularly in vibration and acceleration environments, structural details of the mock-ups of the circuit boards, memory module, and overall package reflect the final design configuration in every detail.

At this point, with the use of development test data, instrumentation is minimized since some correlation data have already been achieved. Fixtures, including hardware used to mount test samples to vibration-, shock-, and acceleration-producing equipment and external heat sinks used in thermal tests will be thoroughly checked-out to assure that the desired environments are accurately transmitted and/or represented by this equipment.

Following completion of the full-scale development test and analysis of the data, any indicated deficiences in the package design will be corrected by incorporating the required design modifications. The necessity for a carefully planned and executed development test program is obvious with respect to providing the package design prior to computer qualification and prior to committing the package hardware to production where changes are extremely costly from both a dollar and schedule standpoint.

ACKNOWLEDGMENTS

The authors wish to acknowledge contributions made by Messrs. K. Harris, J. Kracke, M. Panaro, R. Tomek, F. Price, J. Corcoran, and D. Vullemier.

Standard Package for Microcircuits

N. Shapiro, P. Smith, and M. Genser

General Precision Aerospace
Little Falls, New Jersey

Two attempts to package thin-film circuits are discussed. A major difficulty remaining is the inability to produce active and passive components in a single discipline. In one of the studies described the active elements were introduced in the form of packaged components, in the other they were specially processed to facilitate incorporation in the thin-film circuit.

INTRODUCTION

ATTEMPTS TO DEVELOP some form of standard package for thin-film circuits have been frustrated because of the lack of control of the form factor of the active components available for such circuitry. It does not appear that in the near future it will be possible to produce the active components and the passive components in a single processing discipline as is the practice with silicon monolithic blocks. Tunneling amplifiers do not at this time appear to be sufficiently reduced to practice to warrant serious consideration. Thin-film field effect amplifiers have been built; however, these units do not exhibit the stability of characteristics or the uniformity of production for application in the near future.

We are left with the problem of incorporating some form of semiconductor device as an active element in thin-film microcircuits. The following discussion centers around two attempts to package a thin-film circuit.

In the first study, the active elements are used in the form of packaged components (TO-51, flat packs, etc.). In the second study, the active elements have been specially processed to facilitate incorporation in the thin-film circuit so as to provide the advantages of the monolithic characteristics exhibited by the silicon integrated circuit.

It should be noted that it is quite likely that the silicon monolithic circuit will be predominant for digital applications, while the thin-film circuit will be used primarily in linear applications. There are many and well-known technical and economic reasons for this division of applications which we need not pursue here. However, some degree of packaging standardization is beginning to emerge for silicon circuits for digital applications in the form of the familiar flat packs. These flat packs are assembled by soldering or welding techniques to multilayer boards. One of the techniques to be described permits the fabrication of linear circuits in thin-film format in a package consistent with this type of assembly.

THIN-FILM CIRCUIT MODULE

To best illustrate the package developed in the first study of thin-film microcircuit packaging, we have selected a 5-W servo amplifier. The specifications and photographs of this unit are shown in Table I and Fig. 1, respectively. This unit demonstrates the versatility of thin-film techniques when applied to functions which deliver appreciable power to electromechanical output devices. This amplifier operates Class B at 400 cycles, and was designed to drive a

Kearfott 3.5-W servomotor. The entire amplifier is contained in a copper capsule about 0.3 cu in. in volume. The finished unit weighs about $\frac{1}{2}$ oz.

The amplifier consists of a preamplifier stage driving a power-output stage. The preamplifier consists of a vacuum-deposited thin-film resistor–conductor network. The resistive films are nichrome, and the power-output stage consists of two silicon planar power transistors.

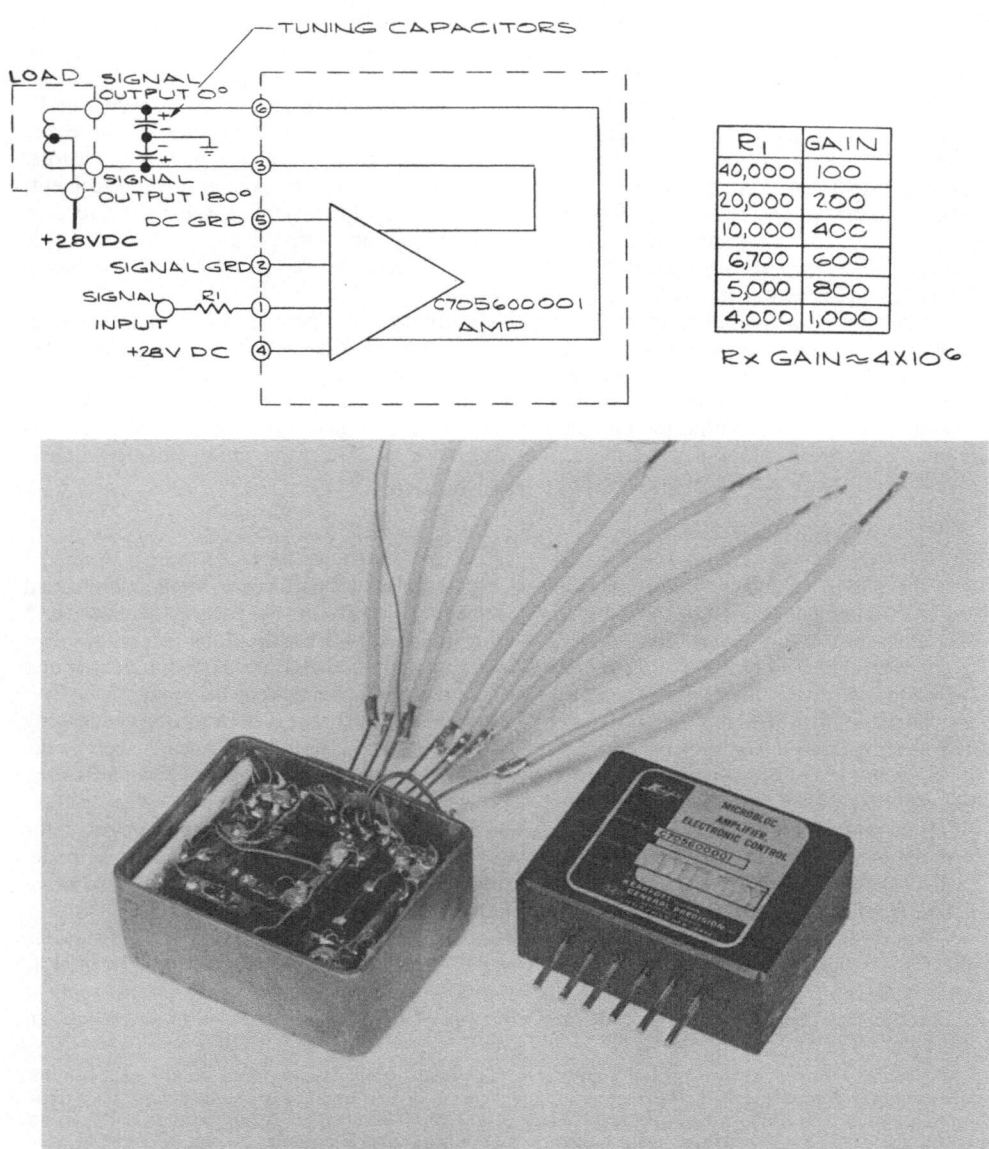

R_1	GAIN
40,000	100
20,000	200
10,000	400
6,700	600
5,000	800
4,000	1,000

$R \times GAIN \approx 4 \times 10^6$

Fig. 1. Microbloc 5-W servo amplifier. Top: block diagram. Bottom: the amplifier before and after final closure.

The power stage, i.e., the silicon transistors, are soldered directly to beryllia chips which are brazed directly to the bottom of the copper can. When the capsule is mounted on a heat sink, the copper case forms an easy path for removal of heat from the unit.

Since the amplifier operates at 400 cycles, the largest discrete components in the unit are four solid tantalum capacitors. These units, suitably insulated, are suspended in a viscous resin in the capsule. The preamplifier substrate, in this case alumina, is also suspended in the capsule in the viscous resin. The viscous resin completely encloses the preamplifier substrate, but does not extend up to the top cover of the copper capsule, leaving an air space for expansion of the resin during thermal cycles. Finally, a top cover plate of copper is soldered in place to hermetically seal the unit.

The amplifier illustrates a method of construction in which an entire function can be packaged in a single universal package. The unit is repairable since the cover can be easily removed, the resin cut away, and the preamplifier tilted up to get at the power transistors. Alternately, the preamplifier stage may be completely removed for repair or replacement. Where necessary, discrete components of appreciable size, when essential, can be included in the package. The copper case lends itself to heat sinking, and some degree of hermeticity is obtained when the lid is soldered shut and a glass-to-metal header is used to communicate electrically with the outside world.

A primary consideration in packaging this amplifier was the thermal gradients set up across the substrate and those from substrate to case. All substrate layouts are measured under load to determine the temperature distribution before they are approved. In addition, approximations are made to establish the most likely heat-flow pattern from substrate to heat sink. The maximum temperature within the module is limited to 150°C due to the use of a potting compound. Data taken show a substrate temperature rise of 27°C when mounted on a quasi-infinite heat sink, and 78°C if no heat sink is used.

SYSTEM PACKAGING

The circuit capsule described above can be employed to package a more complicated system. Figure 2 contains a photograph of a power supply for driving the rotor of a gyroscope.* This unit consists of a crystal-controlled oscillator, integrated circuit flip-flops in a countdown arrangement, and a power-output stage. The individual modules are bolted together and interconnected by an interconnect matrix. The interconnections are welded.

This unit illustrates the versatility of the circuit capsule in that it includes in one packaging scheme such discrete components as a crystal oscillator, silicon monolithic circuits such as the flip-flops in the countdown circuits, and a power stage constructed in a manner similar to the 5-W servo amplifier described above.

It should be noted that the copper capsules again serve as particularly efficient paths for heat removal. For heat sinking, the power stage can be mounted directly to a cold plate. Each capsule moving upward from the heat sink then dissipates less heat and can be more distant from the cold plate without incurring undue thermal stresses.

Figure 3 illustrates maximum substrate temperatures experienced when a 5-up module stacking arrangement is employed. In both cases 17 W are being dissipated from each module stack as shown. The stacks were heat sunk on a 20°C cold plate. Both stacks were identical in all respects except that case No. 1 utilized no copper can but hard expoxied modules with an

---

* The new three-phase gyro-wheel supply provides square-wave voltage drive to the spin motors rather than the conventional sine-wave drive. Equivalent motor performance is obtained, with the exception of an additional 2°C winding temperature rise. The square-wave supply is primarily three transistor switches driven by a 400 cps three-phase input. A 1.2288 Mc signal is provided by use of a crystal in a Pierce oscillator. This signal is divided to 2400 cps by a countdown circuit using a series of integrated circuitry flip-flops. The 2400 cps signal is used as the clock pulse to a ring counter whose output is a three-phase 400 cps square wave, each phase positioned 120° with respect to the other two. Three power switches are driven by the ring counter output; each switch provides a 32-V peak-to-peak square wave. These square waves are applied to the respective phases of the motors. A 32-V peak-to-peak square wave applied to the three phases of the motor provides a fundamental voltage of 26 V rms line-to-line, since the motor is Y-connected.

TABLE I

Characteristics of the Five-Watt Servo Amplifier*

	Electrical Characteristics
Amplifier input impedance:	Approximately equal to gain setting resistor R1 (see Fig. 1).
Signal frequency:	400 ± 20 cps
Maximum signal input voltage:	30 V rms
Voltage input:	28 V dc $\pm$ 10%
Power input:	1.12 W (zero signal input)
	11 W (maximum output)
Voltage output:	40 V nominal in R119-5 motor
Power output:	5.0 W into R110-5 motor
	3.5 W into R119-5 motor
Gain:	1000 maximum (adjustable down by external resistor)
Gain stability:	± 3 db over temperature range
Typical loads:	Kearfott motors R110-5, R119-5
	Mechanical and Environmental Characteristics
Volume:	Approximately 0.25 in.^{3}
Operating temperature: (temperature of mounting base)	-55 to $+71°C$ (at 5 W maximum output)
Mounting:	Unit must be in intimate contact with surface on which it is mounted.
Interconnection:	Interconnecting pins located on 0.100-in. grid to be compatible with standard spacing used on strip wiring and printed circuit connectors. Connections to these pins may be made by soldering or welding.

*The C70 5600 001 amplifier will operate over the temperature range of -55 to $+71°C$ while delivering 5 W of power to a servo load, or -55 to $+100°C$ while delivering 3.5 W of power to a servo load.

aluminum heat-tree arrangement. Case No. 2 shows modules employing copper-case enclosures stacked directly one upon the other. For the given dimensions and assumptions the maximum substrate surface temperatures were 44.0°C and 31.6°C, respectively, as shown.

The unit is repairable, in that individual modules can be readily removed from the stack for rework. Alternatively, depending on the degree of modularization, the module can be thrown away if the module cost is low enough.

FLIP-CHIP† THIN-FILM CIRCUITS

We discussed in the previous paragraphs the design of a packaging scheme to allow the uniform containment in a single enclosure of a wide variety of components, either integrated or discrete, added to a thin-film substrate. An alternative approach to the fabrication of

† General Precision, Inc., Trademark.

Fig. 2. Gyro wheel supply.

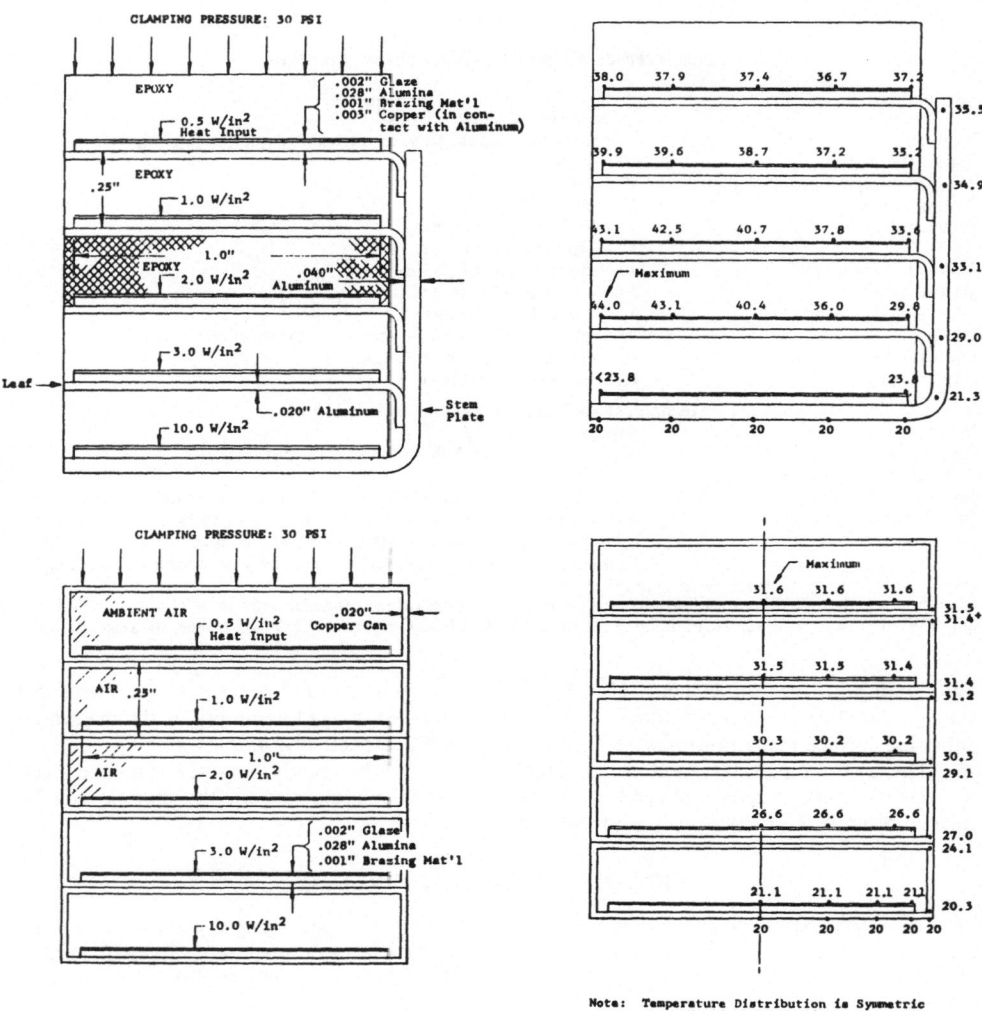

Fig. 3. Maximum substrate temperatures. Top: Case 1; left: heat sink at 20°C; right: resulting steady-state temperatures, °C. Bottom: Case 2; left: heat sink at 20°C; right: resulting steady-state temperatures, °C.

microcircuits in thin-film format is the Flip-Chip approach. This technique, as we will show below, is a method of interfacing silicon circuitry and thin-film circuitry. The Flip-Chip technique solves the problem of building thin-film functional blocks with active elements in many different packages. In this technique, such devices as transistors and diodes are all made in one form factor to facilitate assembly. Secondly, as will be seen, the silicon chips are attached to the substrate and interconnected in a single operation. Finally, no wires or hand soldering operations are involved.

The Flip-Chip technique is a process for incorporating silicon planar devices onto a thin-film circuit. An illustration of the process is shown in Fig. 4. The silicon device is a silicon chip in which contact to the active regions is made by metal film conductors which connect to lands deposited on the periphery of the chip. The device is provided with an inorganic or glassy coating over the active region to provide mechanical protection against abrasion. The

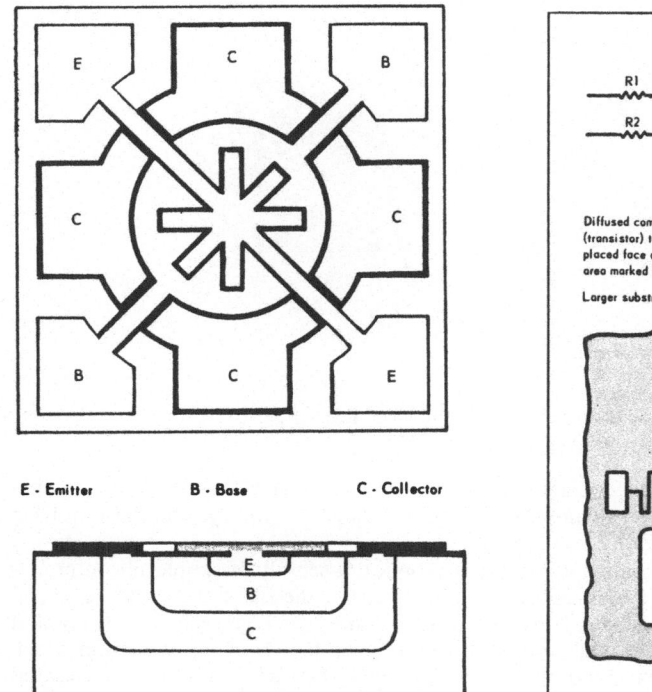
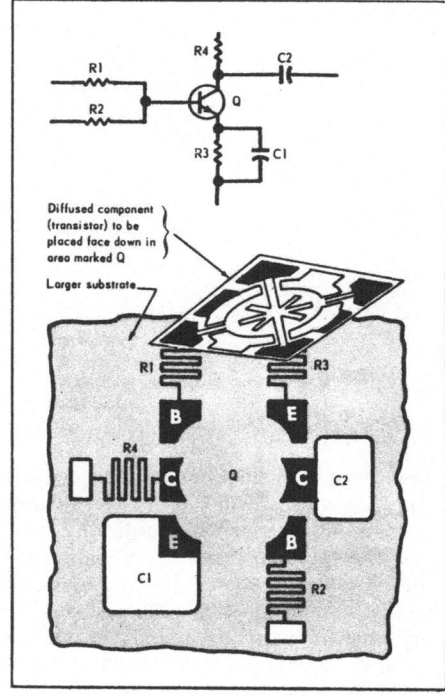

E - Emitter   B - Base   C - Collector

Fig. 4. The Flip-Chip process. Left: example of diffused discrete components. Right: inverted registration.*

lands on the chip register with lands on the substrate. The chip in inverted and suitable alloy preforms are employed to fuse the lands on the chip to lands on the substrate. The name Flip-Chip then refers to the process of inverting the chip to join it to the substrate.

An example of the application of this technique to circuit fabrication is illustrated in Fig. 5. This unit is an AC pre-preamplifier. It contains three Flip-Chip transistors, three Flip-Chip diodes, and a zener diode. These units are all of uniform physical size and are fused to the resistor–conductor substrate. It is apparent that several possible techniques for circuit assembly might be developed. In one process, the chips are loaded onto the substrate in a boat. When they are suitably weighted, the chips are fused to the substrate in a furnace. This process is similar to the methods by which germanium alloy transistor dies are prepared. Alternatively, the chips may be attached to the substrate with a self-heated vacuum pencil. This technique has the advantage that the entire substrate need not be heated.

In addition to the practical advantages of the uniformity of components, Flip-Chip microcircuits approach the monolithic structure of the all-silicon circuit. For example, experimental units have been subjected to accelerations in excess of 5000 g's without removal of any of the chips.

The Flip-Chip process promises to result in low cost microcircuits for another reason. It offers the opportunity of component selection, whereas in the silicon monolithic circuit, circuits might be rejected if there is a defective component resulting in a lowered yield. It is possible to show that microcircuits can be assembled by the Flip-Chip process at considerably

*Diffused discrete components may be silicon integrated circuits, transistors, or diodes. The diffused substrate is passivated with oxide film for isolation. Aluminum interconnections deposited over the oxide film make contact to the silicon through holes etched in the film. Inverted registration of the terminal lands of diffused components with lands of the circuit substrate is shown. The starred pattern can be placed in different positions, allowing best layout of the thin-film components deposited on the substrate.

Fig. 5. Flip-Chip pre-preamplifier.

lower cost than silicon monolithic circuits in quantities up to several thousand. For a complicated circuit for which the silicon monolithic process is limited by poor yields, the Flip-Chip process will be cheaper regardless of volume.

Finally, the process offers a method for the economical assembly of simple silicon monolithic blocks. For example, in the assembly of a shift register, the flip–flops could be silicon monolithic blocks in Flip-Chip format fused to a ceramic substrate containing an evaporated interconnect matrix. An amplifier might be assembled utilizing the Flip-Chip process in which the first stage might be a silicon monolithic block containing two transistors in a differential connection and a third transistor as a current source. The second stage could again be a p-n-p differential pair. The final stage could be still a third chip capable of dissipating considerable power. All of the bias resistors, and frequency-shaping networks would be deposited on the substrate by conventional deposition processes.

In conclusion, we see that the Flip-Chip process offers a low-cost method for microcircuit assembly which can approach the reliability of the monolithic circuit. This process can be extended to the assembly of integrated monolithic blocks.

Semiconductor Circuits and Modular Packaging

ALLAN V. PAINTER AND REGINALD A. ALLEN

The Bendix Corporation
Industrial Controls Division
Detroit, Michigan

A method of packaging integrated circuits on plug-in modules suitable for commercial, industrial, or military applications is described in this paper. Some of the problems that are common to any method of designing plug-in modules for packaging integrated circuits are discussed, and the specific solutions that were developed are described in detail. A cost breakdown of the complete assembled module is given, and the methods for adapting the design for thin-film circuits are also described.

INTRODUCTION

FOR THE PAST DECADE or so, the majority of complex ground and shipborne electronic equipment has been constructed from discrete components, mounted on plug-in modules, and assembled in racks or cabinets. Plug-in modules became popular due to the cost savings and ease of maintainability when compared to unitized construction. The cost saving begins with a reduction in engineering time. Because of the versatility of plug-in modules, they can be used for many functions in several systems that may be in design or production. Cost saving continues into the production area, as the pluggable capability of the modules enables the equipment to be built as a parallel instead of a series program, normal with unitized construction. A parallel program of assembly reduces the assembly space and inventory cost by reducing the assembly time. Down-time of equipment, normally paid for by the manufacturer for leased equipment, is reduced, since a good module can be quickly plugged in place of a faulty one, and the system put back on the air.

When semiconductor integrated circuits first became available in production quantities, they were used in systems for aerospace computers such as A.C. Spark Plugs' "Magic" [1], North Americans' "Monica," and Librascopes' "L-90," where their minute size could be taken advantage of, regardless of cost. In these aerospace systems, microminiaturization was of prime importance, and the equipment was therefore of unitized construction. In the next design phase we saw attempts at modularization, beginning with the Univac "1824" [2] and the Martin-Marietta "Martac 420." These two systems both used modules connected permanently in the equipment. The third phase that brings us up to the state-of-the-art is the packaging of integrated circuits on semipermanent modules, exemplified by the Sippican Corporation [3] technique that uses split wire-wrap posts and wire wrapping to make the connections to the backpanel wiring. Although these semipermanent modules can be removed with care and do save on design and development costs, they are by no means easily removable and the time involved in removing and replacing a module is excessive, compared to a plug-in module.

The initial cost of integrated circuits was very high, and the original techniques for packing them appeared low in comparison; costs of the circuits have now fallen rather steeply and can now be considered for industrial and commercial equipments. We can expect the costs of the circuits to continue to fall. To prevent the proportional packaging and interconnecting costs from rising, all techniques for reducing costs must be investigated.

To take full advantage of the cost saving and reliability improvement aspects of integrated circuits, the packaging philosophy should be based upon the use of the plug-in modules, thus incorporating the advantages of a new technique with an established concept of proven worth. The marriage of these two techniques is fraught with problems but the result of this forced coupling of mature and new ideas is very fruitful.

DESIGN PHILOSOPHY

A packaging system for electronic equipment can be separated into distinct areas, e.g., outer covering (cabinets, cases, racks, etc.), unit assembly (chassis, drawer, book, etc.), the module (printed-circuit card, cordwood pack, etc.), and the module interconnecting method (motherboard, back-panel wiring, etc.). The best packaging systems are a full integration of all these areas since they are all interdependent.

The basic building block of the packaging system is the plug-in module, and it is the area that required the greatest ingenuity. The plug-in module should be the starting place for all design. In this paper we will deal with a module design and its immediate related area, the module interconnecting system (motherboard).

The approach that we took in the design of the plug-in module was to start with the objective that the design should be easy to manufacture and service and that it be versatile for use in many applications. Using these objectives as criteria, the Scamp (Semiconductor Circuits and Modular Packaging) concepts were developed at Bendix and are the basis for the plug-in module described in this paper.

From the above objectives, the following ground rules were established to act as idea parameters:

a. The packaging system will be for ground use; therefore, miniaturization is of secondary importance and will be merely that which is inherent in economically designing with integrated circuits.
b. The total system will be oriented toward the use of automatic manufacturing methods that are usually associated with high-quantity production.
c. The system will be maintainable at the integrated circuit level, i.e., it will be possible to replace the integrated circuit in the field with normal technician hand tools.
d. The circuit will have a standard module configuration that will package 95% of the digital logic circuits and 75% of interface circuits.
e. Costs must be less than those for packaging standard circuits.
f. Maintenance facilities must be included to reduce test, and checkout, and field-service time to a minimum.

In early 1962, Bendix realized that when the integrated circuit program had been in effect for some time, the T.O. 5, T.O. 47 can type of circuit would disappear and the flat packs would emerge as the sole method of mounting semiconductor integrated circuits; therefore, our plug-in module would only have to be capable of mounting the flat packs, and the can types could be ignored. The problems and solution offered are therefore restricted to the flat-pack circuits.

MODULE SIZE

If a plug-in module is considered for optimum size from the human engineering, manufacturing, storage, and maintenance points of view, then the size of the module should be small enough to be handled by one person, yet not so small that it is easily lost. In practice, any module larger than a book of matches and smaller than a telephone book could satisfy this requirement.

The determination of optimum size of a plug-in module involves the optimum electrical size in reference to circuit versatility, reliability, maintainability, and ultimately overall costs.

Increasing the number of integrated circuits in a plug-in module improves reliability, since the number of connections will be reduced. As the number of interconnections between integrated circuits on the plug-in module increases, however, the versatility of the plug-in module in the system is reduced.

For example, suppose a system was to require 1170 integrated circuit packages; these can be mounted on a 13 × 6 matrix on 15 plug-in modules, the method used on Minuteman D37B computer. In this case, the 15 modules are all different because of the complexity of the interconnections between circuit packages. This feature, while undesirable, may be acceptable in a missile application because of the need for maximum density. Furthermore, for installations such as Minutemen where there may be ten or more missiles on a site, storage of one replacement for each of the 15 modules represents only 10% or fewer maintained spares.

A single item of equipment such as a launch computer, however, presents a different situation. The same packaging concept and the same requirement for one spare module of each type would result in 100% spares, equivalent of a spare computer. For a system comprising 1170 integrated circuits, as an example, a requirement for 10% spares can be conveniently satisfied if the total number of integrated circuit packages to be mounted on one spare of every type of plug-in module is approximately 117. Then, the average number of integrated circuits mounted on each plug-in module is 117 divided by the number of module types. This approach can be used to determine an average number of semiconductor packages per plug-in module when the complexity of the end product is established and an estimate of the total number of integrated circuits can be made. This condition does not normally exist since both the mechanical packaging of the circuits and the logic design of the system usually proceed together.

To provide an optimum size module to be used as a standard in many systems of different sizes is a more difficult problem. We approached this problem by considering two production computers of different sizes, the Bendix G15 and G20.

The systems were broken down into logic blocks that were repeatable throughout the computers. During this study it became apparent that the same basic logic had been performed in different ways and that by modifying some logic designs and by standardizing on logic methods, standard counters, shift-registers, adders, etc., could be designed. Further study showed that these standard logic blocks could be split at places in the system where the number of connections were minimum.

The result of this investigation showed that the majority of the repeatable logic blocks in a computer employ between 20 and 30 integrated circuit packages, each package containing a dual, three-input NAND gate, a flip-flop, or an equivalent amount of circuitry. Each of these standard logic blocks requires at least 40 input or output pins plus power supply connection pins. A module containing 30 integrated circuits would have to be at least 3 in. wide and 2 in. long. The smallest preferred module size specified by MIL-STD-736 is 3 × 3 in. We therefore decided that these dimensions would form the general size parameter for the Scamp module. The space required by the circuit packages is approximately 2 × 3 in., leaving an area of 1 × 3 in. for the connectors and test points.

CONNECTOR SPACING

The standard spacings for printed circuit connectors until 1959 were 0.100 in. and 0.250 in. At about that time the high-density packaging achieved by Litton Systems, IBM, and others caused connector manufacturers to reduce the spacing to 0.078 in. This distance between connectors will still provide only 37 connections across the 3-in.-wide module. These closely spaced contacts must be very narrow to allow insulation space between them; thus the contact area is small, and the contact becomes unreliable and likely to disconnect under vibration. One attempt to remedy this problem is to turn the contact area through an angle of 90° and stagger each contact as is done in the AMP-blade. The Elco or Cinch Varicon connector, which is one example of connector staggering but using a patented contact method, have contacts spaced at 0.100 in. These attempts at solving the contact problem have improved the reliability of the connections but the typical allowance of only 29 connections causes plug-in modules incorporating cordwood modules to be pin-limited.

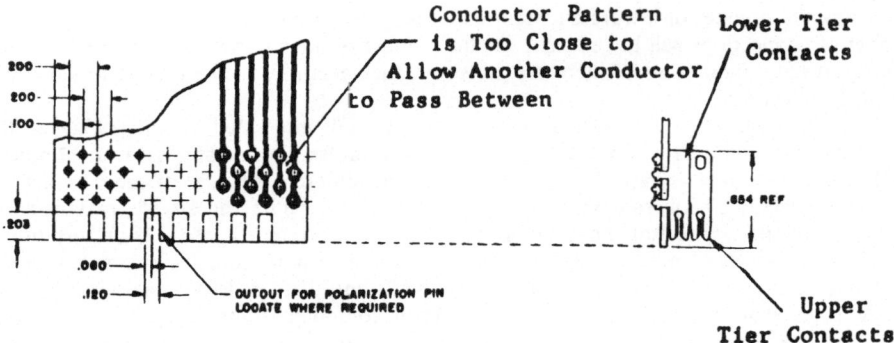

Fig. 1. Pattern dispersion limits the rows of contacts to two tiers.

The AMP-blade, Malco-Wire and Elco Varion connectors all provide two rows of contacts and rely upon passing through the module card and soldering to the pattern as shown in Fig. 1. As can be seen from this illustration, there is no room between the pads for any more conductors.

In the logic design of modules the designer usually finds himself pin-limited. We wished to obtain as many pins as possible within the space allotted. This space was dictated by the 3-in width of the module and, therefore, design investigation for the connector section centered around pin spacing and number of pins. The Automatic Manufacturing Methods ground rule made us give due consideration to the module interconnecting method. Methods considered will be delineated under the "Motherboard" section of this paper; however, those considerations made us settle on the automatic wire wrap.

After visiting Gardner-Denver Company and discussing their automatic wire-wrap techniques with them, it was concluded that the minimum spacing for our purposes is 0.150 in. using a 0.045 in. square wire-wrap pin and 26 gauge wire. To make diagonal wiring more convenient, it was decided to stagger alternate pin rows, i.e., the second row of pins would have to be situated one-half between the first and third rows. Using one of the two row connectors mentioned above would provide 37 connectors. This would certainly give pin-limitations and so we resolved to use three rows of connectors and overcome the pattern dispersion limitation. This we did by extending the lower row of contacts through the molding to the rear of the module, as shown in Fig. 2.

This technique in effect stretches part of a three-row connector so that the module can be considered an odd-shaped connector.

With this pin layout shown in Fig. 3 our module size allowed us a total of 56 pins; however, as one of the pin designs had a possible extraction force of 8 oz per pin, a jacking screw was necessary. The center four pins were discarded to mount the necessary jacking hardware leaving a net of 52 pins. After rechecking our selected standard logic blocks, it was found that 52 pins would suit our purposes admirably.

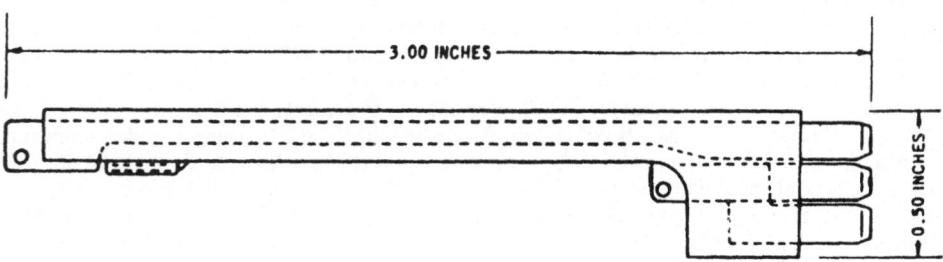

Fig. 2. Pattern dispersion limitation can be avoided by extending third contact tier to the rear of the module.

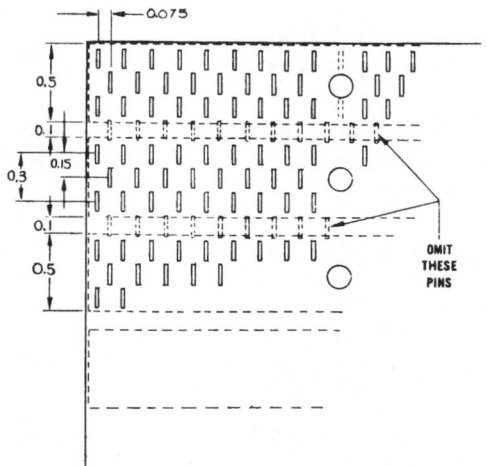

	1	2	3	4	5	6	7	8	9	10
I	LO 2	E2 3	L1 2	E4 2	L2 2	E6 2	L3 2	G8 2	L4 2	H9 2 / C9 1
H	H2 1 / A1 1 / D4 1	H1 2 / E1 1 / I2 1	F3 1	G3 2 / I4 2 / J3 1	E6 1 / G6 3 / E10 1 / D8 1	F5 2 / I6 2 / I5 1	G8 1 / E8 3 / DIO 1	F7 2 / I8 2 / I7 1	I40 1 / CIO 3	I10 2
G	H1 1	H2 2	H4 1 / H3 1 / LI 3	H4 2	H5 1	H6 2	H7 1	H7 2 / E7 1 / F8 1	H9 1	F9 2 / H10 2 / I9 1
F	F2 1 / G1 1 / LO 3	F1 2 / G2 2 / I1 1	E4 1 / G4 3 / F10 1	D4 3	H6 1 / G5 1 / L2 3	D5 2 / C5 2	H8 1 / G7 1 / L3 3	D8 3 / C5 3	G10 1 / G9 1 / L4 3	C3 1 / C4 1
E	F1 1	F2 2	G3 1 / F3 2 / F4 1	E3 1 / F4 1	F5 1	H5 2 / E5 1 / F6 1	F7 1	H8 2	D3 1	C6 1
D	D2 2	D3 2	E5 2 / G6 2	C3 2 / C4 2	C6 2	C7 2	E7 2 / E8 2	B5 2	D7 1	B5 1
C	E3 2 / G4 2	G9 2 / H10 1	D1 1 / D5 1 / E9 2	C5 1	B5 3	D6 1 / D9 2	D7 2	B8 1	F9 1	G10 2
B	A2 2	G5 2 / I6 1 / B7 2 / I8 1	G1 1 / G2 1 / H3 2 / I4 1	B3 1 / B2 1 / C2 1	B6 1 / C8 2	B7 2	B8 2	C9 2 / CIO 2	E7 3 / E8 1	C9 3 / CIO 1
A	B1 1 / A3 2	C1 1	C1 2	A5 1	A6 3 / B4 1 / B4 2 / B4 5	A7 1 / A7 2 / A7 3	A8 / A10 / A9 / B10	XM48	E1 2 / E2 2 / E3 3	G4 1 / E5 3 / G6 1

Fig. 3. The half-inch modular height of the modules gives repeatability of the three rows of connector pins on the module, and permits continuation of the pin matrix on the motherboard.

USE OF TABLE :

THE ROW AND COLUMN INDICES OF A PARTICULAR SQUARE IDENTIFY THE WAFER WHOSE OUTPUT PIN (6) IS CONNECTED TO THE INPUT PINS LISTED IN THE SQUARE

EXAMPLE :

| H6 1 |
| G5 1 |
| L2 3 |

THIS IS READ : "PIN F5-6 IS CONNECTED TO PINS H6-1, G5-1, + L2-3".

As indicated in Fig. 2, the upper row of pins are stepped down so that they protrude through the rear of the connector section level with the center row. These protrusions are used to make the connection between the circuitry and the pin during deposition of the circuitry. The lower row of pins are extended through the integrated circuit mounting section for the same purpose. A key slot of $\frac{1}{4}$-in. diameter was added in one side of the module so that the module could not be reversed in the motherboard. This would not prevent an operator from plugging in modules in the wrong slot; however, as it was intended, it would have the power supply bus connected to the same pin in every module, which meant that if a module was plugged into the wrong slot, no damage could be done.

TEST POINTS

Military requirements for maintainability necessitate the provision of test points that are accessible when the plug-in module is functioning in its normal position in the equipment. Single or multiple test receptacles are available for plug-in modules containing welded or cordwood modules. RCA, Ess Gee, and others also extend the printed circuit conductors to the rear of the plug-in module as test position. None of these methods are acceptable for the narrow conductors and close spacing involved in microelectronic packaging; the use of eyelets or some other device may be feasible. Probably the shape and location of test points may best be determined as development proceeds, provided it is a design criterion from the start.

Our solution to this problem in the case of the Scamp module was to utilize the extended lower row of pins that pass through the center of the molding which forms the integrated circuit mounting section. They are brought up flush with the integrated circuit mounting plane and are also extended 0.250 in. beyond the rear of the molding, as shown in Fig. 2. This fulfills two purposes:

a. By exposing these pins at the rear of the module flush with the mounting plane, it enables us to connect the circuitry to the lower row of pins, obviating the necessity of crossover and wasting valuable surface area for long surface circuit paths, and

b. By extending the pins beyond the rear of the molding we can use these test points, i.e., by carefully selecting the points in a logic block that requires testing and laying out the circuitry so that these are on the bottom row of pins, we have a very convenient set of 18 test points.

Some success has been met in designing a wander plug to mate with these test points and displaying a whole word or logic function system on a switchable register, resulting in considerable savings of built-in maintenance and test and checkout circuitry.

CIRCUIT MOUNTING AREA

This parameter of the packaging design does not have too many problems unless high-density packaging is being sought. However, care must be taken that the surface is flat and not liable to bending or warpage during assembly or use. For the Scamp module the integrated circuit mounting section is an extension of the lower side of the connector. It is 0.250 in. thick in order to withstand the pressure required for applying the circuit. It has a circuit mounting area of $7\frac{1}{2}$ in.^{2} (dictated by the requirements of our standard logic blocks), and will mount 30 integrated circuits—$\frac{1}{4}$ square inch allotted to each circuit, as shown in Fig. 4.

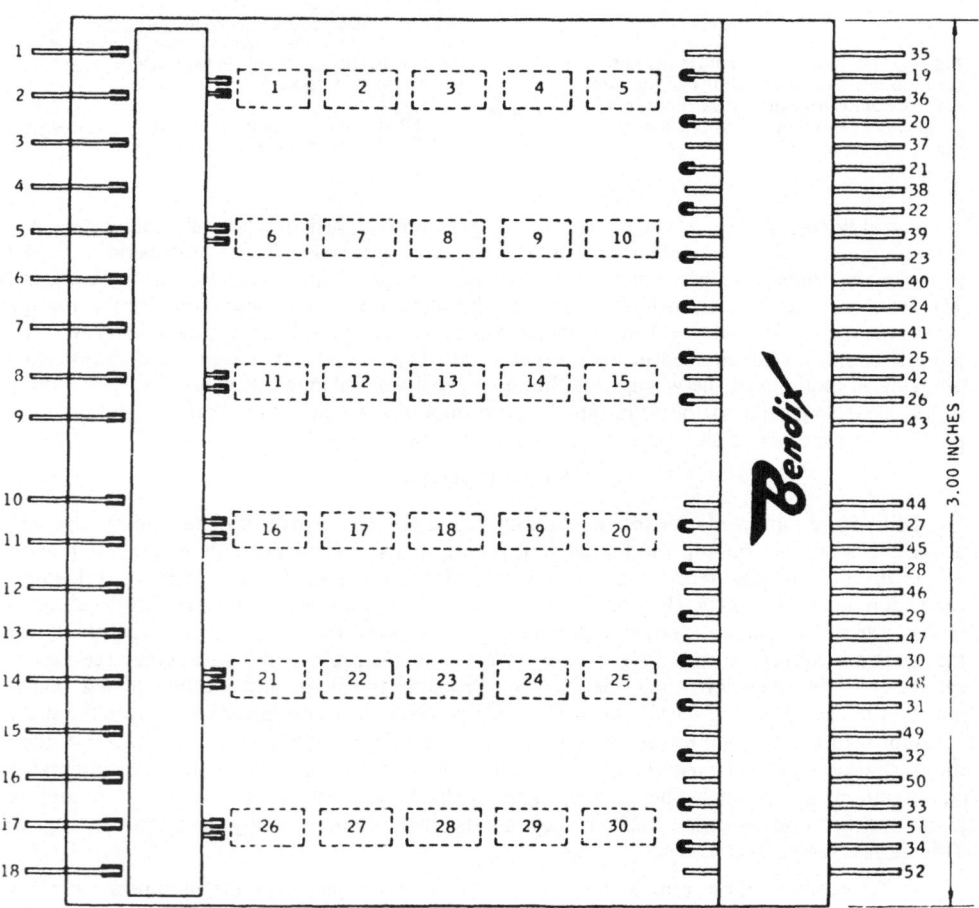

Fig. 4. Master layout of Scamp module.

INTERCONNECTION CONDUCTOR PATTERNS

The copper-clad laminates for packaging of cordwood modules typically use 0.0014-in.-thick copper, $\frac{1}{16}$-in.-wide conductors, with $\frac{1}{16}$ in. minimum spacing between conductors.

When microminiature conductor patterns are laid out, 0.01-in.-wide conductors are found necessary to get sufficient conductors in the small areas. The copper should have at least 0.0028 in. nominal thickness to prevent ohmic losses with such narrow conductors. A characteristic of the photo-etch process is that the etchants cut under the mask a distance of approximately $\frac{1}{2}$ the stock thickness at the surface, tapering to the nominal dimension at the base. This must be allowed for in the design. If the minimum conductor width is maintained at 0.01 in., the spacing between the conductors will be reduced by approximately 0.003 in., thus reducing the number of conductors permitted across any area.

Another problem that occurs with the narrow conductors required for microminiature packaging is that the reduced adhesive area makes it easier for the conductors to lift off the card under vibration or stress.

With these problems in mind we evaluated several techniques in order to ascertain the optimum method for use on the Scamp module. The methods evaluated included:

Vacuum Deposition—Method 1

A copper film was deposited on the phenolic through a stainless steel mask that was the negative of the required circuit pattern. Advantages of this method are its integrity of reproduction, the ease of obtaining very fine circuit paths, and the fact that all connections to connector pins can be made in a single vacuum operation. Disadvantages are: high tooling cost involving the initial cost of the vacuum system, and then, to realize any product rate at all, the stainless steel screens must be duplicated many times and a complicated lazy-susan-type mask transfer mechanism must be designed to operate within the vacuum; the limited thickness of the conductor, involving some fancy circuit load studies; and the copper circuit rubs off quite easily, and damage during mounting of the integrated circuit occurs too frequently to be healthy. (The circuit is fixed by an epoxy spray after mounting the integrated circuits.)

Vacuum Deposition—Method 2

The whole surface area of the phenolic is plated by vacuum methods, and then photo-etched.

Advantage of this method is minor mask cost—all pin connections are made during vacuum process.

Disadvantages are that it is very difficult to obtain consistent thick films, and the film requires protection during handling until fixing after mounting the integrated circuits.

An alternate method is to put a thin film on and electroplate to the required thickness, but it is quite difficult to obtain an even thickness of electroplate.

Copper Shim

A 0.002-in. shim of copper is etched by photo-resist methods and then adhered to the phenolic using a hot cured two-phase epoxy. The circuitry is then gold-plated except for the blanks that held the circuitry together in the shim form. The blanks are then etched off. At this point, the module can be passed through a solder bath to make all the connections to the connector pins.

Advantage of this system is the low tooling cost. Disadvantages are high handling and processing costs, and poor bond strengths due to reverse-side undercutting.

Electroplating

A thin film of copper is put onto the phenolic by chemical methods and then the unit is electroplated to 0.002 in. of copper. The unit is then photo-etched.

Advantage of this method is that it is inexpensive—all pin connections are made during the plating process.

Disadvantages are: Electroplating is slow and blowholes in the original film tend to give high reject rate; bond strength is poor; copper film is of uneven thickness.

Dendritic Powder

Experiments with this type of circuitry construction proved to be most fruitful. The manufacturing methods are quite simple. The base material (phenolic or dial) is sprayed with a single-phase quick-drying epoxy—the epoxy half cures to a nontacky state in about 30 seconds. A two-phase metal powder mixture is sprinkled on the integrated circuit mounting surface. Pressure is then applied to this surface by a hydraulic press, the upper platen of which is negatively etched with the circuit pattern. In the areas under pressure, the two-phase powder changes to a single-phase solid. In those areas not under pressure, the mixture remains a powder and is blown off and recovered. The single-phase solid remaining is eutectoid, that is, it changes directly to the solid at a unique pressure. The powder also forms dendrites on solidifying. It is believed that the epoxy returns to liquid under pressure filling the interstitial spaces of the dendrites, accounting for the very high bond strength experienced.

It is quite simple to set up a production line for this approach. Production rate is limited by press-cycle time. To obtain circuitry 0.002 in. thick, a 15-second cycle with a 3-ton peak is required.

It should be noted that when the powder is applied, small fillets are built up in the inside corners formed by the contact protrusion and the integrated circuit mounting plane. By accurately designing the upper platen of the press these fillets solidify, making reliable connections between the circuitry and connector pins.

By sprinkling the lower platen of the press with the powder, the vertical ground plane is formed during the same press operation. Some success has been met using a pressure-solidifying, age-curing, filled, conductive epoxy instead of the powder. However, this has been costly on materials as the excess must be washed off with a spray solvent and we have not yet found an economical way to recover the fill.

MODULE MATERIAL

Relative production rates, strength (sufficient to withstand the pressure required for the circuit-printing method described above), and module shape made us decide on a filled plastic formed by transfer molding. Specifically, the materials found to be most suitable were long filament glass fiber-filled diallyl phthalate for military applications and mica-filled phenolic for commercial use. Figure 5 illustrates the assembled Scamp plug-in module.

The contacts are gold-plated hard brass and are either flat blade or Elco type, depending on application.

GROUND PLANES

Previous experience with packaging of discrete-component circuits and the constant trend toward lower-power (noise sensitive) and higher-speed circuits implies that we would be negligent by omitting a ground plane from a new design, and to provide one, the reverse surface of the Scamp module has a copper coating. This coating is connected to the ground pin of the module and forms a vertical ground plane, effectively reducing crosstalk between modules.

The copper has been applied by several means but the impressed-dendritic-powder method was found to be the most convenient.

LOGIC INTERCONNECTIONS

One of the major factors that produce low reliability in a machine is the failure of connections. Semiconductor integrated circuits increase the system reliability by reducing the number of connections between components and by the use of low-failure-rate joints. The reliability of an equipment can be further increased by making the connection between logic functions on the module with the conductor pattern, rather than through the two connections required when using back-panel wiring. Economically, it is also desirable to make the logic interconnections on the modules, for the cost of each back-panel interconnection is of the order of 40 cents.

Logic connections on the module shorten the length of the conductor paths, thus reducing capacity and signal-propagation time. This is an important factor at clock rates in excess of 3 Mc/s.

The first layout of integrated circuit packages described in this paper is for an arrangement loosely designated Universal Logic. This type of module-configuration performs no specific

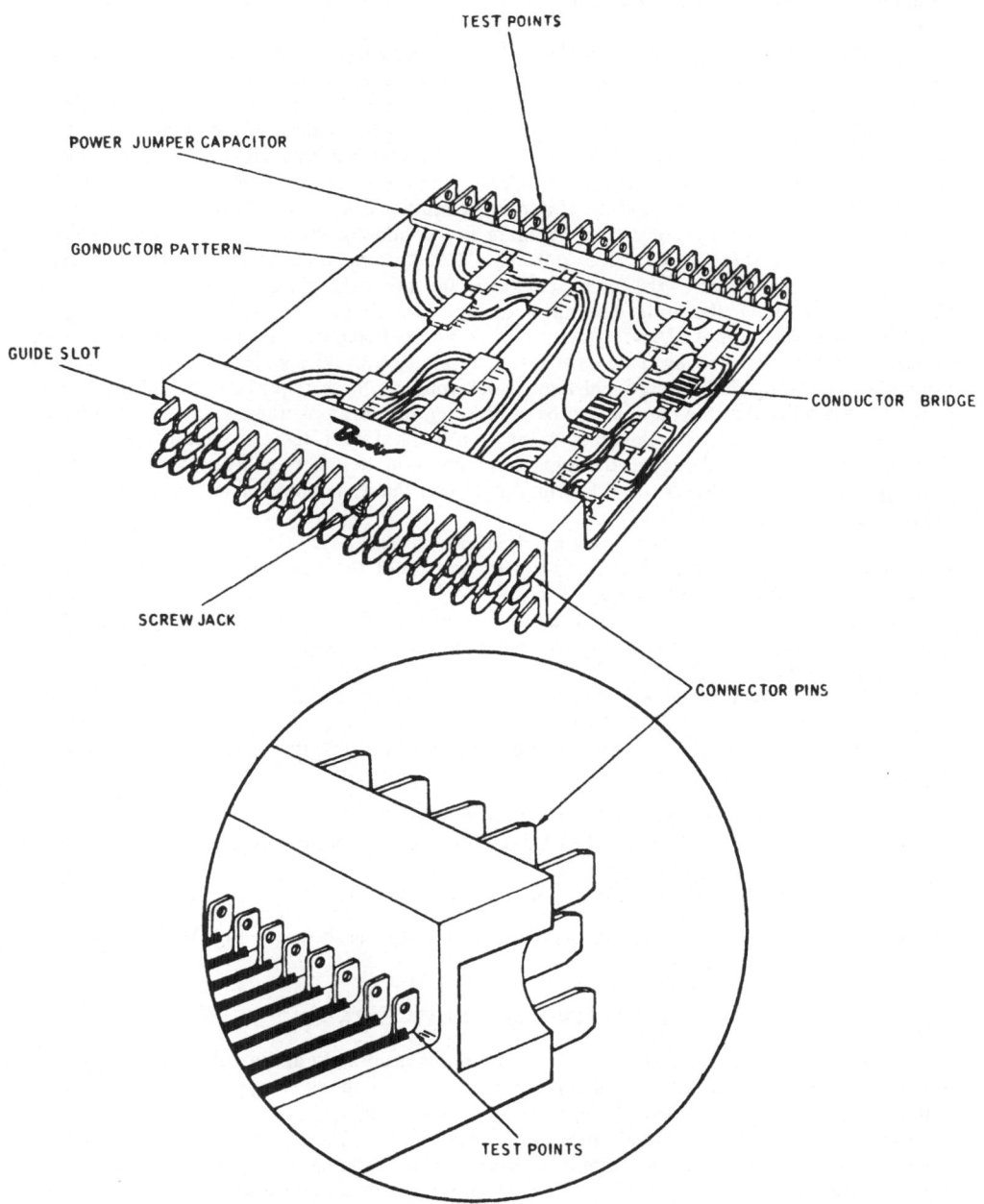

Fig. 5. View of a typical Scamp plug-in module.

function such as a decoder or shift register, but consists of flip-flops and gates, and is used in various places of a system where no repeatable logic circuits occur.

The Universal Logic is connector-pin-limited, less reliable, and tends to be more expensive than functional-logic-block configurations. However, although the majority of the logic on a machine will be built from repeatable-logic-block modules, certain functions will remain that can be handled only or more economically with a Universal Logic module.

Mechanization of the Universal-Logic configuration was made using eleven TRW semiconductor microcircuits, but could just as well have been made from any of the other manufacturers' standard dual three-input gates, as shown in Fig. 6.

The PC number in the circuit blocks of Fig. 6 represents the TRW part number for that circuit, and the number shown in the upper right-hand corner of each circuit block defines the location of that circuit on the module, with reference to the standard layout shown in Fig. 4.

No difficulty was experienced in laying out the conductor pattern for the Universal Logic, and no attempt was made to obtain the simplest conductor patterns. In Fig. 7, the layout of the pattern shows that it is sometimes necessary for the output of one circuit package to cross the power-bus and connect to the inputs of another circuit package. To do this a special bridge was designed. This bridge consists of molding five conductor strips 0.285 in. long, 0.010 in. wide, and 0.003 in. thick into a 0.125 × 0.250 × 0.035 in. plastic block. The bridge is therefore similar in size to one of the circuit packages and takes the place of a circuit package on the module. In Fig. 7, the bridges occupy locations 12 and 19.

With dual three-input NAND gates in up to 30 locations, this is a maximum of 60 logic functions that can be packaged on one Scamp module, with 49 signal input or output pins available. The design of a functional logic block must fit these requirements and also be of sufficient versatility that the module can be used in ten or more positions within an equipment.

The design of functional logic blocks that can be used with high repeatability within one or many systems is a challenge to the logic designers. The designers will probably require the development of new methods of performing logic and the acceptance of redundancy in some module applications.

The example of a functional logic block shown in Fig. 8 is a four-stage counter with a two-phase clock. The signals flow from top to bottom of the diagram, while the clock and "carry" signals flow from left to right.

The four-stage counter was chosen as an example for this paper because this is about as complicated as you can get and is a test of our theory that all logic interconnections could be made on a single layer.

By the use of two flip-flop configurations and four bridges, the pattern was successfully laid out as can be seen from Fig. 9. Part of the success in interconnecting the 25 circuit packages with one layer of conductors is due to the use of the power distribution capacitor. This special capacitor has three functions:

a. Filters high-frequency noise from the power line.
b. Acts as a power-distribution bus by transferring the supply and ground lines to each row of integrated circuits.
c. Acts as a bridge over the conductors that connect to the lower row of connectors.

The method of fabricating this power distribution capacitor is shown in Fig. 10.

MOUNTING THE FLAT PACKS

A template is used to locate the integrated circuits accurately and a rubber base adhesive is used to apply the semiconductor integrated circuit. If the epoxy were used, it would adhere to the base epoxy making it virtually impossible to remove these items.

Experimentation with several methods in making the connections was made. These include parallel-gap welding, programmed electron-beam welding, soldering, and conductive epoxy, all with the same relative success. Most thinking was done in the area of programmed electron-beam welding. As the integrated circuits are in fixed positions on a 5 × 6 matrix,

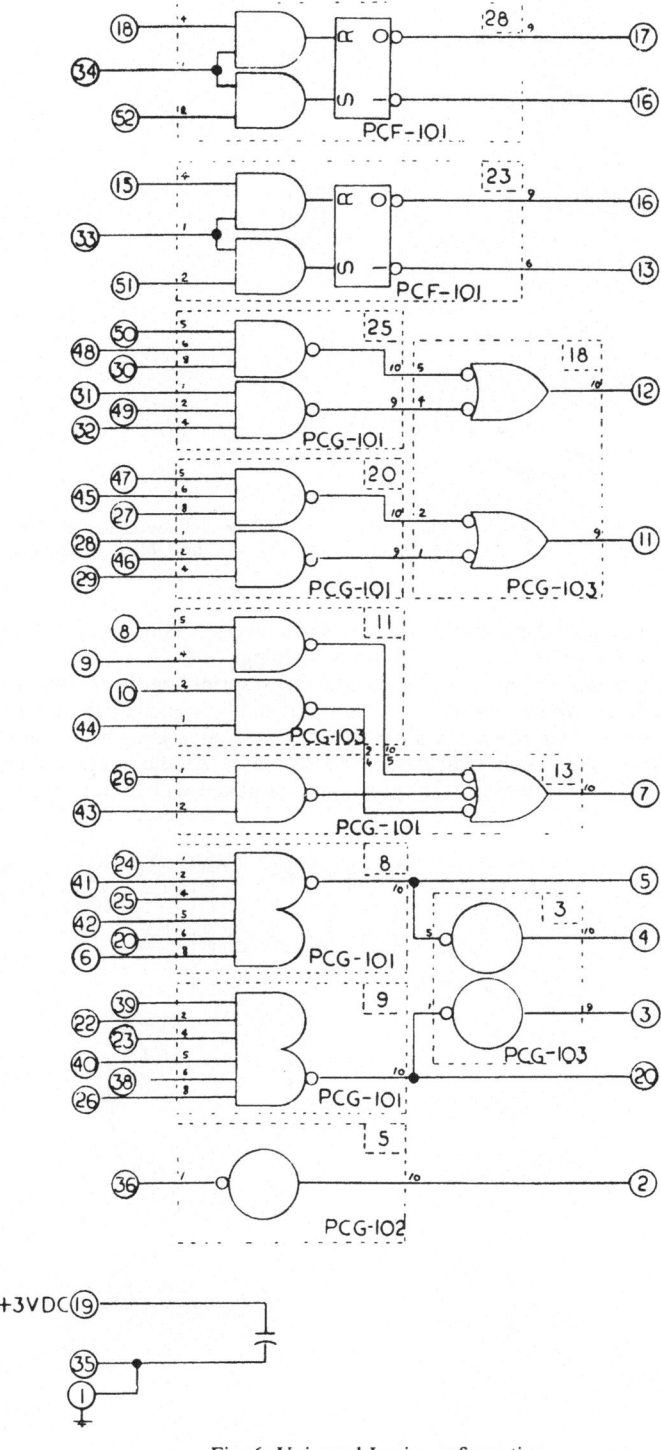

Fig. 6. Universal Logic configuration.

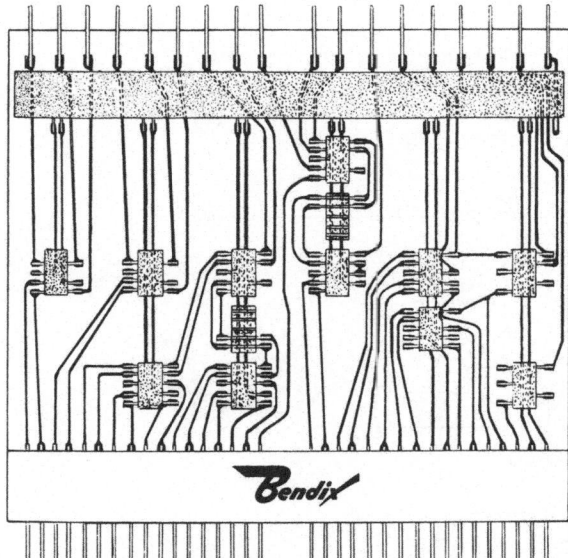

Fig. 7. Conductor pattern layout of the Universal Logic module.

a single program would make the welds for all modules of every circuit configuration, eliminating the necessity of costly one-at-a-time soldering or welding.

However, a much better way has been developed. By resprinkling the module with powder and running it through the press again, all connections are made at once by the solidifying alloy. This type of platen has 30 clearance cavities for the integrated circuits with protrusions around the platen face in those areas where connections are required. As the integrated circuits (and therefore their connection tabs) are in fixed position, one platen will suffice for all circuits.

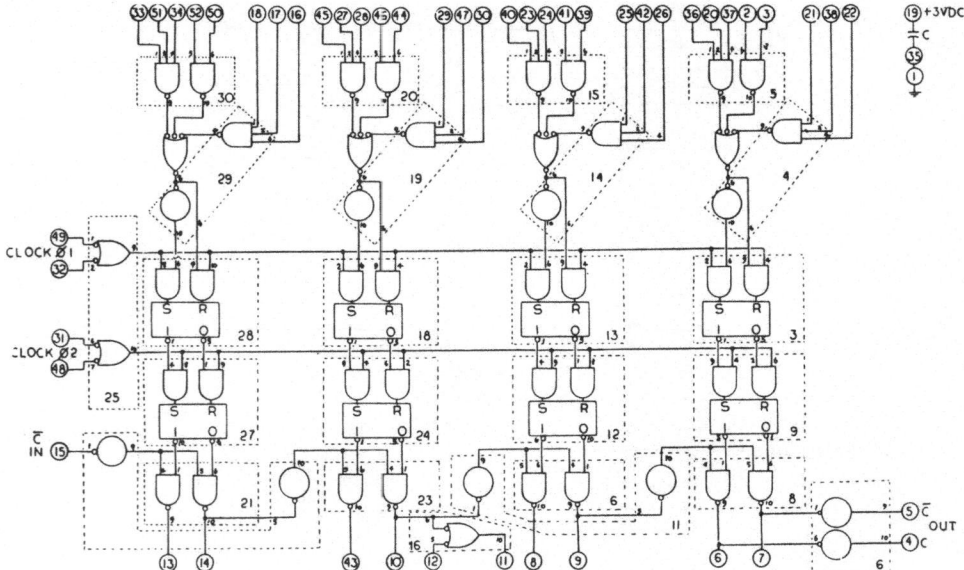

Fig. 8. Logic diagram of a four-stage counter with two-phase clocks.

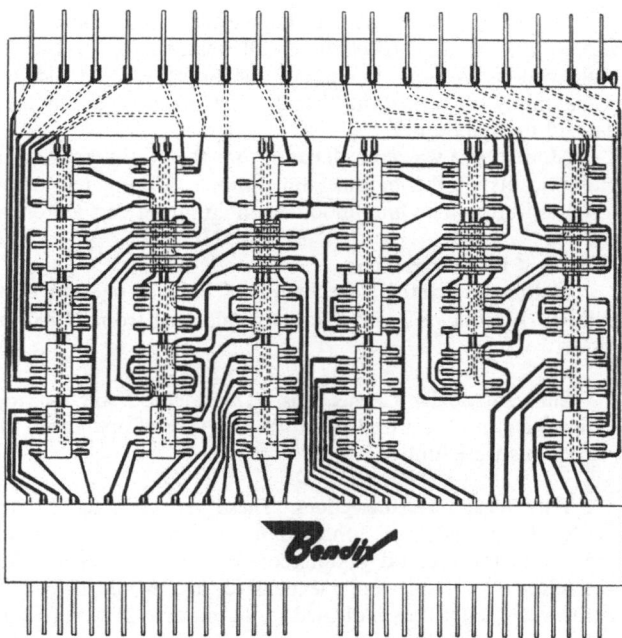

Fig. 9. Pattern layout of a
four-stage counter.

To replace the integrated circuit, the tabs are cut through with an exacto knife or surgical scissors, and the new circuit is connected by rewelding on top of the existing tab stubs or using solder or conductive epoxy.

THE MOTHERBOARD

The motherboard for integrated-circuit-type equipment assumes far greater importance that it did for discrete components inasmuch as it encompasses whole logic sections rather than a small logic block. For example, a small central processor can be packaged on one motherboard. It can be seen, that because of this greater importance, more external factors influence its design. Some of the factors that were considered are as follows: size, method of interconnections, quantity requirements, orientation of module, cooling methods, and ground planes.

Size of Motherboard

The size of the motherboard is always controlled by several conflicting considerations. First, there is convenience of accommodating the wiring method. Electronic engineers would like to have them as large as possible to reduce the number of disconnectable connections, and also sized to reflect their function. Manufacturing would prefer to have them of uniform size to facilitate tooling and fabrication.

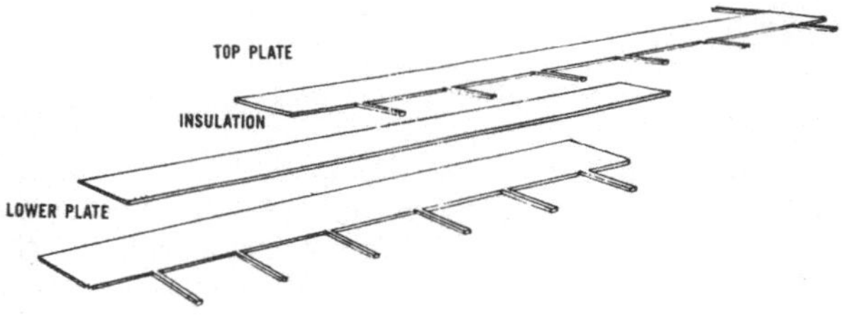

TOP PLATE

INSULATION

LOWER PLATE

Fig. 10. Exploded view of noise
supression and power distribution
capacitor.

Naturally, it was decided to make the motherboards of uniform size, and to ascertain this size, a large central processor (G-20) was analyzed to ascertain the optimum magnitude for a logic section. We were looking for the largest size of logic section that would give us reasonable repeatability. It turned out that the best size for a motherboard was one that would accommodate 16 to 20 modules.

Module size was considered in that the motherboard dimensions should be approximately multiplicants of the module dimensions. Keying requirements made a positive number of rows desirable. The final module configuration decided on was two rows of ten modules.

Method of Interconnections

Multilayer Printed Wire Boards. This was rejected because of base cost and nonrepeatability. (Multilayer wiring boards would cost from $1000 to $1500 with approximately $1500 for artwork.)

Standard Solder Connections. This method was considered from the cost point of view as our factory was set up to perform this type of operation. It was rejected because cost savings projections were limited.

Wire-Wrap Interconnections. These were considered as the projected product improvement for reliability and cost appeared to be attractive. After due consideration, wire-wrap was selected as the method of interconnections, as it would lend itself to interconnections and was relatively reliable and inexpensive, and would fit our standard shop practices. The pin layout was finalized as indicated in the module section of this paper.

Quantity Requirements

The estimated quantities for this motherboard dictated the manufacturing method. The two methods investigated were: (a) a punched aluminum plate containing the National Connectors Corporation "Rap-Lok" type of terminal and (b) a transfer-molded diall or phenolic block.

The perforated-plate type of construction was selected for our prototype as it is best suited to low and moderate quantity production. These have the advantage of accurately locating the wire wrappings for purposes of matching the tolerances in the automatic wire-wrap machine. The perforated aluminum plate also formed the horizontal ground plane that isolated the back-panel wiring from the logic on the modules. The cost is approximately $200 per board. The molded motherboard required quite accurate tooling. It was expensive but in high quantity it turns out the motherboard at approximately $50.

Orientation of the Module

The keying requirements of the module required a $\frac{1}{4}$-in.-diameter boss between the two rows of modules. As the keying slot is on only one side of the module, this meant that alternate rows of modules were reversed. This alternating of the module rows prevents the long-fin effect when the modules are cooled by a cross draft, i.e., the boundary layer is prevented from forming, and air in contact with the integrated-circuit mounting surface remains turbulent, giving a superior coefficient of heat transfer.

The incoming power is on pin 1 of the module and ground is on pin 52. This means that the power bus travels around the module and a single ground bus up the center of the motherboard, thus preventing logic interwiring crossing the power bus, eliminating extraneous noise on the power bus.

SCAMP-MODULE COST

Most of the cost of the Scamp module is in the semiconductor circuit packages, and it follows that the cost of any module will be proportional to its complexity.

Although the cost of a semiconductor circuit package is still high at the present time compared to a similar logic function of discrete components, there are numerous cost reductions

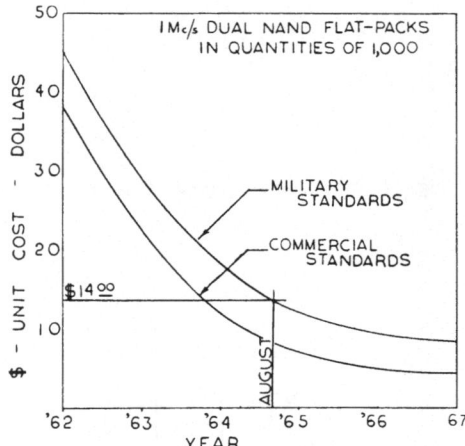

Fig. 11. Cost prediction curves for 1 Mc/s dual
NAND gate integrated circuits.

that accrue with the Scamp plug-in modules. For example, the blank Scamp molding costs $3.89 in small quantities; this is complete with the 52 connector pins and replaces some printed circuit connectors and copper-clad board.

Other reductions include the large savings in back-panel wiring costs due to the amount of logical interconnections that are made on the printed circuit pattern, and further large savings in cabinet and rack cost.

In order to forecast the cost of typical modules, a price prediction curve for the most commonly used integrated circuit package was made and is shown in Fig. 11.

Based on the military standard curve shown in Fig. 11 the following breakdowns of a Scamp module can be estimated (see Table I).

INTERFACE CIRCUITS

At the present time the number of integrated circuit types is somewhat limited, and on most equipment it is necessary to perform level changing or power amplification at the interface with another equipment, and these circuits are usually constructed in welded, cordwood, or thin modules. Our solution to this problem can be illustrated by considering the line-driver circuit shown in Fig. 12. The purpose of this circuit was to convert from the 2.0-volt integrated circuit level to a 28-volt output level into a 1000-ohm transmission line.

TABLE I

Cost Breakdown for Universal Logic Module

Part or operation	Cost, dollars	
	Manufacture August 1964	Manufacture December 1967
Blank module (quan. of 2000)	1.83	1.50
Eleven integrated circuits	154.00	88.00
Manufacture of a conductor pattern and mask amortized over 75 modules	3.88	3.88
Adhering and bonding flat packs	3.27	3.00
Power-jumper capacitor	10.00	7.00
Inspection and test	10.95	12.00
Total cost of Scamp module	183.93	115.38

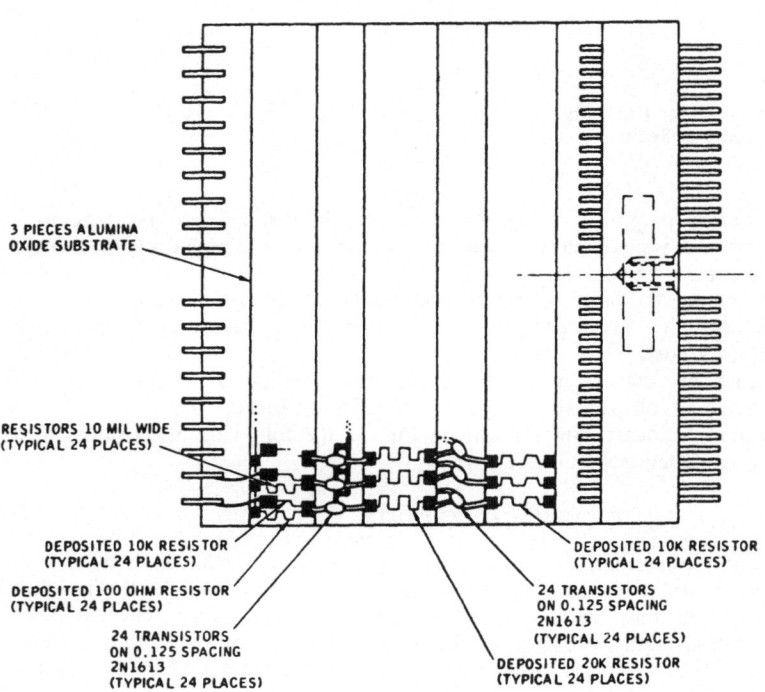

Fig. 12. Schematic of level converter and output driver.

Fig. 13. Thin-film resistors and chip transistors on Scamp-module form interface circuits.

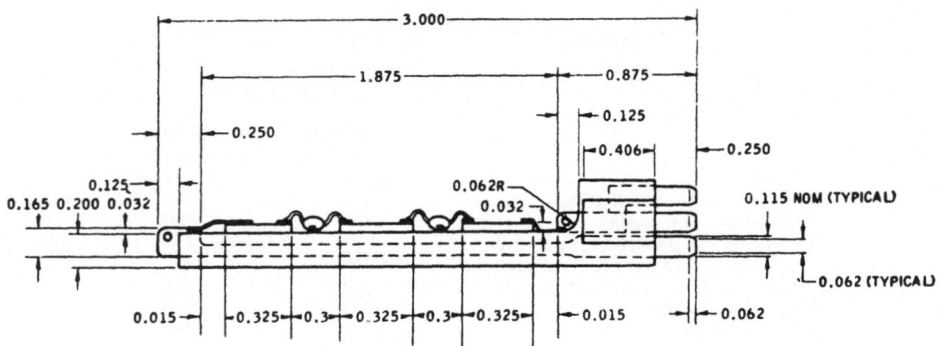

The interface circuit was assembled upon the blank Scamp module, using the 2N1613-epoxy-encapsulated chip transistor, and resistors evaporated onto 0.324 × 0.32 in. thick slices of alumina. The resistor slices and chip transistors are fixed in place with a rubber-base adhesive and the leads parallel-gap welded to the conductor pattern as shown in Fig. 13.

By using this thin-film and chip-transistor technique we can package 24 interface circuits on a Scamp plug-in module. By maintaining the same Scamp configuration when the interface circuits are available in semiconductor integrated form, redesigned plug-in modules can replace the existing ones without any wiring changes.

CONCLUSIONS

We have listed the problems that occur in packaging integrated circuits on plug-in modules, as we see them, namely:

1. Choice of optimum electrical and mechanical size module.
2. Provision of sufficient connectors.
3. Conductor pattern dispersion.
4. Adequate number of test points.
5. Low fabrication cost.
6. Serviceability.
7. Versatility of logic function.
8. Manufacture of high-grade conductor pattern.
9. Provision of ground-plane and supply-noise decoupling.
10. Automatic methods for fabricating the back-panel wiring.

In the presentation of the Scamp concepts, we have suggested solutions to these problems, showing that the advantage accruing from integrated circuits can be enhanced by retaining the plug-in-module concept in the equipment design.

REFERENCES

1. A. H. Faulkner, F. Gurzi, and E. L. Hughes, "Magic—An Advanced Computer for Spaceborne Guidance Systems," Proceedings of the Spaceborne Computer Engineering Conference, Anaheim, October 30, 1962, pp. 83–94.
2. Philip J. Klass, "Microcircuit Use Slashes Computer Size," *Aviations Week and Space Technology*, December 24, 1964, pp. 43–46.
3. W. H. Ayer and T. E. Kirchner, "Integrated Circuit Packaging and Interconnections," 1963 Western Electronics Show and Conference, San Francisco, August 1963.

A Packaging Evaluation of a Five-Bit Adder

E. F. UBER, W. D. FULLER, AND A. J. DOMENICO

Research Laboratories, Lockheed Missiles & Space Company
Palo Alto, California

A comparative evaluation has been made of the packaging of a five-bit adder which originally was designed as an assembly of discrete component modules. These modules were later directly translated into individual thin-film integrated circuit components. Recently, the five-bit adder was repackaged using thin-film integrated components containing eight circuits each. The eight-circuit component is a developmental step which adds flexibility and versatility to the original packaging concept by placing more functions in each integrated component. This paper compares and evaluates the merits and shortcomings of each packaging design.

INTRODUCTION

THE DEVELOPMENT of an equipment technology to meet the specialized requirements of data processing in aerospace projects poses interrelated problems in logic design, component selection, package design, and manufacturing methods. Cost is one of the primary constraints in each individual data processing system, which will not be mass produced but must be designed to perform a specific task in a special environment, be packaged in a minimum volume, have high reliability, and be immediately available. The latest electronic technologies must always be used to maintain a competitive business position.

A technology satisfying these requirements must have two basic parts: a standard logic function and a computer program for logic design; and a simple package design which lends itself to high rates of fabrication.

The development of such an equipment technology over the past several years can be illustrated in the packaging and fabrication of a five-bit adder. The basic logic function is a Resistor–Transistor Logic (RTL) circuit, as shown in Fig. 1, which may be directly connected to other RTL circuits in various combinations to form flip-flops, logic chains, and amplifier–buffer combinations. The computer program is termed LAMP (Lockheed Automatic Mechanization Program), and converts Boolean logical equations directly into wiring lists for the interconnection of RTL circuits for the prescribed data-processing system. Package design

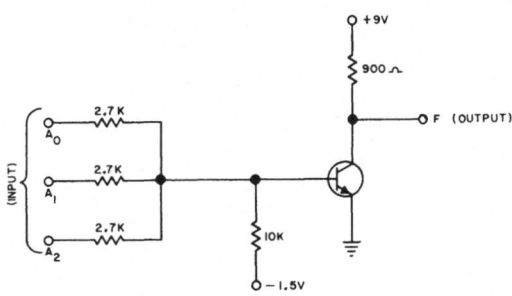

Fig. 1. Nand-nor RTL circuit.

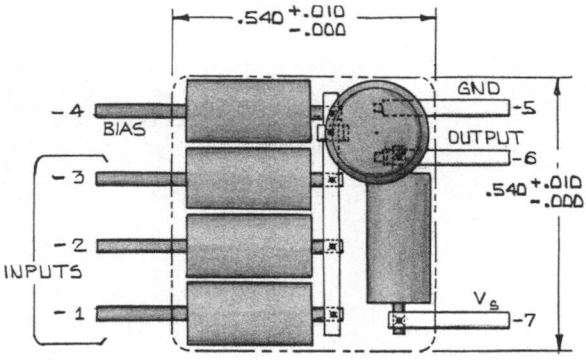

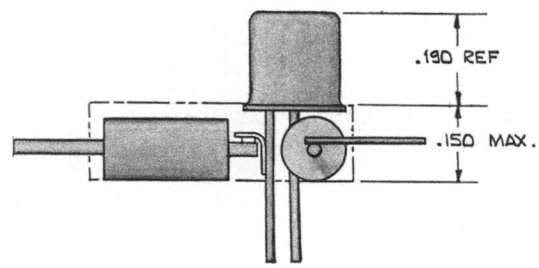

Fig. 2. RTL wafer assembly.

and manufacturing methods have been continually up-dated to take advantage of advances in the state of the art throughout this period. They have progressed from welded electronic (W.E.) techniques to hybrid thin-film circuit configurations.

WELDED ELECTRONICS ASSEMBLY

The RTL circuit module for the W.E. assembly was fabricated by welding components together with short lengths of nickel ribbon, as shown in Fig. 2. The welded structure was then encapsulated to produce a wafer-shaped package 0.540 × 0.540 × 0.150 in. thick, with the transistor case protruding an additional 0.190 in. A working model of a five-bit adder with an integral clocking generator was then assembled from 90 of those RTL modules, using the computer-generated wiring instructions (shown in Fig. 3). The 90 RTL modules were mounted on an aluminum plate which was embossed in a pattern of pads and grooves to increase its stiffness. The RTL wafers were cemented to the pads, and the interconnecting wires were routed along the grooves with all connections made by welding. The aluminum plate served as a heat sink for the wafers and supplied an electrical ground plane for the circuits in addition to its structural function. The welded electronic assembly is shown in Fig. 4.

THIN-FILM CIRCUIT ASSEMBLY

With development of the LMSC titanium thin-film technology, the ability to directly translate component-type circuits into thin-film circuits was demonstrated by fabricating the RTL circuit as a titanium thin-film circuit. The configuration of this thin-film circuit was required to be as nearly identical to the W.E. module as possible (as shown in Fig. 5). This thin-film circuit measured 0.540 × 0.540 × 0.080 in., and had the leads perpendicular to the

	1	2	3	4	5	6	7	8	9	10
I	L0 2	E2 3	L1 2	E4 2	L2 2	E6 2	L3 2	G8 2	L4 2	H9 2 / C8 1
H	H2 1 / A1 1 / D4 1	H1 2 / E1 1 / I2 1	F3 1	G3 2 / I4 2 / I3 1	E6 1 / G6 3 / E10 1 / D8 1	F5 2 / I6 2 / I5 1	G8 1 / E8 3 / D10 1	F7 2 / I8 2 / I7 1	I10 1 / C10 3	I10 2
G	H1 1	H2 2	H4 1 / H3 1 / L1 3	H4 2	H5 1	H6 2	H7 1	H7 2 / E7 1 / F8 1	H9 1	F9 2 / H10 2 / I9 1
F	F2 1 / G1 1 / L0 3	F1 2 / G2 2 / I1 1	E4 1 / G4 3 / F10 1	D4 3	H6 1 / G5 1 / L2 3	D5 2 / C5 2	H8 1 / G7 1 / L3 3	D8 3 / C5 3	G10 1 / G9 1 / L4 3	C3 1 / C4 1
E	F1 1	F2 2	G3 1	F3 2 / E3 1 / F4 1	F5 1	H5 2 / E5 1 / F6 1	F7 1	H8 2	D3 1	C6 1
D	D2 2	D3 2	E5 2 / G6 2	C3 2 / C4 2	C6 2	C7 2	E7 2 / E8 2	B5 2	D7 1	B5 1
C	E3 2 / G4 2	G9 2 / H10 1	D1 1 / D5 1 / E9 2	C5 1	B5 3	D6 1 / D9 2	D7 2	B8 1	F9 1	G10 2
B	A2 2	G5 2 / I6 1 / G7 1 / I8 1	G1 2 / G2 1 / H3 2 / I4 1	B3 1 / B2 1 / C2 1	B6 1 / C8 2	B7 2	B8 2	C9 2 / C10 2	E7 3 / E8 1	C9 3 / C10 1
A	B1 1 / A3 2	C1 1	C1 2	A5 1	A6 3 / B4 1 / B4 2 / B4 3	A7 1 / A7 2 / A7 3	A8 / A9 / A10 / B9 / B10	XM48	E1 2 / E2 2 / E3 3	G4 1 / E5 3 / G6 1

USE OF TABLE :

     THE ROW AND COLUMN INDICES OF A PARTICULAR
SQUARE IDENTIFY THE WAFER WHOSE OUTPUT
PIN (6) IS CONNECTED TO THE INPUT PINS
LISTED IN THE SQUARE

EXAMPLE :

    H6 1 / G5 1 / L2 3   THIS IS READ : "PIN F5-6 IS CONNECTED
                     TO PINS H6-1, G5-1, + L2-3".

Fig. 3. Wiring instructions for five-bit adder.

Fig. 4. Welded electronic assembly.

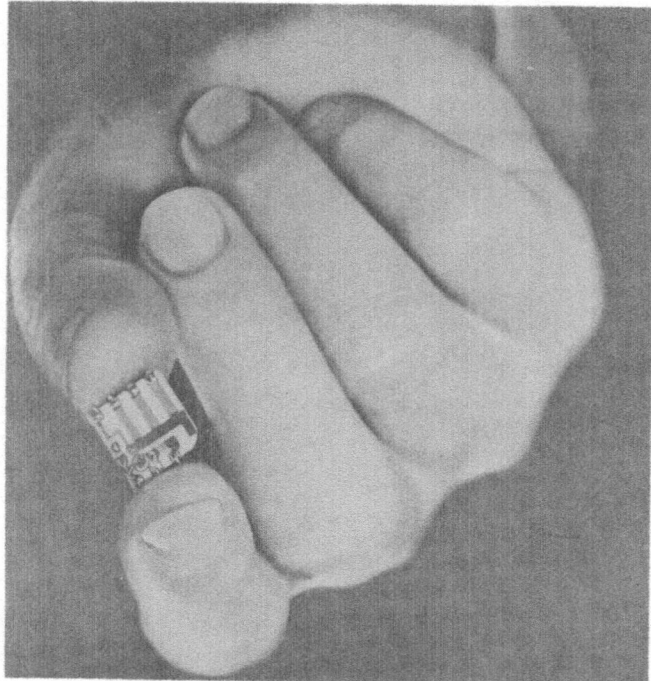

Fig. 5. Thin-film logic circuit.

wafer, around the edge. The fabrication yield of the research titanium thin-film circuit pilot line exceeded 80% in the fabrication of this RTL circuit in the quantity required for the five-bit adder.

The thin-film circuit wafers were then mounted on a printed-circuit-board assembly that afforded three wiring planes for the interconnections. The final assembly is shown in Fig. 6.

DESIGN INTERRELATIONSHIPS

Since this particular assembly did not illustrate the inherent advantages of thin-film technologies in size reduction and interconnection minimization, the five-bit adder was redesigned by a team consisting of personnel skilled in packaging and personnel skilled in the titanium thin-film fabrication processes.

In the application of integrated-circuit technologies, the package designer needs a greater amount of information than ever before concerning electrical and physical characteristics of the circuit materials—the limits and tolerances in all phases of the fabrication processes, acceptable interconnection mechanisms, and compatible hardware. The process engineer in turn must have greater knowledge of the configuration requirements in order to recommend process changes to simplify the package design.

TITANIUM THIN-FILM TECHNOLOGY

The titanium thin-film technology is based upon an economical series of chemical processes for the fabrication of hybrid microcircuits. These sequential steps (as shown in Fig. 7) involve: molten-salt metallizing of ceramic substrates with a uniform film of titanium; electroless plating for a nickel overlay; photoetching for pattern formation; electrochemical conversion of

Fig. 6. Thin-film circuit assembly.

titanium into a high-resistance material; and electrochemical conversion of titanium into titanium dioxide for the dielectric of capacitors. A dynamic testing method is used for in-process control of the circuit-element valves. Welding, resistance soldering, and dip soldering may be used for lead attachment to the circuit pattern.

The initial titanium coating process is uniquely able to metallize simultaneously all surfaces of the substrate, including holes and cavities in the substrate. Metallized holes are used for side-to-side connections of circuit patterns on substrate or as pin terminations when connections are made by dip soldering.

Stable resistive films in the range of 1 to 2000 ohms/square are produced using the molten-salt method. The range of 1 to 50 ohms/square results from the time-temperature control of the molten-salt deposition process, while the range of 50 to 2000 ohms/square is produced by anodic conversion of the lower resistance films. Tolerances of $\pm 1\%$ or $\pm 3\%$ are achieved as specified by continuous monitoring of the resistor during conversion. The characteristics of titanium resistive elements are shown in Table I, and the results of accelerated life tests are shown in Fig. 8.

Circuits are usually designed for 1000 ohms/square, which allows the use of line widths of 0.020 inches while still maintaining a resistance density of 2.0 megohms/in.^{2}. Those comparatively wide lines in the continuous titanium circuit pattern minimize the effects of random pinholes on resistor stability and circuit yield. The wide circuit-pattern lines in conjunction with the dynamic testing technique minimize the necessity for extremely precise control in the circuit-pattern formation process as well as in the initial titanium deposition thickness. The conversion process is terminated when the specified resistor value is reached, rather than at a specified sheet resistance.

FABRICATION PROCESSES

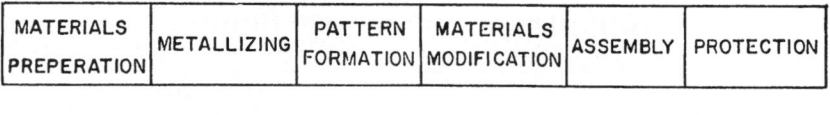

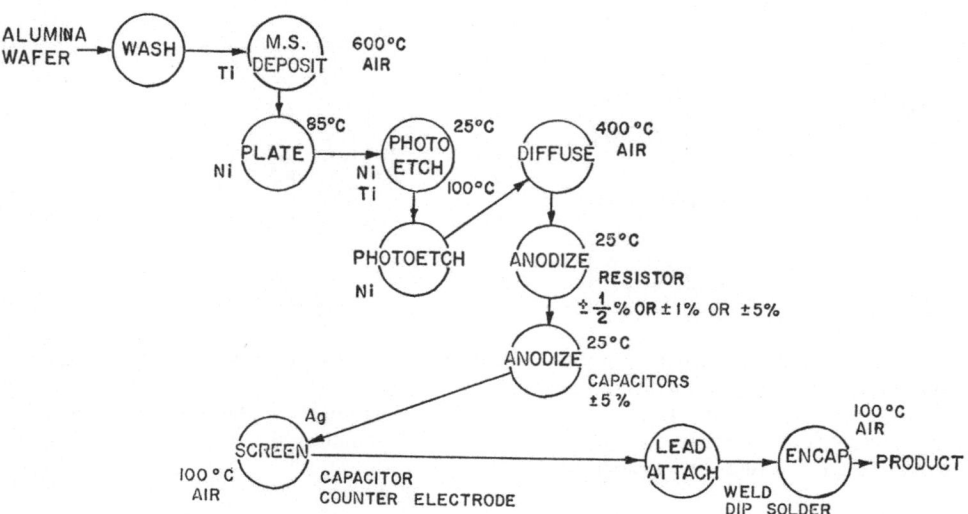

Fig. 7. Titanium thin-film fabrication processes.

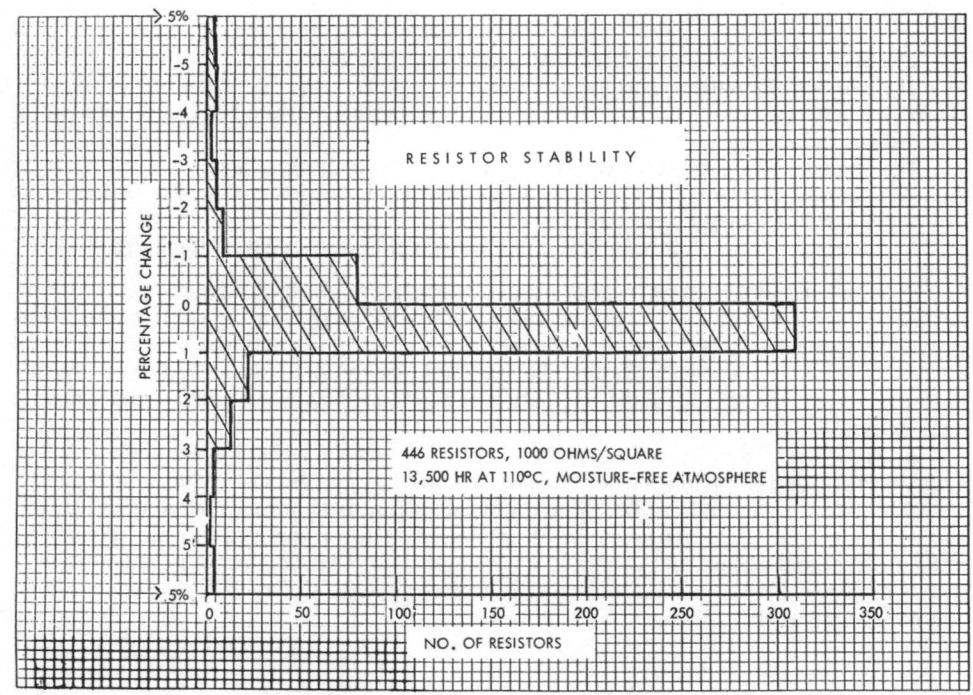

Fig. 8. Stability of titanium resistors.

TABLE I

Characteristics of Titanium Resistive Elements

Characteristic	Ohms/square (alumina substrate)			
	100	500	1000	2000
Temperature coefficient				
($-55°C$ to 85°C), ppm/°C	130	120	50	-100
Voltage coefficient, ppm	0	180	425	1050
Aging (%/1000 hr at 1000 hr)				
No load (110°C)	0.5	1.6	1.7	2.3
R. T. (5 W/in.²)	-1.8	0.6	-0.3	-4.5
Aging (%/1000 hr at 15,000 hr)				
No load (110°)	0	0.02	0.08	0.1
Radiation effects (% change)				
Gamma (5×10^8 R)	—	—	0.3	—
Neutron (2.6×10^{16} nvt)	—	—	1.0	—

Although thin-film capacitors produced by this process have advantageous characteristics, no capacitors were used in the five-bit adder.

A significant product characteristic of thin-film technologies is the ability to fabricate many interconnected circuits on a single substrate. This significantly reduces the number of interconnection junctions in a system assembly.

WAFER CONFIGURATION

The wafer is the result of compromise between the requirements of the system assembly and circuit fabrication, such as:

The number of circuits per wafer versus manufacturing yield.

The number of circuits versus the wafer-size limitations of the manufacturing process.

The number of circuits per wafer versus the allowable temperature rise due to total power dissipation.

The number of external leads allowed by the available wafer-lead area.

The number of interconnection wires and crossovers in the external wiring device versus the number of circuits per wafer.

Resistance to physical environment versus wafer size and assembly configuration.

With this in mind, a design was originated which placed 8 Nand circuits on a 1-in. substrate-sized reduction of 50% over a similar quantity of single ($\frac{1}{2}$ in. square) circuits. This procedure coupled with plated-through holes allows (1) many interelement connections to be formed at the substrate level, (2) the elimination of many intercircuit connections by the integral substrate wiring, and (3) the elimination of much assembly work by use of dip-soldering techniques.

At this stage of development, it was also seen that the eight-Nand wafer could be made a very flexible device by making all interconnections between Nands at the substrate level available (but not connected). The addition of an intermediate wiring matrix would permit any of the possible 256 logic functions of two variables to be formed, and also permit the wafer–matrix combination to be handled as a single unit. This combination, therefore, was termed a universal logic function wafer.* The prototype model of this unit is shown in Fig. 9, and the circuit pattern is shown in Fig. 10 for clarity.

* "A Universal Logic Function Wafer," E. F. Uber and A. J. Domenico, IEEE International Conference, March 1964.

Fig. 9. Prototype of universal logic function wafer (top), and its interconnection wiring matrices (bottom).

Several schemes were considered for packaging the required substrates for the five-bit adder. Basically they fell into two groups: (1) stacking arrangements and (2) flat cards. A study of the stacking arrangements indicated that the chief value of the arrangement is increased packaging density. The required twelve wafers, plus intermediate wiring matrices, could be fitted into a one-inch cube. This cube was exclusive of riser wires or other interconnection schemes. There were several disadvantages which precluded use of the design. (See Figs. 11, 12, and 13 for typical stacked assemblies.) One of these is the necessity of bringing all interconnection leads out to the edges of the board. This would be inconvenient using the previously

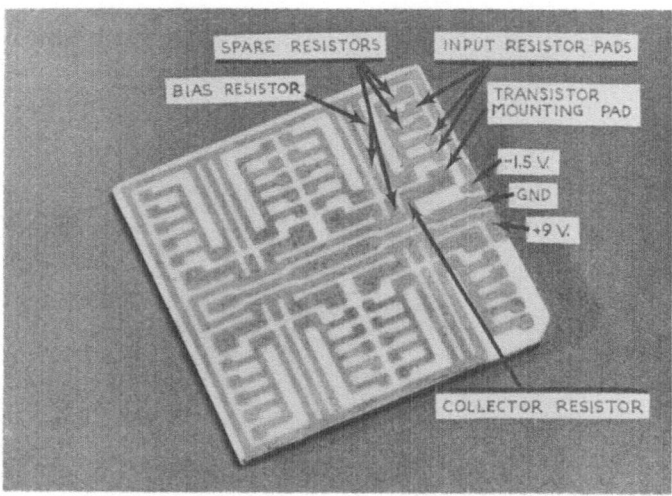

Fig. 10. ULFW circuit pattern.

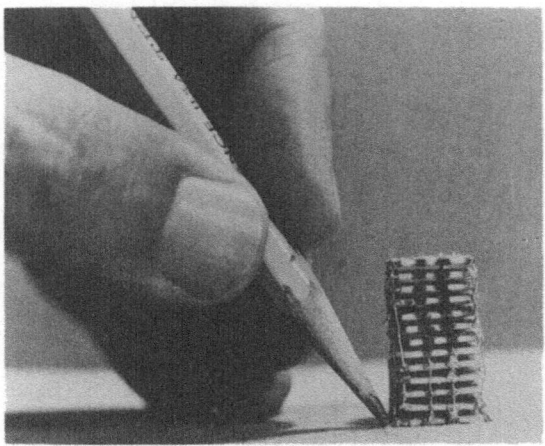

Fig. 11. Stacked-wafer assembly.

designed logic wafer. It would at least entail a complete redesign of the logic wafer or intermediate wiring matrix or both. Another disadvantage was that once placed in such an arrangement, the logic wafers are virtually inaccessible for repair or replacement, particularly if riser wires are placed on two or more edges. It is also difficult to replace or repair parts even if all interconnects are on one edge, as was learned from a previous design. Another important disadvantage of this technique is that it vastly increases the heating problems. This is a particularly relevant problem since one-half of the Nand circuits in this design are "on" at any one time. The disadvantages mentioned indicate that a different packaging scheme would be better than the stacking arrangements.

From the standpoint of packaging sophistication, flat cards are not very elegant. On the whole, they do not utilize space particularly well. They are very simple arrangements, however, and this is, perhaps, their chief advantage. By their simplicity of arrangement, they skirt

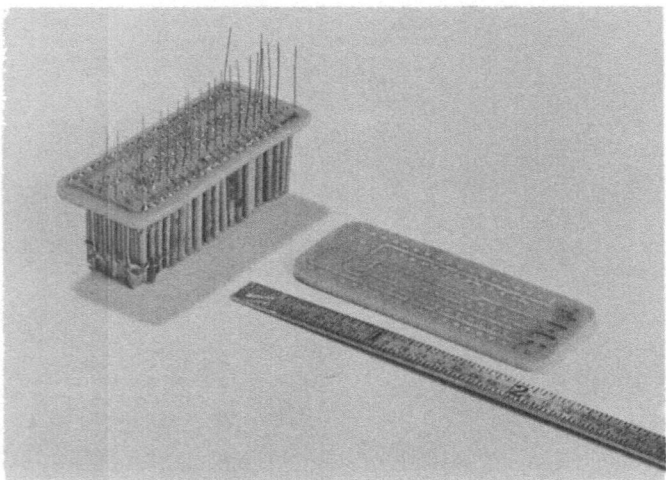

Fig. 12. Stacked-wafer assembly.

Fig. 13. Stacked-wafer assembly.

problems posed by the stacking schemes. Since substrates are arranged in orderly fashion on a master (mother) board, they are completely accessible for replacements of entire substrates, repair of substrates by replacement of transistors, repair by resistor substitution (with other resistors on the substrate), and for necessary troubleshooting in event of failure. The master board offers sufficient space for the wiring necessary, and can be prepared as a multilayer printed circuit board if more than two layers are necessary or convenient. Dip soldering is possible and desirable for reliability. The most apparent drawback of the flat-mounting technique is that the cards are rather large in length and width (though not in thickness) and therefore can cause rather awkward mounting of the package. To improve this an alternate flat mounting was devised. This packaging technique requires the use of a "folded" master board. This folded board is actually two printed circuit master boards, mounting six substrates each, connected by a flexible cable and joined together rigidly at the other end by a connector. This packaging technique efficiently utilizes space, allows repairability by substitution of substrates, substitution of active components at the substrate level, substitution of resistors by jumping, and provides troubleshooting accessibility.

Having established the packaging technique, work was started on the master-board layout, holed substrates were ordered, and art work completed for the basic eight-Nand substrate. The master-board layout quickly showed an additional advantage of the system. Since many of the interconnections are made at the substrate level, the master board is a relatively simple layout. It was discovered that a multilayer board was unnecessary, that the wiring needed only two layers and, indeed, much of the wiring could be accomplished on one layer. While awaiting delivery of the holed substrates, prototype circuits were processed on blind (holeless) substrates, through the titanium thin-film pilot line. Producibility of the unit proved to be excellent. Due to the inclusion of spare resistors, virtually no rejects of entire wafers occurred, despite the large number of resistors (40) necessary to make an acceptable substrate. These parts contained the circuit pattern only. Due to the absence of holes no preliminary interconnect pattern was made on the reverse side. Leads could not be dip soldered and were therefore hand soldered, using a fixture which spaced the leads to comply with the hole pattern. The parts produced were checked-out electrically and placed on life test.

The fabrication of the prototype universal logic function wafers revealed some difficulties with the holeless substrates. These substrates did not permit the utilization of the reverse side for wiring. To supply the wiring which was designed for the reverse side of the logic wafer it was necessary to fabricate intermediate wiring matrices. The circuit side of the substrate on which the transistors were mounted was obscured when the substrates were mounted on matrices or a master board. This was due to having the exiting wires soldered to the circuit side of the substrate, which was necessitated by the holeless substrates. These exiting wires caused other difficulties. It was necessary to make them flexible to allow some latitude in mounting. The flexibility, however, presented a problem in itself. It is difficult to insert 40 flexible leads into a printed circuit board. Holeless substrates eliminate or reduce these problems. By having heavier wires positioned by soldering into the holes, and exiting from the reverse side of the substrate, the handling problems are reduced, and the circuit side is exposed for testing and repair. The use of the holed substrates also reduced the number of external connections beyond those shown in Table II by utilizing the plated-through holes to connect the circuits to the integral wiring on the reverse side. The holed substrates also obviated the necessity of using intermediate wiring matrices. In this design, they are superfluous because the intercircuit wiring is more easily accomplished on the master board.

Even the prototype substrates exhibited a significant reduction in the number of interconnections between modules. Table II shows a reduction of 234 interconnections. It is interesting to note that this was accomplished simply by providing integral connections for $B+$, $B-$, and ground on the substrate level rather than to each circuit from the master-board level. A further reduction in the number of interconnections could have been made by fabricating them as integral wiring on the substrate. It was decided not to do this, however, because it would require fabrication of 12 unique substrates, and would destroy the concept of universality of substrate manufacture. Further, should a substrate require replacement, it would be necessary to replace it with a unique part rather than a universal part, which complicates stocking of spare parts. When the substrates with holes were received, production of the actual parts needed for the five-bit adder were started. Several substrates were fabricated using the holed substrates, but no attempt was made to fabricate the master boards due to a reduction in project funding.

CONCLUSION

Several conclusions were drawn during the design of the five-bit adder:

1. Where repairability is a factor of large importance, the use of flat cards supporting substrates is preferable to stacking.

TABLE II

Comparative Parameters—Five-Bit Adder

Section	Component assembly	T/F assembly	ULF assembly
Assembly			
Length (in.)	8.0	6.0	3.5
Width (in.)	7.0	6.0	2.35
Thickness (in.)	0.34	0.30	0.375
Volume (in.^{3})	19	10.8	3.1
Components (No.)	541	181	103
Module			
Length (in.)	0.54	0.54	1.0
Width (in.)	0.54	0.54	1.0
Thickness (in.)	0.34	0.08	0.05
Volume (in.^{3})	0.1	0.02	0.05
Number	90	90	12
Connections			
Internal	720	270	270
External	630	630	396

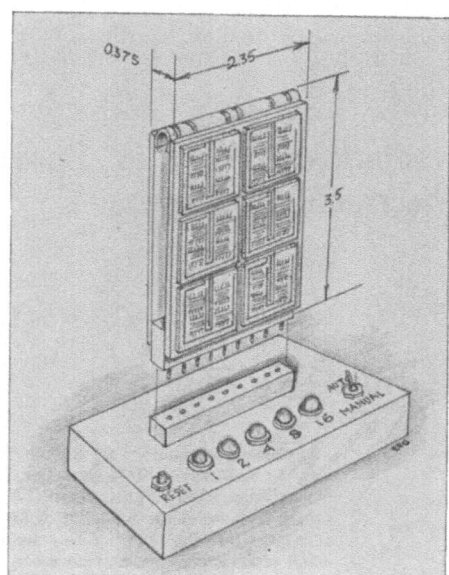

Fig. 14. Planar assembly for five-bit adder.

2. Flat-mounted substrates allow easy replacement of the substrate, simplified trouble-shooting, and limited repairs on the substrate without removal.
3. Stacked substrates are difficult to repair, but the resultant assemblies are more compact. They also tend to become pin-limited.
4. Using holed substrates, speeded assembly techniques, allowed transistors to be placed on the outside of the package, and eliminated many interconnections (by plating through). They slightly complicated fabrication by necessitating more critical placement during photoetching, and also restricting of mask placement for anodizing.
5. Designing spare elements on the substrate made virtually every substrate an acceptable piece.

The consideration of packaging requirements and manufacturing constraints has resulted in the development of a thin-film, universal-logic, function-wafer assembly which meets the many data-processing requirements of equipment to be used in the specialized aerospace environment. The design of a planar assembly to perform the function of a five-bit adder is shown in Fig. 14. The evolution of the five-bit-adder package has kept pace with the evolution of microelectronic technologies, and the change in physical characteristics of the package is vividly illustrated in Table II, which shows a 6:1 reduction in package volume, a 7:1 reduction in substrate count, and a 35% reduction in interconnections, while maintaining all the functional versatility of the original equipment technology.

ACKNOWLEDGMENT

The development of the RTL circuit, the LAMP design program, and the titanium thin-film technology were funded by the Lockheed Independent Development Program. Many personnel have contributed to these developments, but particularly the efforts of R. A. Quinn, T. Lawrence, E. Poe, G. T. Uber, B. Leitner, W. M. Holmes, G. J. Whyte, and M. Lipanovich must be acknowledged in the evolution of equipment design techniques for thin-film circuits.

Radiation Effects in Epoxies Used as Encapsulants for Electronic Packages

F. F. STUCKI

Research Laboratories, Lockheed Missiles & Space Company
Palo Alto, California

Scotchcast-type epoxies are among those encapsulants which are being considered for applications in outer space. It is expected that these epoxies will be exposed to a considerable dosage of radiation when flown in space. At LMSC we radiated Scotchcast-type epoxies in a cobalt source and tested the influence of this radiation to the epoxy. A special microtransducer was embedded into the material as a sensor, which allowed monitoring the behavior of the sample under test. It is shown that radiation leads to an after-curing effect, which decays exponentially with time. After a dosage of approximately 10^6 roentgens, the material is completely cured. Besides these curing effects, a shift of the secondary transition point from +10 to +25°C was experienced. This shift of the secondary transition point from a low temperature into the room-temperature range can lead to difficulties for the circuit designer. The results of the test indicate that electronic components embedded into Scotchcast-type materials are not too greatly affected by changes in the epoxy due to radiation.

INTRODUCTION

HIGH RADIATION DOSES are known to have adverse effects on epoxies which are used as encapsulants for electronic packages. This is important for any electronic package to be sent into outerspace because a high-radiation environment exists in outerspace. Until recently, it was difficult to investigate the effects of radiation on encapsulants such as epoxies because no sensing element was available which could be embedded into the test sample without disturbing the overall stress pattern in the test sample. The Microsystems Electronics Lab of the Research Laboratories at Lockheed Missiles & Space Company developed a small magnetic transducer which can be embedded into the material. This is shown in Fig. 1, where two transducers and two glass diodes are embedded into a translucent epoxy; compared with the stresses introduced by the diodes, the stress pattern around the transducer is negligibly small. The stress distribution in this package is made visible by the polarization of the light source.

Throughout our investigations a 3M-241-type Scotchcast was used as the test sample. The main aim of these investigations was to demonstrate the capabilities of the test system. The results should enable the selection of a proper epoxy, which could withstand radiation effects as they exist in outerspace.

THE MICROTRANSDUCER

The microtransducer, as shown in Fig. 2, is based upon a ferrite core, 0.050-in. in diameter, which has been specially processed to have a high magneto-damping characteristic under hydrostatic pressure. The core is wound with two coils, a drive coil and an output coil, and operated as a high-repetition-rate pulse transformer, where the amplitude of the output pulse

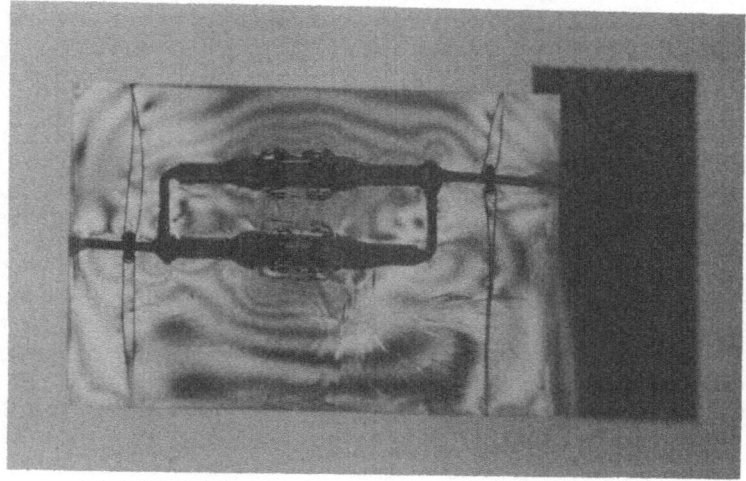

Fig. 1. Two glass diodes and two pressure transducers encapsulated in a translucent epoxy.

varies inversely with the hydrostatic pressure. The microtransducer is operated at a pulse rate which generates 300,000 samples per sec with a resulting pressure response bandwidth of 30,000 cps. With a standard current drive of 800 mA-turns on the core, an output signal of 50 mV per turn is generated at zero pressure, and 15 mV per turn at 20,000 psi. The usable pressure range of the microtransducer is from 0–20,000 psi. The microtransducer has shown no change in characteristics for exposures up to 10^{16} roentgens, and one of its important applications has been the monitoring of the stress changes in semirigid epoxies exposed to

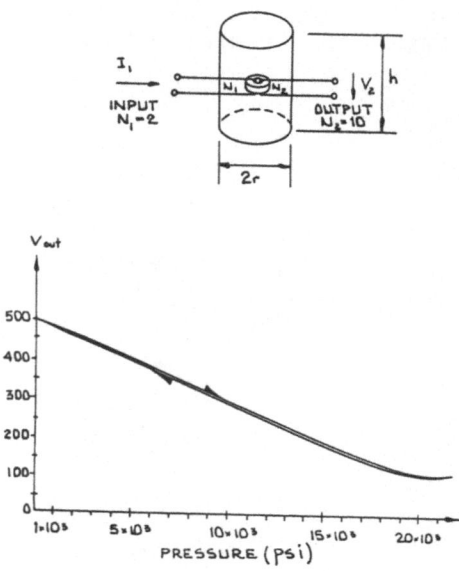

Fig. 2. Ferrite pressure transducer embedded in epoxy (top) and calibration curve $V_{out} = f$(psi) (bottom).

gamma radiation. The primary advantages of this microtransducer are its small size, high frequency response, and high sensitivity, which allow it to be embedded in a wide variety of materials to explore localized stress characteristics of the material configuration.

THE TEST SYSTEM

To monitor the radiation effects in the epoxy, a small electronic system was developed such that even slowly changing pressures in the epoxy could be monitored and recorded on a strip-chart recorder. In Fig. 3 the block diagram of the system is shown. The pressure transducer is activated by a modified blocking oscillator which acts as pulse generator. The rise-time of the current pulse is 0.5 μsec, and up to 1.5 A are available to drive the transducer. The pulse width has a duration of 2 μsec. The pulse driver is terminated into 10 ohms, and at the output leads, which are terminated into a 175-ohm resistor, an output signal of 500 mV can be detected. This signal then passes through a simple, single-stage amplifier, followed by a high-speed integrator, and is again amplified by a high-input impedance dc amplifier. At the output of the dc amplifier an analog signal of 5 V is available to activate a strip-chart recorder. The electronic circuitry operates from a +28-V dc source.

THE PREPARATION OF THE TEST SPECIMEN

A calibrated microtransducer is selected and properly wound with the desired input and output leads. This transducer is then cleaned and potted with epoxy in a mold, half an inch deep and half an inch in diameter. Scotchcast-type 241 is a semirigid epoxy resin, a filled version of type 235 resin [1]. The curing agent is a proprietary anhydride, used in a ratio 1:2 with the resin. The curing cycle was 16 hr at 160°C. The cured material had a hardness of about 65 Shore D. The samples were then monitored for stresses through a temperature range of 100°C to −40°C (Fig. 4).

DOSIMETRY

The samples were irradiated in the LMSC cobalt-60 facility. This source provided a dose rate of 2 × 10^5 R/hr gamma rays with energies of 1.17 and 1.33 Mev. The average energy of this source is comparable to the average energy of the reactor-produced gammas and, since the potting compound of interest was organic in nature, there was no need for neutron irradiations [2]. The gamma dose rates were based on ferrous sulfate dosimetry, and converted from roentgen to rad (carbon) on a 1:1 basis.

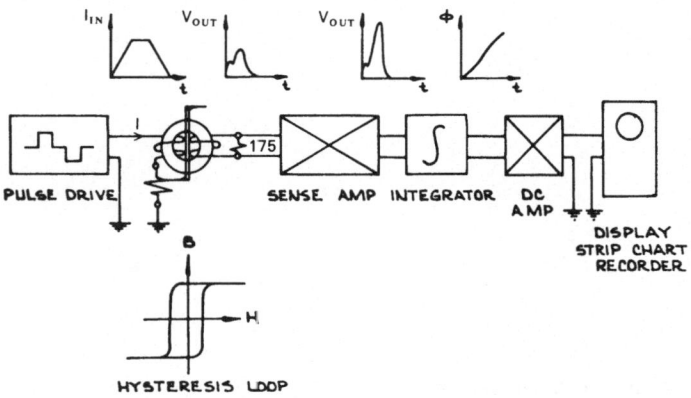

Fig. 3. Test-circuit diagram of ferrite transducer.

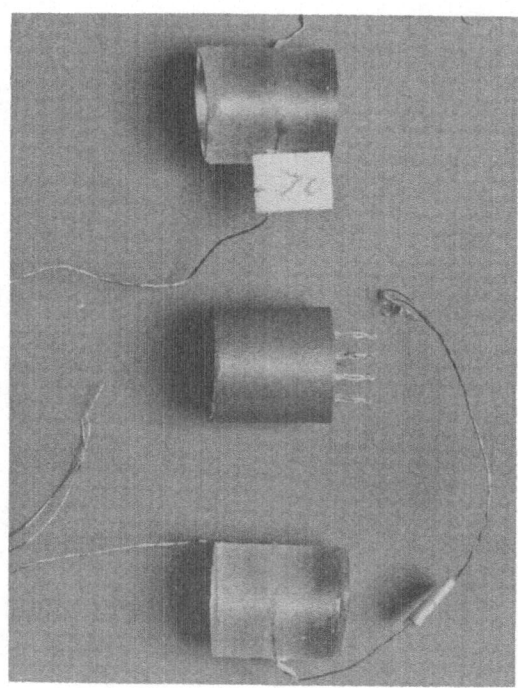

Fig. 4. The type 3M 241 (center) and two other test samples of epoxy materials. (A special connector is used to simplify the test procedures.)

Both passive and active tests were performed [3]. The active tests were performed in order to study the stress relaxation properties at various doses. In these tests a dummy cable was used to assure that radiation-induced cable transients were not causing erroneous readings in the output signals. This was confirmed by dropping the cobalt-60 source while taking measurements on samples. Tests were carried out to a maximum dose of 3×10^6 rad (C).

DISCUSSION OF RESULTS

Passive Irradiations

The internal stresses in the epoxy material increased with the radiation dose. The increase was evident at all temperatures between -20 and $+100°C$ (Fig. 5). The increase at all temperatures was nonlinear, increasing rapidly at low doses, and then approaching a plateau at the highest dose of 3×10^6 rad (C). The increased stress values result in a shift in the secondary transition toward a higher temperature range.

The radiation-induced increases in the internal stresses are believed to be due to the cross-linking process, and could be referred to as a post-cure of the polymer. However, it is possible that all or part of the pressure is due to gas formation from radiation degradation of the polymer. An epoxy will generate approximately 3×10^{-4} liters (STP) of H_2 gas per 10^6 rad (C) of energy absorbed. The pressure resulting from this gas is related to the geometry of the sample and to the diffusion-rate constants; it is impossible to calculate whether a measurable internal pressure results from this dose. The plateau in the pressure-dose curve could be due either to a gas diffusion equilibrium or to the formation of a fully cross-linked three-dimensional network in the epoxy polymer structure.

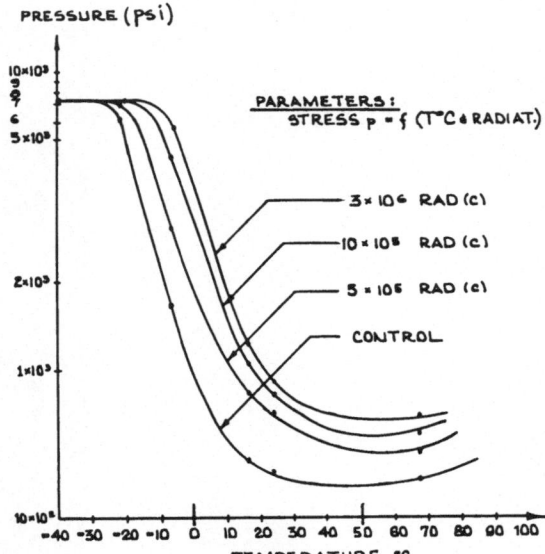

Fig. 5. Stress increase in epoxy (type 3M 241 Scotch-cast) due to radiation and subsequent temperature cycling.

Active Irradiations

In the active irradiations, it was possible to periodically drop the radiation source and observe the stress values over a short period of time. Early in the test, before the pressure–dose curve reached its plateau, removal of the source resulted in a stress relaxation (Fig. 6). The relaxation process amounted to about a 30 psi decrease, and resembled the stress relaxation which is well known and fairly well understood in elastomers. Since the epoxy is not elastomeric in its physical structure, it cannot be assumed that the same mechanisms are responsible for the relaxation process. As the dose was increased, the amount of stress relaxation became less each time the source was removed. When the exposure reached 3×10^6 rad (C), and the epoxy appeared to be fully cured, no further relaxation occurred upon removal of the source. The stress-relaxation data tend to favor the highly cross-linked three-dimensional hypothesis rather than the gas-pressure hypothesis as the explanation for stress changes in the epoxy during irradiation.

SUMMARY AND CONCLUSIONS

A simple, pressure-sensing electronic system which allows the monitoring of radiation effects in epoxies was discussed. A new microtransducer was used as pressure sensor. The gamma radiation led to an after-cure effect which could be detected by an increase in pressure in the radiated epoxy. The results can be summarized as follows: The possibility of embedding into the epoxy a small sensor enabled us to follow any static and dynamic change in the stress pattern of the material under test. The resulting shift in the secondary transition point should be the main concern of the design engineer because the sharp increase in internal stresses is shifted from $+10°C$ to close to room temperature. Since the electronic components embedded in an epoxy are mainly under hydrostatic pressure, an evenly distributed temperature change should have little effect on the package. A marked temperature differential between different areas in the same package will result in a pressure differential which can damage the encapsulated components. To minimize the existence of stress gradients, the size of an electronic package should be kept small.

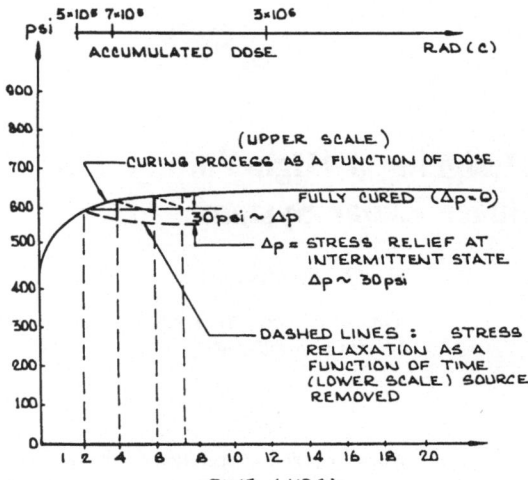

Fig. 6. Stress relaxation in type 3M 241 epoxy.

ACKNOWLEDGMENT

The help of R. D. Carpenter, who was instrumental in the test procedures and the subsequent testing, was greatly appreciated.

REFERENCES

1. J. Dallimore, F. F. Stucki, and D. Kasper, "Measurements of Internal Stress in Electronic Encapsulating Resins with a Small Solid-State Transducer," Reg. Conference of the Society of Plastic Engr., Syracuse, New York, April 1963.
2. R. W. Roos, C. F. Kooi, and M. E. Baldwin, "Neutron and Gamma Irradiation of Some Square-Loop and Microwave Ferrites," Transactions of AIEE Communication and Electronics, Paper 60-1015, September 1961, pp. 1–5.
3. J. P. Nichols and E. D. Arnold, "Shielding Isotopic Power Sources for Space Missions," *Nucleonics*, February 1964, pp. 52–56.

Designing a High-Voltage Power Supply for a Space Radar System

H. G. FRANKLAND

Ryan Electronics, Ryan Aeronautical Company
San Diego, California

When the period of operation of a power unit in a space environment is limited to a few minutes, and weight is at a premium, it is necessary to resort to thermal storage to reduce hot spots. If the same unit has to undergo prolonged periods of testing, a different approach has to be made to the cooling technique. Add to these requirements a vibration-test input of 18 *g*'s and there is a challenging design task. This paper describes the steps taken to meet these requirements in the packaging of the awkwardly shaped parts needed for high-current and high-voltage application.

THE LANDING RADAR SYSTEM, of which the high-voltage power supply formed one of the four major assemblies, is a radar altimeter and doppler velocity sensor. A line diagram (Fig. 1) shows the relationship and connections among these four assemblies.

Early in the design program it was decided to locate the klystron transmitting tubes in the same enclosure as the power supply. Since a modulator was directly linked with the altimeter klystron tube it was decided to package this also in the same container. This electronic assembly was called a klystron power supply and modulator.

The reasons for accommodating the klystrons with the power supply instead of mounting them on the transmitting section of the antennas were:

1. To obviate the necessity for high-voltage-system interconnecting cable and the attendant problems with high-voltage connectors in a hard-vacuum environment.
2. To reduce the number of velocity-sensor transmitting tubes from two to one in order to conserve weight. The transmitted microwave energy was piped through a splitter in the microwave tubing to the separate antennas.
3. To enable the power-supply structure to serve as a thermal capacitor for the energy dissipated by the klystrons. These were the highest dissipators by a large margin. The altimeter klystron used 37.5 W and the velocity-sensor klystron dissipated 110 W.

The following outputs were required: 55 mA at −2200 V ± 50 V dc with ripple not exceeding 1 mV at 4 kc, to a VA515 klystron tube; 65 mA at −500 V ± 10 V dc with ripple not exceeding 3 mV, to a VA246 klystron tube; reflector power at −800 V dc with a sawtooth wave superimposed on it by the modulator; 1.1 A at 6.3 V, regulated to ±0.1 V, to the filament of each klystron.

Because the cathode and filament are connected in the transmitting tube, the 6.3-V filament circuits are at −2200 V and −500 V below ground and had to be treated as high-voltage circuits. The primary power was from a 28-V battery, so the electronic design was a dc-to-dc conversion type. The output stepped up to 2800 V and rectified, filtered, and regulated to the various dc loads.

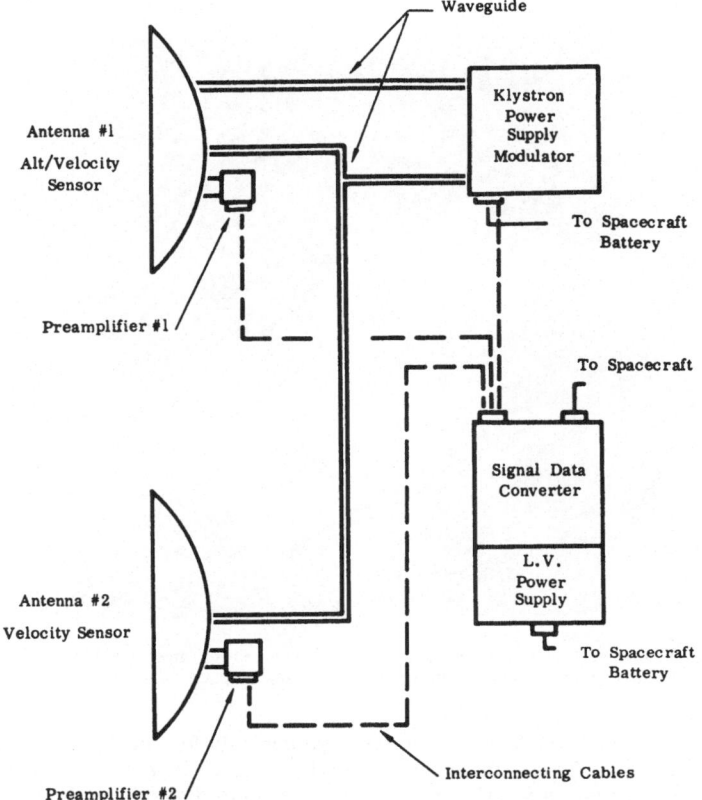

Fig. 1. The relationship between the high-voltage power supply and the other units in the landing radar system. The waveguide is rigid rectangular tubing. The electrical connections to the power supply consist of both unregulated power from the spacecraft battery (28 V, 365 W dc), and regulated power from the low-voltage power supply.

The operating requirements were for a 30-sec warm-up and 5-min operating time in the landing-descent stage in a hard vacuum (10^{-8} mm Hg) environment. The mode of cooling was to be by a passive rather than an active method. The prime contractor would not allow any of the change of stated methods of cooling, due to their unpredictable nature. Convection cooling was not possible, and any appreciable loss by radiation was negligible because of the short duration of operation. Heat transfer by conduction alone would be effective, and sufficient heat transfer media and thermal capacity had to be designed into the equipment where it was necessary.

One of the first tasks of the packaging engineer was to work with the klystron-tube manufacturer to develop suitable mechanical configurations to meet thermal requirements.

Figure 2 shows how thermal capacity was built-in to one of the klystrons, and how by arranging a large area of interface and an adequate number of attachments, efficient heat transfer to the mating structure could be arranged. Every electronic part in the unit had to be evaluated to ensure that it alone or with added thermal capacity would not exceed the reliable temperatures at the end of $5\frac{1}{2}$ min of operation in the space environment.

With weight at a premium, it was a great challenge to have to design this unit for continuous operation at normal pressure during prolonged testing, without adding extra weight permanently to the flight hardware. The chassis arrangement and the design and placement of parts, especially those of high heat dissipation, had to be consistent with this requirement in addition to all others. This paper will show later on how this was achieved.

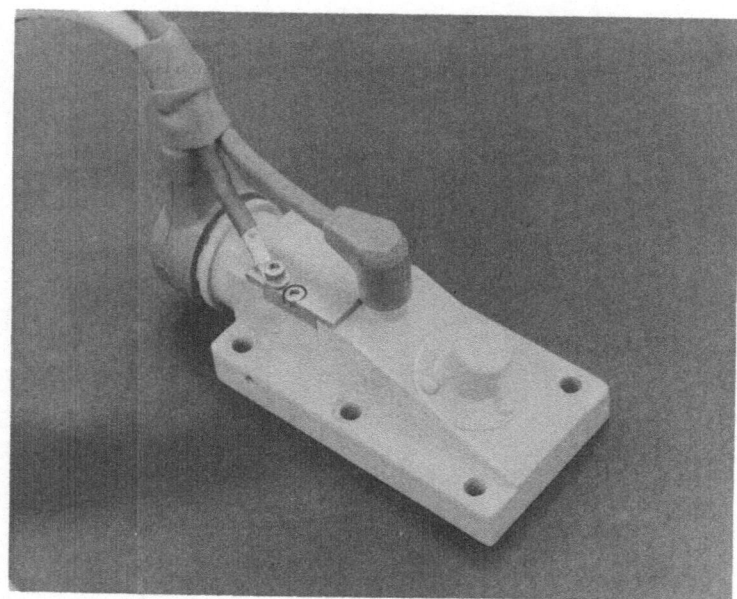

Fig. 2. A special klystron transmitting tube developed to meet electrical and thermal requirements. The large flange is for thermal capacity and efficient heat transfer.

An attempt to summarize the vibration-test requirements for the whole landing radar system is shown in Fig. 3. This is the vibration spectrum which it was anticipated would be "seen" by equipment during retrorocket firing. The maximum dynamic loading due to sinusoidal vibrations was 18 g between 20 and 125 cps, zero to peak. At 125 cps the sinusoidal loading was reduced to 4 g over the range 125 cps to 1500 cps. The random noise inputs were from 100 to 1500 cps. The worse vibration range then was between 100 and 125 cps where 18 g sinusoidal and random noise were superimposed. To avoid excessive loading within this range it was decided to design the structure so that no major- or secondary-mode resonant frequency was below 200 cycles.

MOUNTING ON THE SPACECRAFT

The overall thermal requirements for the spacecraft dictated that certain details of the thermal characteristics be imposed on the design of this unit, such as:

1. The assembly was thermally isolated from the spacecraft structure. This was required to ensure that no heat transfer by conduction took place to or from the structure itself.
2. The location and attitude of the unit was such that a maximum area is exposed to the sun on the descent part of the mission. By suitable surface treatment of this area, an accurate solar absorptivity-to-emissivity ratio can be arranged. The turn-on temperature most suitable to the unit was thus arranged for minimum weight and maximum reliability of the unit.

This latter requirement set the pattern for the attitude and position of the assembly and its mounting provisions. Figure 4 is an outline sketch of this mounting, showing the flight hardware and the cooling apparatus for long-duration testing mounted to it. The attachment-bracket interface was designed to span approximately two thirds of the length of the power supply. To take the loads directly from the power-supply chassis, the basic structure was designed with two diaphragms through the whole section, to line up with the bolts in the

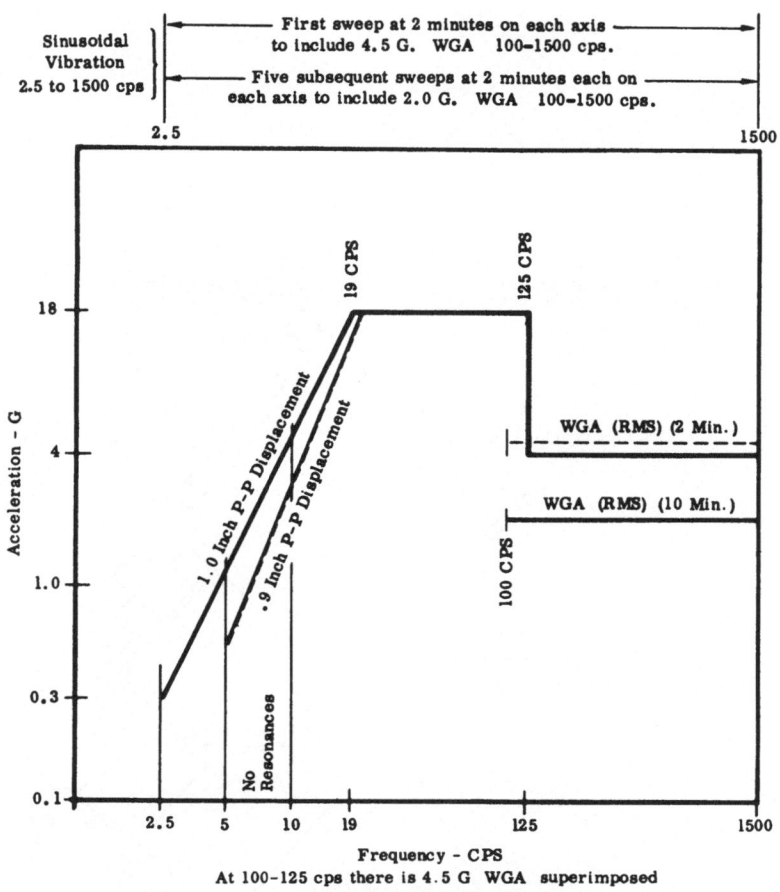

Fig. 3. Vibration-input spectrum (both sinusoidal and random) between 5 and 1500 cps. A waiver was obtained to reduce the low-frequency displacement slightly, and to commence testing from 5 cps instead of from 2.5 cps. This was done to enable existing vibration-test equipment to be used.

attachment bracket (Fig. 5). These diaphragms subdivided the box volume into three parts. By not continuing the attachment bracket to the ends of the unit, it was possible to arrange the microwave-output waveguide from each klystron on each side of the bracket. This determined the positions of the transmitting tubes. Ideally, they were positioned remotely from each other, with two separating walls of structure which served also as electrostatic shields. Members at right angles to these diaphragms insured that loads were taken directly from two mutually perpendicular members into the supporting bracket.

COOLING ARRANGEMENTS

It had been decided by thermal evaluation and testing on thermal mock-ups that it was necessary to have cooling air in direct contact with the dissipators. Air must be blown right through the unit. Plenum chambers were then arranged to be built into the structure, through the center and running the depth of the box. All high heat dissipators excepting klystrons would be mounted to, or in good thermal contact with, the walls of the plenum chambers.

The task now was to decide how the double-cruciform structural arrangement should be built. A permanent structure would be very difficult for assembly of electronic parts and for wiring.

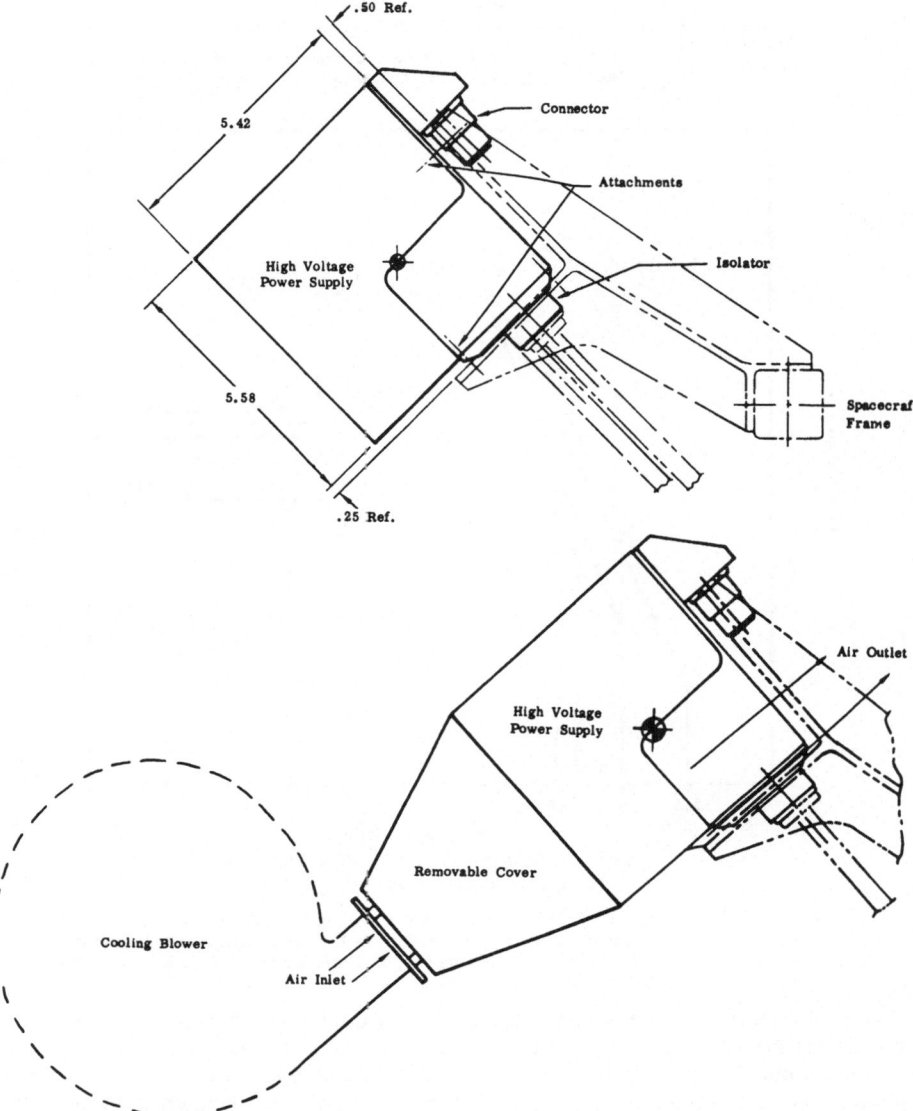

Fig. 4. The position of the unit on the spacecraft mounting structure. The orientation is with the axis of the spacecraft vertical. The upper figure shows the hardware as mounted for flight; the lower, the hardware and cooling blower assembly installed for test purposes.

BASIC STRUCTURE

Several structural arrangements were considered, as were different materials. Requirements were good thermal conductivity, light weight, high specific heat, adequate strength, and good workability. Aluminum alloy sheet 6061 T4 was the best available material to satisfy all these requirements reasonably, especially since dip brazing was chosen as the means of joining the smaller elements together. The structure was divided into four separable parts. Figure 6 shows the four basic structural parts as they are separated. The screwed joints would not be considered as vital to the thermal design because the three sections of plenum chambers each had its own cooling airflow, and the thermal capacity for $5\frac{1}{2}$-min operation was calculated on the

Fig. 5. The structure assembled. Some stud-mounted diodes, the two diaphragms, and the plenum chambers are visible. The flared half tubes are for intercompartment wiring.

basis of no heat transfer to structure other than that for directly mounting the electronic part. The stud-mounted power transistors and diodes are mounted through a wall of the plenum chambers, which resulted in the most efficient cooling for both service and test operations.

BUILT-IN SHIELDING

It was fortuitous that the opposite walls of these parts completed the electrostatic shielding between the assembly sections. In fact it was double shielding. The subdivisions of the assembly were arranged to house (1) the preregulator section; (2) the switching circuits, H.V. transformers, rectifiers, and filters; (3) the filament transformer and converter; (4) altimeter klystron

Fig. 6. The four basic structural parts as they appear when separated. Each subassembly is made from 6061 T4 aluminum alloy with joints dip brazed. Some of the transistors with high heat dissipation are mounted on the plenum-chamber wall.

post regulator (beam-current and filament) and modulator; (5) V.S. klystron post regulator (beam-current and filament); and (6) high-voltage bleed assembly and high-voltage capacitors. Each section was shielded from the others by the structure and the covers which fitted over it.

Weight and assembly complexity were minimized by making all major fabricated parts, structural members, contribute to thermal capacity and conduction, and perform an electrical function—shielding.

DETAIL THERMAL DESIGN

Thermally, the design had to be good for $5\frac{1}{2}$-min operation, only a small fraction of the time which would be required to reach steady-state conditions. During this time, the loss of heat by radiation, even from surfaces of high emissivity, was negligible. Every part had to be considered as a thermal storage element. If any electronic part did not have sufficient thermal storage capacity it had to be mounted on an element of adequate thermal capacity, having a high specific-heat-to-weight ratio, ensuring that the thermal junction was of low resistance. Generally the aluminum-alloy structure was used, additional material being brazed-on where the structure itself was insufficient.

Figure 7 is a nomogram which was designed as a rapid means of checking each semiconductor used, to see if it had sufficient inherent thermal capacity to keep the surface or junction temperature below the reliability figure calculated. It is based on parts having an average specific heat of 0.1 BTU/lb · °F.

The thermal mass is obtained from the relationship

$$Qt = MC_p\Delta t$$

where Q is the dissipation in BTU's; t is the time in hours ($= 5.5/60$ for this design); M is the mass (or weight) of the parts and leads; C_p is the specific heat of the materials from which the part is made; Δt is the temperature differential between the turn-on temperature of the unit and the maximum allowable operation temperature in °F.

From this

$$M = \frac{Qt}{C_p\Delta t}$$

If the effective mass of the part was less than this figure, additional thermal capacity had to be provided. The method of using the nomogram in Fig. 7 is explained at the end of this paper. For other parts, such as axial lead resistors, it was necessary to prepare special derating data. Manufacturer's or Mil Spec derating curves could not be used, as they are based on the steady-state operation of these parts under normal temperature and pressure conditions. On many occasions it was necessary to request a change of parts in the circuit because the part specified in the electronic design did not have sufficient thermal capacity to meet temperature requirements in vacuum. The weight and specific-heat data were not easy to get from the component manufacturers. They are not very interested in the small quantities used in special applications such as this. In addition, they are not anxious to divulge proprietary data concerning materials of construction, needed so that specific heats can be obtained. Nevertheless parts were sectioned, basic material weights were obtained, and the specific heat of each element was found. An average figure of specific heat for each component was obtained as shown in Fig. 8.

MECHANICAL DESIGN TRANSFORMERS

Placement of the heaviest parts such as the transformers, inductors, klystrons, and high-voltage capacitors was carefully arranged. The objective was to have as many of the heavy parts as possible located within the triangle of the attachment points. This would reduce the overall moment of inertia of the unit. A potted inductor for the preregulator section weighing 1.6 lb was 15% of the overall weight. This had to be located with its attachments right inside the vee of the attaching bracket.

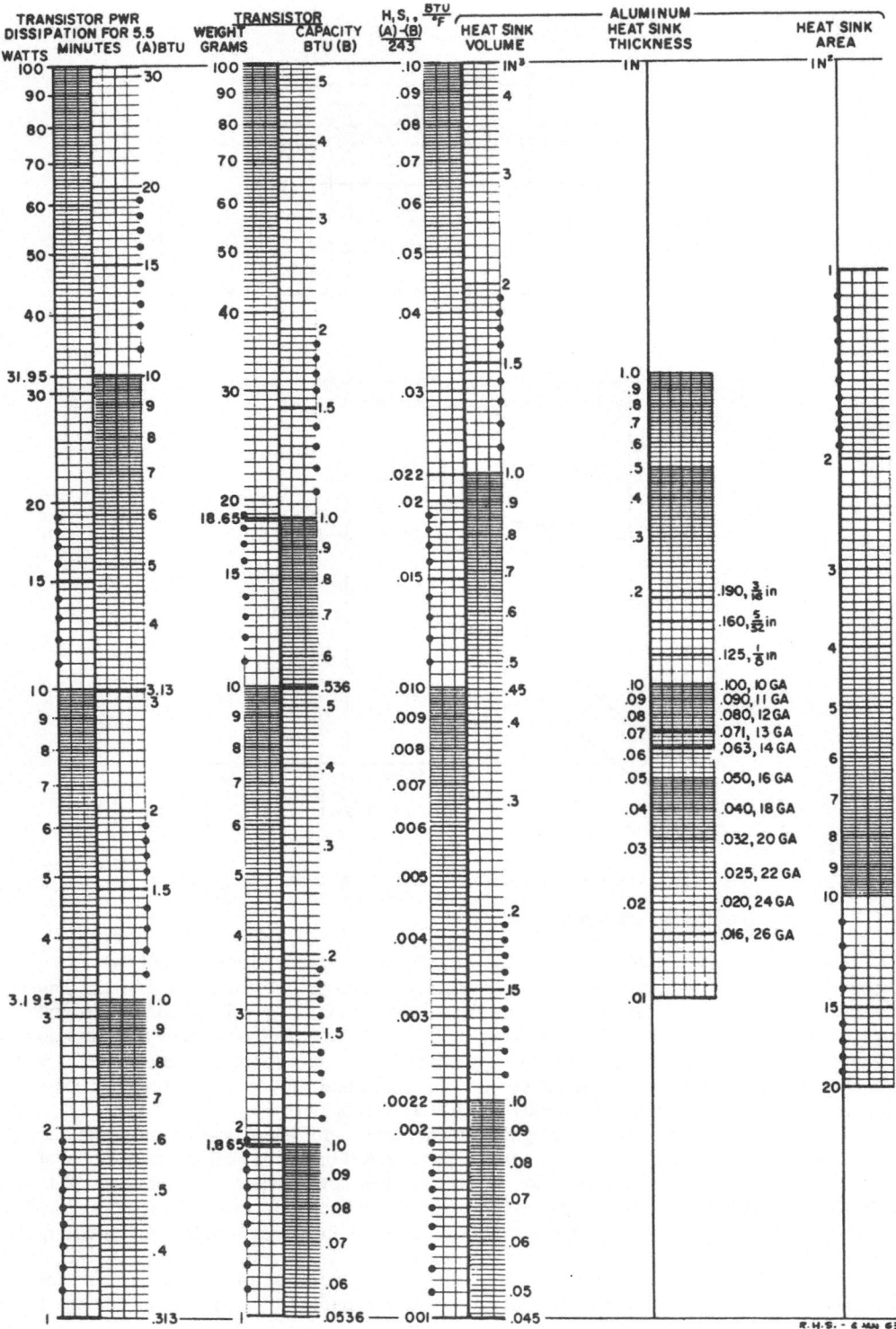

Fig. 7. Nomogram used for checking thermal capacity.

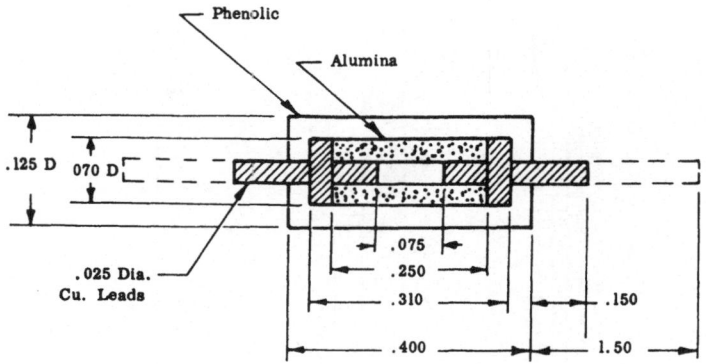

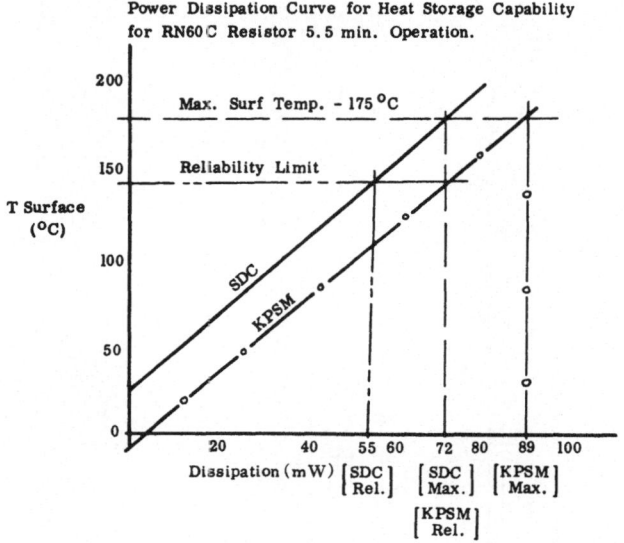

Fig. 8. A section through a typical RN60C-type resistor (top), and a derating curve based on a specified surface temperature for a given dissipation (bottom figure).

The natural frequency of each element of the structure as loaded with parts and the natural frequencies of panels and cover elements were calculated to ensure that no f_n would be less than 200 cps. A vibration mock-up was made, and the weights and C.G.'s of all individual parts and circuit boards were simulated. A resonant survey on the shake table showed that there was no natural resonant frequency below 240 cps. At a point where one of the heaviest masses was attached an excessive amplification showed up. An additional support at the center of mass altered the resonant frequency, and shifted the Q to a tolerable level.

It is the writer's opinion that with flight electronic equipment, where weight is at a premium, every part and item of hardware should perform an electrical or mechanical function. Some of the greatest contributors to excessive weight are the dust covers which go over a complete assembly. This hardware should be made to work. In this design, the external cover was part of the structure. The flanged ends of the two outer-plenum chambers are unstable until the covers are bolted down, then a stiff structural box is formed.

Figure 9 shows the cover arrangement. The design was arranged so that the cover was in two parts. A fixed part which was only removable when the unit was dismounted was assembled

Fig. 9. The two-part cover: the heat-sink blocks are dip brazed on the fixed part; the larger section is removable.

between the mounting bracket and the inner structure, and was clamped by the attaching bolts, and a removable part which was larger in size than the fixed part. Removal of this enabled access to be obtained to all test points and for changing the resistors in various circuits. This part had to be interchangeable with the cooler assembly so that testing could be accomplished on the spacecraft and the cooling cover replaced by the flight cover with a minimum of disturbance to the remainder of the assembly. This is shown in Fig. 4.

The klystron tubes had to be mounted to the fixed cover so that the microwave tubing runs need not be disturbed for normal access to the unit. Choke grooves were to be machined

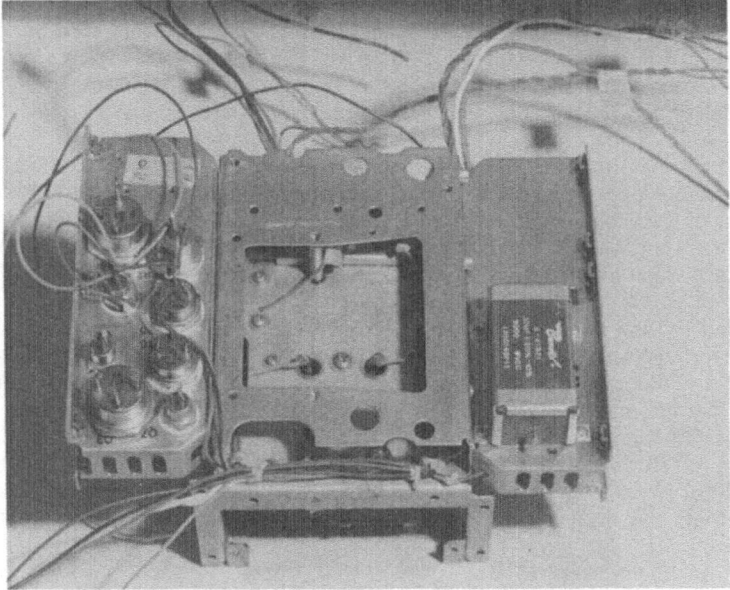

Fig. 10. The partly-assembled structure showing the mounting provisions for the high-voltage transformer through the wall of the center plenum chamber. Stud-mounted transistors and diodes are mounted on a thick heat sink. One of the high-voltage capacitors was located as shown. The ground connection is toward the outside.

at the interfaces. One electrical requirement was that the tubes be electrically isolated from the chassis. This could not be done with an isolating plate such as teflon, because any appreciable thickness interposed between the tube and mounting face would cause microwave leakage. The only insulating method which would meet the requirement was hard anodizing, and this was done directly to the inside of the heat-sink mountings which formed the two corners of the fixed cover.

Within the limitations of the electrical and magnetic design, mountings of transformers and inductors were arranged for low C.G. and good thermal paths. Three examples of how this was achieved are:

1. The transformer was bolted through holes in the magnetic core directly to the structural heat-sink plate. The core face had to be flat and free from impregnants and encapsulants. Figure 10 shows the hole cut in the plenum-chamber plate to allow the potted winding to protrude into the air stream for direct cooling during testing. The great difficulty with this design was in preventing gaps from opening up between the core and potting in the cylindrical cup.
2. Aluminum-alloy brackets were made to straddle and be in contact with the core inside the potting cup. Flanges on these brackets carry the nut plates, which are the attachments to the structure. By this means the attachment is not dependent upon potting material for its strength or its thermal conductivity which at best is poor.
3. When potting-cup materials are even poorer thermal conductors than acceptable potting materials, mounting through the cup is inefficient. A mounting base on the larger area, made from aluminum alloy, will serve the purpose of strengthening the mounting and reducing its thermal resistance.

HIGH-VOLTAGE PRECAUTIONS

All output sections of the assembly were at high voltage, even tube-filament power. The 2200-V output had to be very carefully considered. Fortunately, the unit did not have to operate at the critical altitude shown in the Paschen Curve. Operation at sea level and in hard vacuum (10^{-8} mm Hg) necessitated a gap of 0.5 in. between exposed high-voltage points and parts at or about ground potential. Small printed-circuit subassemblies were made for the high-voltage-regulator output circuits, and these were mounted back-to-back for structural rigidity. Very careful layouts were made to ensure that the spacing was at or greater than the minimum required for breakdown. Exposed needle points were avoided to reduce corona effects.

It is not very often that the packaging engineer has to consider the maximum allowable voltage which can be applied across a resistor. Each high-voltage regulator in this design had to have a string of precision low-temperature-coefficient resistors across the outputs. When these resistors are tightly packaged in this situation, the breakdown voltage has to be considered rather than the power dissipation. It was necessary to request component callout changes to cater to this requirement. Across the 2200-V output a series string of 12 resistors was needed.

Wiring and routing of wires were closely controlled. It was not economical weightwise to use all high-voltage wiring, so the routing of lower voltage wires had to be carefully watched to ensure that they did not pass too close to components at high voltages. It was realized early in the design that internal hookup wiring would be somewhat of a problem with an assembly having six shielded compartments in this geometrical layout. Wiring channels were designed into the structure. Routing and bunching of wires was decided in conjunction with the circuit design. It did not require much coupling to get a fraction of a millivolt injected on the high-voltage dc outputs and make the ripple exceed specification, causing a degrading of klystron output.

The klystrons themselves had four layers of 0.005 conetic magnetic shielding, totally enclosing them except for the microwave window. Spray coating the magnetic shielding directly on the klystron external surfaces was tried without achieving the degree of attenuation required.

MATERIAL LIMITATIONS

Most metallic materials, except for zinc, cadmium, and tin, are suitable for short-duration space applications. Alloys in which zinc was present were avoided. It was not necessary to use cadmium plating, since certain types of stainless steels were approved. Tin became a problem when it was discovered that solid wire to specification QQ-W-343 could be supplied with commercially pure tin. Special orders for wire coated with a 70/30 tin–lead alloy were placed, and the use of tinned wire was forbidden.

The most serious material problem was in the use of elastomeric materials. In hard vacuum these materials outgas condensibles which recondense on lenses and mirrors. There would be sufficient outgassing to render inoperative such things as television-camera lenses. Additionally, the vacuum pump would not be able to pump down to the required degree of vacuum for test purposes on the unit, if there was any appreciable outgassing from materials. Silicone rubber was the only approved elastomer and it was used in three places: the power connector; the cap and leads from the klystron; and in small antivibration mounts on the wave-guide runs. It was relatively easy to change the silicone in the connector for an approved material, diallyl phthalate. Manufacturing difficulties prevented a changeover in the klystrons. An all-metal vibration mount with the same natural frequency used to replace the antivibration mounts on the waveguide runs became impractical to use. After some research it was found that a thermovacuum pretreatment for certain elastomeric materials would enable them to be used (silicone rubber to aircraft materials specifications; Viton A—DuPont).

Other materials of nonmetallic form are so numerous that tests had to be carried out on each type before approval for use could be given. Generally, teflon, mylar, fiber glass and diallyl phthalate are the more common approved materials. Also, epoxies with aromatic-amine hardeners were approved.

SPECIFIC ELECTRONIC PARTS

Generally, tubes, relays, variable capacitors, and variable resistors are not permitted in space electronic systems due to their low reliability figures. Of these, only the omission of variable resistors proved to be a handicap to the packaging of the unit. Provision had to be made for the factory adjustment of resistors in some circuits. Terminals or solder pads were brought out to accessible positions. If the adjustment could be made in the subassembly stage the location of the solder joints was not critical. If the adjustment had to be made at unit test, it was necessary to make the part accessible and easy to change when the removable section of the cover was off.

There were 26 stud-mounted semiconductors used in this design. Manufacturers usually supply hardware with these parts. Until the component manufacturers get down to producing

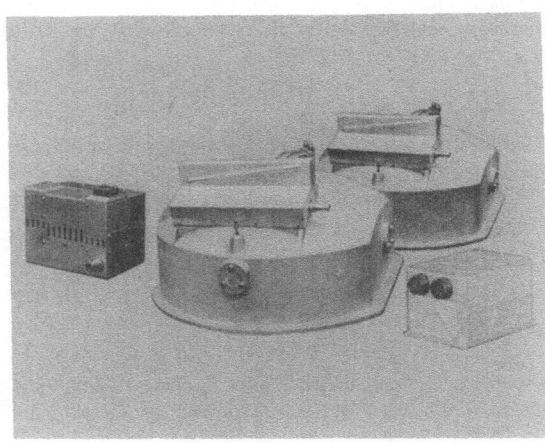

Fig. 11. The complete flight unit showing the waveguide mounting pads. (A wave-guide isolater is attached to the altimeter output.)

good, usable hardware, it is advisable to dispense with what they send, and purchase specification parts, or make special parts. Some of the troubles experienced were:

1. Roughly-stamped washers with one side having a sharp edge often cut through the mica insulators and caused grounding.
2. The locking washer was too clumsy.
3. The plating flaked off nuts and was found inside the unit.
4. Dimensions, especially lengthwise, had very high tolerances. The amount of heat conducted to a heat sink through the stud and nut of a stud-mounted transistor is negligible in comparison with that conducted directly from the flange of the part to the upper side of the heat sink.* Consequently, a better-insulating washer can be used on the underside.

The optimum hardware arrangement was found to be as follows, assembled in this order from the top downward.

A thermally conductive washer, e.g., beryllium oxide or boron nitride.
A teflon insulating bushing—thickness equal to or smaller than that of the heat sink.
A fiber-glass insulating washer.
A standard AN or NAS plain washer.
A self-locking nut or standard thin nut installed with loctite.

Torque-set stainless-steel screws were used throughout. In a small package such as this, number two screws had to be used in some locations. Not a single screw was broken in installation or test. The use of titanium screws resulted in a 30% weight saving for the same strength.

The power supply described successfully passed all thermal and vibration tests. Not a single solder joint or wire joint failed during the whole test phase. This is unusual for units using hookup wire with such high-level inputs as the vibration spectrum required for this space landing radar system.

NOMOGRAM FOR CALCULATING STUD-MOUNTING TRANSISTOR AND DIODE HEAT-SINK REQUIREMENTS†

The procurement specification lists case temperature at power-on time at $-20 \pm 10°C$. The maximum allowable heat-sink temperature has been established as 125°C. Therefore minimum heat-sink temperature is from -10 to $+125°C = 135°C$ or $\times 1.8 = 243°F$ rise. This nomogram is based on a 243°F temperature rise of semiconductor and heat sink. Actually, the semiconductor will have a greater rise, which will be dependent upon the thermal resistance of the mounting. The two columns on the left are conversion scales. The other three columns are alignment scales.

The watts column converts to BTU(A) for 5.5 min.

$$[\text{watts}] \times 3.413 \frac{[\text{BTU/hr}]}{[\text{watt}]} \times 5.5 \frac{[\text{min}]}{60[\text{min/hr}]} = \text{BTU(A)}$$

The grams column converts to semiconductor heat capacity in BTU(B) for a 243°F rise, and an estimated specific heat of 0.1 BTU/lb · °F.

The specific heats of some semiconductor materials are: silica, 0.19; solder, 0.04; glass, 0.16; steel, 0.12; copper, 0.094 (all in BTU/lb · °F).

Semiconductor weight is given by

$$[\text{grams}] \times 2.205 \times 10^{-3}[\text{lb/g}] \times 0.1 [\text{BTU/lb} \cdot °F] \times 243°F = \text{BTU(B)}$$

---

* J. E. A. John and J. J. Hilliard, "Heat-sinking techniques for power transistors in a space environment," *IEEE Trans. on Space Electronics and Telemetry*, vol. SET-9, pp. 45–51, June, 1963.

† This section refers to the nomogram in Fig. 7.

The starting location on the center column is found by subtracting BTU(A) − BTU(B), and dividing by 243. This value is located in the (A − B)/243 column, and is the required heat-sink capacity in BTU/°F. The required heat sink BTU/°F is converted directly into the amount of aluminum required in the heat-sink-volume column. From the heat-sink-volume column simply use a straight edge to convert to an acceptable package configuration of area and thickness.

For example, given 2N1936 dissipating 12 W, its weight is 30 g plus an estimated 3 g for mounting, hardware, effective wiring, insulation, and solder; therefore, the total weight is 33 g. The power dissipation of 12 W converts to 3.77 BTU(A). The transistor weight of 33 g converts to 1.75 BTU(B), then

$$\frac{(A) - (B)}{243} = \frac{3.77 - 1.75}{243} = \frac{2.02}{243} = 0.00831 \text{ BTU/°F}$$

which converts to 0.377 in.[3] If due to packaging configuration a 2 × 3 area (6 in.[2]) is required, then the required thickness is 0.063 in.